Liaison Functions

Presentations at the 2010 AIChE Annual Meeting

Salt Lake City, Utah, USA
7-12 November 2010

ISBN: 978-1-61782-163-9

Printed from e-media with permission by:

Curran Associates, Inc.
57 Morehouse Lane
Red Hook, NY 12571

Some format issues inherent in the e-media version may also appear in this print version.

Printed by Curran Associates, Inc. (2011)

For permission requests, please contact AIChE
at the address below.

AIChE
3 Park Avenue
New York, NY 10016-5991

Phone: (203) 702-7660
Fax: (203) 775-5177

www.aiche.org

Additional copies of this publication are available from:

Curran Associates, Inc.
57 Morehouse Lane
Red Hook, NY 12571 USA
Phone: 845-758-0400
Fax: 845-758-2634
Email: curran@proceedings.com
Web: www.proceedings.com

TABLE OF CONTENTS

(64a) Food and Fuel: The Technical Potential of Biofuels in the United States Using Land Efficient Animal Feeds

Bryan Bals[1], **Bruce E. Dale**[2], Seungdo Kim[3] and Pragnya Eranki[3], (1)Michigan State University, East Lansing, MI, (2)Department of Chemical Engineering and Material Science, Michigan State University, East Lansing, MI, (3)Department of Chemical Engineering, Michigan State University, Lansing, MI

Despite the potential for biofuels to displace petroleum in the transportation sector, several concerns remain regarding land use issues. It is believed that there is not enough land available for both food and fuel production, and therefore large-scale biofuel production would cause adverse land use change such as deforestation. We introduce a separate approach, wherein current cropland is intensely managed to either produce both food and fuel or to increase food production, allowing land to be freed for biofuels. Cover crops can be grown on the same land as corn or soybeans, increasing the amount of stover removed and providing a fiber source for ruminants. In addition, protein recovery from early harvest hays can displace soybean production, while pretreating switchgrass or corn stover with ammonia pretreatments can displace traditional forages in cattle diets. Likewise, biofuel production can be performed on idle land or Conservation Reserve Program land.

In this study, the theoretical maximum amount of biofuel produced on current cropland without affecting overall food production is considered. Current fiber, energy, and protein needs for human consumption, animal feed, and exports are met. This study suggests that over 50% of current US gasoline consumption can be displaced by biofuels grown on current cropland in the United States. The large volume of biofuel produced is due to intensive harvesting of corn land, which includes grain, stover, and cover crops. New animal feeding technologies also contribute to increased biofuel production. This result is consistent across multiple sensitivity scenarios, with the largest changes being due to changing crop yields and the loss of corn land or cover crops. Net soil organic carbon increases in this scenario; however, nitrate leaching is greater than if little to no biofuels are produced. Greenhouse gas emissions are reduced by 670 Tg CO_2-eq/yr, which is over 10% of current U.S. emissions.

See more of this Session: Sustainable Biomass Feedstock Production and Supply for the Emerging Biorefinery Industry
See more of this Group/Topical: Sustainable Engineering Forum

(64b) Optimal Design of Sustainable Cellulosic-Based Advanced Biofuels Supply Chains

Fengqi You[1], Ling Tao[2], Diane J. Graziano[1] and Seth W. Snyder[1], (1)Argonne National Laboratory, Argonne, IL, (2)National Bioenergy Center, National Renewable Energy Laboratory, Golden, CO

Fuels made from renewable resources, such as cellulosic plant materials, have been recognized as important sources of energy that will reduce the nation's dependence on petroleum and have a positive impact on the economy, environment, and society [1]. In final rule-making of the Renewable Fuel Standard (RFS) mandated by Congress, the EPA has set a renewable fuels requirement of 36 billion gallons by 2022 of which 16 billion gallons must be cellulosic biofuels (defined as yielding a minimum 60% reduction in greenhouse gas emissions relative to a 2005 baseline gasoline or diesel) [2]. Current cellulosic biofuels capacity is less than 0.1 billion gallons [3]. The rapid expansion of biofuels production needed to meet the RFS will require optimized design and management of new, sustainable, and regionally dependent supply chains [4]. Novel decision tools and systems analysis approaches are needed to facilitate supply chain development that minimizes the cost and negative environmental impact while maximizing the benefits to regional economies [1].

In this work, we consider the optimal design and operations of cellulosic ethanol supply chains under economic, environmental, and social criteria. A mixed-integer linear programming (MILP) model is developed that takes into account the main characteristics of cellulosic ethanol supply chains, such as seasonality of feedstock supply, biomass deterioration with time, geographical diversity and availability of biomass resources, feedstock density, conversion technology and performance, fuel price, infrastructure compatibility, spatial distribution of fuel stations, regional economic condition, tax subsidies, and policy. Techno-economic models based on ASPEN Plus for the conversion processes of potential feedstocks with possible biochemical and thermochemical pathways are linked to the MILP optimization model for detailed process economics of biorefineries. The MILP optimization model integrates decision-making across multiple temporal and spatial scales and simultaneously predicts the optimal network design, facility location, capital investment, production operations, inventory control, and logistics management decisions.

In addition to the economic objective of minimizing the annualized net present cost, the MILP model also integrates with life cycle assessment (LCA) and economic input-output (EIO) analysis through the multi-objective optimization scheme to include another two objectives: the environmental objective measured by life cycle greenhouse gas emissions and the social objective measured by regional economic impact. This allows the model to establish trade-offs between the economics, environmental impact, and social performance of the cellulosic ethanol supply chains in a systematic way. The multi-objective optimization problem is solved with ε-constraint method and produces a Pareto-optimal surface, which reveals the tradeoffs between the three objectives.

We apply the modeling and optimization framework to study cellulosic ethanol production in the state of Illinois and the Midwest area. Three major cellulosic biomass feedstock sources (corn stover, switchgrass, and woodchips) with biochemical and thermochemical conversions are considered. County-level results will be presented that provide regionally-based insight into transition pathways and consequent energy, economic, environmental, and social impacts of biomass production and conversion. Armed with this insight, we identify economically viable strategies for expanding biofuel production to benefit regional economies while minimizing negative impacts on the environment including land, water, and energy use.

References:

[1] Growing America's Fuel - An Innovation Approach to Achieving the President's Biofuels Target; Biofuels Working Group, U.S. Whitehouse, February 2010

[2] National Renewable Fuel Standard Program for 2010 and Beyond; U.S. EPA, February 2010

[3] Biomass Program Multi-Year Program Plan 2010; EERE, U.S. DOE, March 2010.

[4] National Biofuels Action Plan; Biomass Research and Development Board: U.S. DOE & USDA, October 2008.

See more of this Session: Sustainable Biomass Feedstock Production and Supply for the Emerging Biorefinery Industry
See more of this Group/Topical: Sustainable Engineering Forum

(64c) Establishing a Dedicated Energy Crop Supply Chain: The Tennessee Switchgrass Experience

Samuel W. Jackson, Office of Bioenergy Programs, University of Tennessee, Knoxville, TN

The University of Tennessee Biofuels Initiative (UTBI) is Tennessee's answer to the global challenge of securing sustainable, renewable, affordable energy while advancing the local economy and protecting the environment. The Biofuels Initiative is a farm-to-fuel business plan developed by State of Tennessee, UT Institute of Agriculture, and Oak Ridge National Laboratory researchers. Tennessee has made an unparalleled commitment to lead the transition to an advanced biofuels economy with a commitment to support the development of a dedicated bioenergy crop supply chain and the construction of a 250,000 gallon per year demonstration cellulosic ethanol facility. Working with private industrial partners, UT aims to establish a dedicated energy crop (switchgrass) supply chain, demonstrate and improve the technologies used to create cellulosic ethanol, reduce the costs of production, and ultimately commercialize the technology across the state. Tennessee has the potential to produce over a billion gallons of ethanol each year, replacing 30 percent of the state's current petroleum consumption. This effort will serve as a model for commercialization for cellulosic biofuels as well as other biobased energy and products. Multiple research programs involving dozens of faculty are underway in support of the biofuels program. Researchers with foci ranging from agronomics to biofuels conversion to bioproducts are leading the way. One of the key elements of the integrated Biofuels Initiative is a demonstration-scale cellulosic ethanol biorefinery in East Tennessee, supplied by local biomass. For this endeavor, Genera Energy, LLC has been established. Genera Energy is a for-profit limited liability company formed in 2008 by the University of Tennessee Research Foundation as a vehicle to carry out the cellulosic biorefinery activities and capital projects of the UTBI. Genera Energy was specifically created to provide the commercial flexibility needed to develop collaborations and partnerships with private entities with technology or other resources to contribute to the UTBI. Genera Energy, which is managed by a Board of Directors, collaborated with DuPont Danisco Cellulosic Ethanol to construct a pilot-scale cellulosic ethanol biorefinery. The biorefinery commenced operations in December 2009. When running at full capacity, the biorefinery will produce 250,000 gallons per year of cellulosic ethanol from switchgrass. The Tennessee Biofuels Initiative takes a farm-centric approach to feedstock development, working with local farmers to develop a program that provides direct payments to farmers for switchgrass production as well as one-on-one technical assistance through UT Extension and wide-ranging research related to all aspects of the feedstock supply chain. Approximately 6000 acres of switchgrass have been planted on over 60 private farms in the region. The UT Biofuels Initiative has the largest planting of switchgrass on private farms in the United States. Switchgrass produced on participating farms will be utilized in the pilot scale biorefinery to produce ethanol as well as other products. An established feedstock supply of this scale, provides significant opportunity for research and development related to feedstock production, harvesting, storage, transportation, and preprocessing, and environmental sustainability. The University and Genera Energy have made significant progress in these areas, integrating new and existing technologies to improve the overall efficiency and sustainability of the feedstock supply system. Ultimately, the development of the cellulosic biofuels supply chain, from biomass to fuels and products, will have a significant impact on the energy future of the state and nation. These new systems, technologies, and products will provide increased economic opportunities for rural economies, increased energy security for the nation, and improved environmental impacts of energy production/utilization.

See more of this Session: Sustainable Biomass Feedstock Production and Supply for the Emerging Biorefinery Industry
See more of this Group/Topical: Sustainable Engineering Forum

Techno-economic feasibility of Miscanthus as biomass feedstock: Analysis using BioFeed model

Yogendra Shastri, Alan Hansen, Luis Rodriguez & K.C. Ting

Energy Biosciences Institute
Department of Agricultural and Biological Engineering
University of Illinois at Urbana-Champaign
1304 W. Pennsylvania Avenue,
Urbana IL - 61801.

Abstract

The success of the bioenergy sector depends significantly on ensuring efficient and sustainable biomass feedstock production and provision, which requires a comprehensive systems theory based approach. BioFeed is a system level model that has been proposed to optimize the feedstock production and provision activities, and has been applied in the past to study switchgrass production in Illinois. This work presents recent additions to the BioFeed model to enable a more accurate representation of various biomass production activities for the energy crops. While maintaining the original model framework that focused on farm level design and operational issues in addition to storage and transportation logistics, new biomass packing and size reduction operations such as pelletization and grinding have been added. The selection and operation of biomass handling equipment such as loaders, unloaders and infield transportation equipment have also been incorporated. The addition of these new operations creates the challenge of ensuring the logical validity of the operational sequence during model simulation. A super-structure of all possible operational sequences is developed and the biomass form at the output of every equipment is tracked to ensure appropriate equipment selection. The model is then applied to a case study of Miscanthus production as the energy crop in southern Illinois. The results show that the optimized delivered cost based on existing technology is about 45 \$/Mg. Biomass packing and storage are important components of the total cost distribution. The potential alternatives to reduce the delivered cost include using a single-pass mowing and baling operation, increasing the packing throughput capacity, and extending the harvesting window.

1 Introduction

The importance of biomass feedstock production and provision in the success of the biomass based energy sector has increasingly been highlighted in recent times [1, 2]. Shastri et al. [3] emphasized the importance of taking a systems perspective and propose the use of systems informatics and analysis techniques to overcome these obstacles. A generic systems analysis framework is a powerful tool because it is applicable to any situation subject to the availability of data. BioFeed is an optimization model which has been developed as a first step towards the development of such a framework [4]. It is a mixed integer linear programming (MILP) model that incorporates various biomass production activities such as harvesting, packing, transportation, storage and handling and determines the optimal system level configuration. An important unique feature of the model is to develop the operational blueprint that can be used by farmers and managers to efficiently operate the whole system. The model has been applied to study switchgrass production in southern Illinois [4].

This paper presents the latest developments in the extension of the BioFeed model. The major additions are in the post-harvest biomass processing operations. In addition to baling, the new model can simulate biomass chopping, grinding and pelletization. In the first version of the model, biomass handling (loading and unloading) and infield transportation operations were approximated. These operations are modeled in greater detail in the new version of the model. This includes the selection of the appropriate handling and infield transportation equipment for a farm as well as optimizing the operating schedule of that equipment. New biomass storage options, particularly covered storage, reflecting different ways of constructing a storage facility are added. The addition of new biomass processing options such as chopping and grinding create the challenge of developing a sequence of operations that is logical. The output form of an equipment therefore must correspond to the input form of the subsequent equipment. This is ensured by first developing a pre-defined super-structure of operational sequence and then matching equipment attributes to determine the operations sequence. These additions significantly enhance the optimization capability of the model since all possible combinations are explored for each farm. A new computational scheme, called the Decomposition and Distributed Computing (DDC), has been developed to solve the BioFeed model in a computationally efficient manner and has been successful in reducing the simulation time by an order of magnitude without significantly compromising the optimality. This scheme has been used to solve the new version of the BioFeed model. The new version is then applied to a case study of Miscanthus production in southern Illinois.

The article is arranged as follows. The next section presents a brief overview of the first version of the BioFeed model, its important features and assumptions. Section 3 presents the new enhancements in the BioFeed model, including modeling of various packing options, modeling of farm handling and infield transportation equipment operations, and selection of the appropriate equipment using a numeric scheme. It also explains the creation of a superstructure of possible scenarios to enable logical scenario construction during model simulations. Section 4 presents the results of the application of the model for a case study of Miscanthus production in southern Illinois. The final section summarizes the work and presents the importance conclusions.

2 BioFeed model background

Shastri et al. [4] developed and presented the first version of the BioFeed model. It covers the feedstock production and provision activities between farms growing energy crops and the biorefinery. The model assumes that standard crop-establishment techniques will be used that will result in standing biomass on a farm, which constitutes one of the boundaries for the model. For each farm under consideration, the production activities include harvesting, raking, packing (e.g. baling), on-farm storage and/or ensiling. The biomass can remain on farm post raking if a packing machine is not immediately available. The packed biomass can either be directly transported to the refinery, or stored in one of the three storage options, namely, on-farm open storage, on-farm covered storage, and centralized storage which is shared by all the farms. The transportation activities are carried out using a set of trucks that is independently owned. Each compartment is modeled using a set of linear algebraic equations that reflect the mass balance as well as the equipment capacity and availability constraints. The decision variables for the optimization model include equipment selection and their operating schedule for all activities on the farm, biomass distribution among various alternatives, on-farm storage selection and sizing, centralized storage selection and sizing, transportation fleet size selection and utilization of the fleet (logistics), and the sustainable biorefinery capacity. It is a mixed integer linear programming (MILP) model, where the equipment and storage site selection decisions constitute the integer decisions while the operating decision such as harvest scheduling and storage sizing constitute the continuous decisions. The design decisions such as equipment selection are one time decisions while the operational decisions are optimized for every time-step during the simulation horizon. The simulation horizon consists of one year. Thus, biomass harvested during the harvesting horizon, which typically lasts for 2-4 months, must be stored and supplied to the refinery during the non-harvesting horizon. The smallest simulation time-step is one day, and user can specify a larger time-step consisting of multiple days. Although a smaller time-step is desired for greater accuracy, reducing the time-steps increases the solution time significantly. Each operation is associated with capital as well as operating costs. The capital cost includes the total purchase cost which is amortized over the expected life of the equipment, while the operating cost includes labor, fuel, maintenance and other associated costs. The goal of the model is to optimize all decisions simultaneously so that the best system level configuration can be determined. The objective function, therefore, is the maximization of the total system profit by assuming a certain refinery gate price for each unit of biomass. The division of the total profit among various stakeholders is not considered. It is possible to pre-specify the type and number of various equipment to enable the user to build more specific scenarios. In such cases, the corresponding variables are converted into parameters and hence not included in the optimization decision making process. A comprehensive discussion of the model is omitted here for brevity and interested readers are referred to Shastri et al. [4].

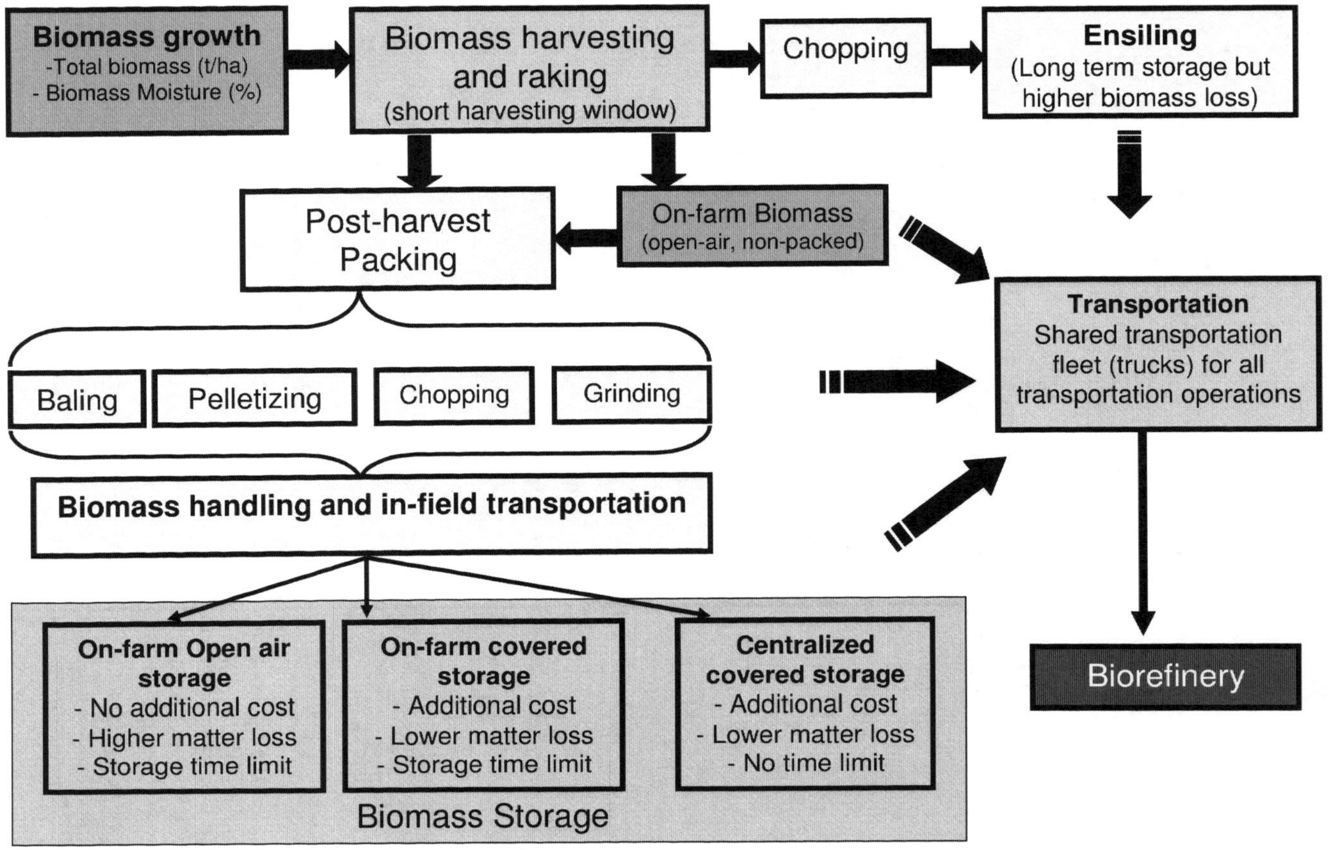

Figure 1: BioFeed model: Version 2

3 BioFeed model enhancements

Figure 1 shows the new version of the BioFeed model, with additional features for a more comprehensive representation of the feedstock production activities. The important modifications are described in the following sections.

3.1 Farm equipment data

The new version of the BioFeed model includes a number of different operations such as pelletization and chopping, as well as more detailed modeling of biomass handling and infield transportation. However, successful modeling of these features necessitates modifications in the equipment database. The typical equipment data required for model simulation include the capital and operating costs, throughput rates, efficiency, fuel consumption, weight capacity and more. These data will be stored in the form of attributes for a particular equipment in MySQL database using an object-oriented format [5]. However, with the addition of biomass handling and infield transportation models, it is necessary to expand the set of attributed beyond these typical ones. This is because the selection of a particular type of equipment for these operations is often a function of the physical form of biomass being handled (such as baled or chopped biomass). For example, a telescopic bale loader can only be used to load/unload bales and cannot be used to handle chopped biomass. Incorporating this logical selection into the optimization model is a challenging task. Hence, in BioFeed, a convention has been developed, as shown in Table 1, to describe each physical form of biomass using numeric a value. This convention is also used to describe additional attributes for an equipment, such as the biomass input form and output forms of the equipment. It is also necessary to know whether a particular equipment is stationary or mobile, since this information is required for modeling the infield transportation operation. Hence, each equipment also includes two more attributes 'Stationary' and 'Mobile', which are binary attributes. Thus, if Stationary=1 for a particular equipment then it means that the equipment is stationary and vice versa. This information is used in the modeling of biomass handling and infield transportation equipment selection. The new model also includes additional harvesting options such as integrated mower-baler and integrated mower-chopper. In such cases, the post-harvest biomass distribution will be a function of the output form/characteristics of the biomass. This is modeled in BioFeed by assigning attributes to the harvester that define what are the possible options for the output from that equipment.

3.2 Biomass packing

The first version of BioFeed considered baling (square and round) and loafing as the only possible post-harvest biomass packing options. However, there are additional forms of biomass packing, such as pelletization, that can potentially be utilized on a particular farm. This version of the model, therefore, incorporated three additional forms of post-harvest biomass packing, namely, pelletization, chopping and grinding. It must be noted that the term packing was used to reflect the form of biomass that could be stored and transported for delivery to the refinery and did not

Table 1: Numeric representation of various biomass forms

Form	Numeric representation
Standing crop	0
Harvested	1
Baled	2
Chopped	3
Ground	4
Pelletized	5

necessarily reflect the physical operation of packing. Thus, biomass diverted to packing operation can further be divided into the four available options. It is important to correctly model the distribution of biomass among these different alternatives, and ensure that the form of biomass delivered for packing is suitable for the type of packing option selected. This is modeled by using the 'output form' attribute of the harvester that is selected for a given farm. The model also enforces the constraint that only type of packing option can be selected by a farm, although different farms can select different options. The primary reason for enforcing this constraint is to simplify the selection of the appropriate biomass handling and in-field transportation equipment. If different packing options are selected, then optimization of these equipment selection is difficult. However, it is expected that this simplification will not significantly affect the model accuracy since an individual farm typically would not invest in multiple packing equipment since the capital costs for these equipment can be significant. Moreover, each farm can select a different packing option without being affected by the selection on another farm. It is also possible that the user might want to eliminate certain packing options from consideration to create a more specific scenario. Therefore, BioFeed incorporates a user-selection option where the user can exclude a particular option by making the binary selection parameter equal to zero. The equations to model each of the four packing options are similar to those for the first version of BioFeed presented in Shastri et al. [4], and hence not included here.

3.3 Farm biomass handling

Biomass handling tasks such as loading, unloading and stacking are important farm operations. As mentioned earlier, the type of handling equipment that can be used depends on the form of the biomass being handled, and hence, on the packing equipment selected for a particular farm. The first version of BioFeed did not model equipment selection decisions for biomass handling. Rather, per unit biomass handling cost from literature was used. The revised model eliminates this approximation and models the selection of biomass handling equipment for each farm. The model also optimizes the operation of these handling equipment based on the capacity of each equipment.

The numerical scheme explained in section 3.1 is used to ensure the selection of the appropriate equipment. Here, only those handling equipment with the biomass input form equal to the output form of the packing option selected are available for selection. This ensures the selection of the appropriate equipment. The model also determines the number of hours for which the selected biomass handling equipment is used. This is calculated by using the total biomass that requires handling, the total weight capacity per trip for the selected equipment, and the total round trip time for the equipment. Once the total time is known, the total cost and fuel consumption is determined by using equipment attributes such as operating cost and fuel efficiency. These equations are not included here for the sake of brevity. Each farm can select only one type of handling equipment. Moreover, similar to the packing machine selection, the user can pre-define the type or number of handling equipment of a particular type for a given farm.

3.4 Infield transportation

Infield transportation of biomass includes primarily the roadsiding operation to move the harvested or packed biomass to the edge of the farm. From there, it can be transported or kept in the on-farm storage facilities. The objective of the BioFeed model is to optimize equipment selection and operation for each farm. However, similar to the biomass handling operation, the equipment selection decisions are based on the form of biomass that needs to be transported. Moreover, the selection also depends on whether the packing equipment is stationary or mobile. Thus, balers are often mobile, which means that the output form of the packing equipment needs to be transported. On the contrary, pelletizers are often stationary and located at the edge of the farm, which means that the input form of the packing equipment needs to be transported. Thus, the selection of the infield transportation equipment is more complex than the selection of the biomass handling equipment. The numeric scheme described in section 3.1 is again used for equipment selection decisions. After the selection of the equipment, the number of hours the equipment is used is again calculated by using the weight capacity of the equipment, the round trip time, and the total biomass being moved. It is again possible for the user to pre-define the type and/or number of a specific type of equipment for a given farm. Those equations are omitted here for the sake of brevity.

3.5 Selection of the sequence of operations: Super structure creation

The long term objective of this systems analysis work is to use the BioFeed model as a web-based decision support tool. This decision support environment will allow the users to build specific scenarios, provide expert knowledge for building relevant scenarios, and then optimize the scenario to determine the operating blueprint. There exist a number of different scenarios (defined here as a specific sequence of operations) that can be followed to deliver biomass at the refinery, many of which have already been described in literature [6, 7, 8, 9, 10]. From modeling perspective, it must be ensured that all the individual operations and the scenarios are encompassed in the modeling framework. However, a critical challenge then is to ensure that the selected scenario is logically correct. For example, the modeling framework must eliminate the possibility of sending the output of a baling machine to the input of a harvester. This is also important for the user interface development, since it must not allow the user to build physically non-realizable scenarios.

In BioFeed, this is achieved through the development of a super-structure of all scenarios, i.e., all possible connections between different operations. Once the individual operations are modeled, all possible connections between these operations are then defined. A subset of these connections, based on the scenario to be constructed, is selected through an input file that can be modified by the user. This input file consists of binary selection parameters pertaining to each connection defined in the super-structure. If the selection parameter is 1, then the corresponding connection becomes valid and can be used by the optimization model. On the contrary, if the selection parameter is 0, then that particular connection is not considered in the scenario. This type of set-up has two advantages. First, since the super-structure contains only the connections that are logical, the possibility of the user or the optimization algorithm selecting an infeasible connection is eliminated. Second, since the connections can be activated and deactivated using an input file, it gives the user the flexibility to easily create the desired scenario. In future, when the web-based interface is developed, this input file will be defined based on the selections made at the graphical interface. The critical aspect of this approach is the creation of a comprehensive super-structure. However, it should not be a major concern since it is a one time exercise. When any new operation is added to the model, all connections involving the new operations will have to be added to the super-structure.

4 Case study: Miscanthus production in southern Illinois

Shastri et al. [4] had applied the first version of BioFeed to study switchgrass production in thirteen counties of southern Illinois. This work considers the same geographical location in Illinois for a case study application, and considers the production of Miscanthus as the energy crop in this region. Figure 2 shows the farming and collection area of about 5540 square kilometers which consisted of 284 farms distributed over 13 counties. A hypothetical refinery is assumed to exist at Nashville (IL), while a centralized storage location is assumed to exist at Okawville (IL), which is about 20 kilometers for the refinery. It is assumed that the 284 farms are uniformly distributed over the 13 counties. The transportation distances between various farms and the refinery as well as the centralized storage location are calculated using Google maps. Please refer to Shastri et al. [4] for a comprehensive discussion related to farm location and distance calculations. The focus of the work presented here is on evaluating various technology options available for Miscanthus production. This is particularly important since the performance of various technologies/equipment for Miscanthus is still being evaluated, and hence the results from this work should shed light on the critical components of the whole system.

4.1 Data gathering

Heaton et al. [11] have reported typical Miscanthus productivity along with the growth profile based on previously published literature. The values reported by them have been used in this work. Miscanthus yield in Illinois should be maximum around September 1, after which the standing crops start to dry. Miscanthus recycles nutrient back to rhizomes after full growth [12]. To take advantage of this mechanism and reduce fertilization requirement for the next growth season, harvesting of Miscanthus is expected to begin in early spring in Illinois. Therefore, this work assumed that the harvesting season began from January 1 and lasted till April end (about four months) when the new growth season begins. Post senescence biomass loss rate for the standing crop was taken from Heaton et al. [11]. Based on the typical moisture variation ranges proposed by Yu et al. [13] and Kumar and Sokhansanj [8], the BioFeed model assumed that the plant moisture content at peak biomass value was 60% and started to drop linearly beyond this point due to gradual drying up to a minimum moisture content of 15%. The short term variations in the moisture content due to weather variability were ignored, but could be included in future model extensions.

One of the major challenges in modeling Miscanthus production was to collect data related to equipment use for Miscanthus. This included not only the cost data for an equipment, but also the performance data such as harvesting capacity, efficiency and fuel consumption. Most of the studies on Miscanthus have been conducted in Europe and there are very little data specific to conditions and equipment in the US [6, 14, 15, 16, 17, 9, 18]. This work, therefore, adapted the performance data published for the European studies, assuming that the performance of those equipment would be replicated for the US. Sokhansanj et al. [7] have used the the ASABE machinery standards [19, 20] to determine the energy consumption, fuel consumption and cost related data based on the machine horsepower and the purchase cost. This work, therefore, utilized those guidelines to generate data for equipment used in the European studies. The greenhouse gas emissions depend on the source of energy used to operate the equipment. For equipment running on diesel fuel, USEPA reports the greenhouse gas emission factor to be 10.1 kg of CO_2/gallon of diesel (http://www.epa.gov/oms/climate/420f05001.htm) which is used to translate fuel consumption into greenhouse gas emission. For equipment running on electricity (such as stationary pelletizer), USDOE has published greenhouse gas emission factors for electricity in various states [21]. The emission factor for the state of Illinois is 0.668 Kg CO_2 per KWh which was used to convert the

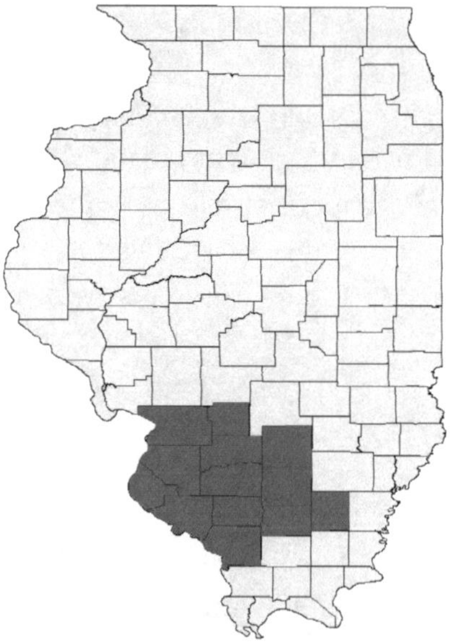

Figure 2: Energy crop farming and collection area in southern Illinois

energy required into corresponding greenhouse gas emissions. The data for typical farm equipment such a loaders, fork lifts, and trailers that are not crop-specific were taken from the IBSAL studies and data sheet [22]. The model assumed that on-farm open storage in on gravel pad, on-farm covered storage is in the form of a shed without walls and the centralized storage is in the form of an enclosed facility without any ventilation, unless otherwise specified. The next section reports the results from various simulation studies.

4.2 Scenario modeling and comparison

The new version of the BioFeed model was first used to compare different biomass packing options, namely, baling, grinding, chopping and pelletization. Four different scenarios were modeled where each farm was forced to select the particular packing option, i.e. the type of packing was pre-decided and not optimized by the individual farm. Within the given packing option, the farms were allowed to select the suitable number and type of equipment from those available for selection. The post-packing options, such as the storage options and availability of the centralized storage facility, were same for all the scenarios.

Figure 3 compares the final total delivered cost, while Figure 4 compares the total energy consumption per unit of biomass delivered for various packing options. Biomass baling is currently practised for hay, and hence is expected to be the most likely packing option utilized for energy crops in the near future. The total delivered cost for baling was about 48 \$/Mg. Figure 5 shows the distribution of the total cost amongst the various operations. It was observed baling was the most cost intensive operation (44%) due to the high cost of baling equipment and relatively low throughput capacities. This was followed by storage, which accounted for about 23%. It is an important result since this highlights the role of biomass storage which has often been ignored in previous studies. It is also interesting to note that when BioFeed model was used to simulate switchgrass production for the same case study, storage accounted for only 13% of the total cost. Thus, higher productivity of Miscanthus, which is desirable for higher biofuel production, increased the importance of storage in the overall value chain. A transportation fleet of 153 trucks was required to support the transportation operation, but the average fleet utilization was only about 43%. The farms had the option of buying a baling machine or renting the baling machine from a contractor. It was observed that only 56 farms, typically large farms, opted to buy the baler, while the rest opted to rent it from the contractor. The model did not limit the total number of balers available for renting, which could be an important factor if renting becomes the predominant alternative. The total energy consumption was 866 MJ/Mg, and harvesting, packing and infield transportation operations were the major components with 32%, 23% and 32% contributions, respectively.

Figure 3 shows that chopping was the costliest biomass packing option, with the delivered cost of 63\$/Mg. The primary reason for the increased cost was the low bulk density of 100 kg/m^3 for the chopped biomass. The low bulk density biomass required higher biomass storage volume and a larger transportation fleet (191 truck). Moreover, the transportation operating cost, both for infield and long-distance transportation, increased since more trips were required. Since chopped biomass was eligible for ensiling, biomass ensiling was extensively used as the storage option. The total energy consumption for chopping scenario, however, was lower than that for baling. This was due to the higher throughput capacity of the chopper which resulted in lower number of operating hours for the equipment.

Pelletization of biomass led to the final delivered cost of about 51.5\$/Mg. Although the difference between the total cost for baling and pelletization was not significant, the distribution of the cost was significantly different. For pelletization, lower throughput capacity of about 4.5 Mg/hour caused the total packing cost, particularly the packing operating cost, to increase significantly. However, the higher density of the pellets (650 kg/m^3) caused the storage and transportation cost to reduce. Thus, only 116 trucks were required in the transportation fleet (average fleet utilization of 43%) and the storage cost component reduced to 5% of the total cost. Pelletization, though, is an energy intensive operation and hence led to the highest energy consumption among all the packing options. The energy consumption related to packing itself went up by almost 100% as compared to baling. However, the higher density of pellets reduced the transportation energy consumption by about 50%, which resulted in a net overall increase of about 13% in energy consumption as compared to biomass baling. Another important observation was the significant increase in the greenhouse gas emission, which was calculated to be 128 kg CO_2 eq. as compared to 63 kg CO_2 eq for baling and 58 kg CO_2 eq for chopping. This was because the pelletizer was considered to be stationary and operating on electricity. Greenhouse gas emissions for the electricity generated in Illinois were higher than those from diesel fuel for the same output energy. Hence, using pelletization as the packing option caused higher greenhouse gas emissions.

Biomass grinding scenario resulted in the lowest delivered cost for Miscanthus, which was due to two reasons associated with the grinder attributes. The model considered tub grinder, hammer mill and rotary knife mill as three possible selection options, with bulk densities ranging from 0.125 kg/m^3 to 0.20 kg/m^3 depending on the output particle size. The throughput capacity for tub grinder was significantly higher than the other two as well as that of the baler and pelletizer. However, the hammer mill and rotary knife mill resulted in smaller output particle size, which increased the bulk density of the ground biomass. The results showed that large farms, which required to process large quantities of biomass, selected tub grinder, thereby taking advantage of the higher processing capacity. On the contrary, the smaller farms selected hammer mill since the lower capacity was not a limiting factor and the higher bulk density reduced the storage and transportation costs. Thus, overall, the smaller and larger farms both were able to select the appropriate grinding equipment, thereby, leading to an overall cost that is slightly lower than the cost while using baling. It was also observed that biomass handling incurred higher cost (17% of the final cost) for ground biomass compared to other packing options. The higher throughput capacity also led to a lower total energy consumption, as shown in Figure 4.

The grinding scenario assumed that all storage options are available for each farm. However, on-farm open storage of ground biomass is difficult since ground biomass will be highly susceptible to weather disturbances such as high wind and rain. Moreover, the on-farm covered storage may also require a closed storage facility in place of a simple shed, as assumed in the previous scenario. Hence, a new scenario for grinding was modeled where on-farm open storage of ground biomass was not permitted, and the on-farm covered storage facility had to be enclosed (shed with four walls and without ventilation). The delivered cost for this case increased to about 51\$/Mg, making it more expensive than the baling option. The ability to store ground biomass at low costs is, therefore, critical to the grinding scenario. It must, however, be acknowledged that the refinery might value ground biomass higher than other forms of biomass since it will eliminate preprocessing at the refinery to get the biomass ready for conversion.

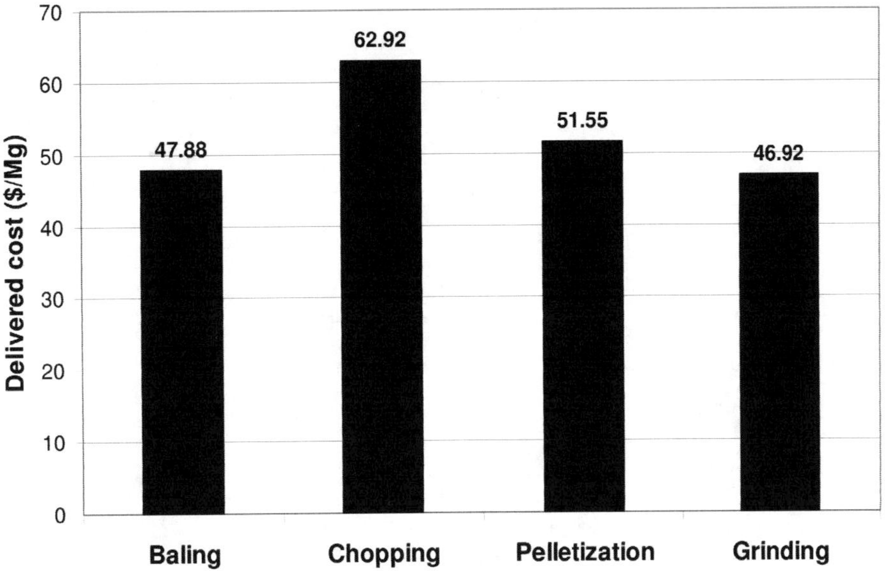

Figure 3: Comparison of Miscanthus production system costs for different biomass packing options

4.3 Technology development sensitivity analysis

New technologies are being developed with the objective of making energy crop production and provision more efficient. If technologies with a significantly impact on the efficiency can be identified, that can direct the research focus. While doing this, it is important to ensure that the impact on the whole system rather than the individual operation is considered. BioFeed model, being a system level model incorporating all major production activities, has the capability of identifying those technologies by conducting a sensitivity analysis.

Table 2 lists the attributes of selected technologies/equipment that were considered for the sensitivity analysis. The model assumed that each of those attributes improve by 25% at the same cost (capital as well as operating). The model simulation results were then compared with the simulation results for the original scenario to quantify the improvement. The improvement was calculated in terms of reduction in the delivered cost as well as energy consumption. The results are documented in Table 2. It was observed that higher throughput capacity of pelletization, grinding and baling had a significant impact on the system efficiency. It led to a reduction in not

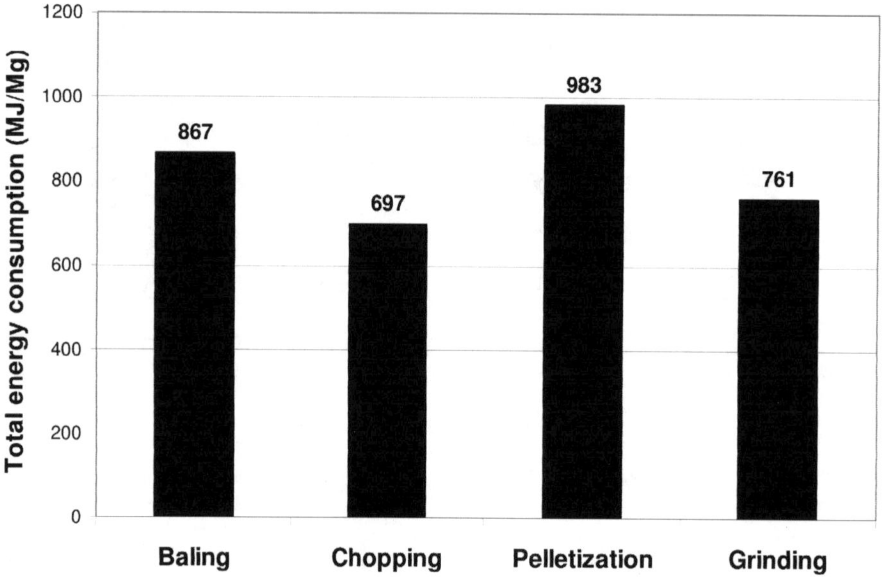

Figure 4: Comparison of Miscanthus production system energy consumption for different biomass packing options

Table 2: Sensitivity analysis results

Technology and attribute	Cost reduction (%)	Energy consumption reduction (%)
Baling density	5.9	0.3
Chopping density	3.8	5.0
Grinding density	8.2	1.7
Baling throughput	8.5	5.1
Grinding throughput	7.8	5.4
Pelletization throughput	11.7	9.8
Truck capacity (weight and volume)	2.3	1.7

only the final cost, but also to the total energy consumption. Both these effects were attributed to the fact that higher throughput capacity resulted in lower operating hours, which led to lower operating cost and fuel/electricity consumption. Improvement in pelletization, in particular, was highly beneficial. If the operating costs associated with pelletization can be reduced, the significant advantages in the storage and transportation operations can then be availed. Increasing the output density of the biomass by packing equipment reduced the delivered cost but did not have a major impact on the energy consumption. Increasing the carrying capacity of a transportation truck did not have a major impact on the overall system.

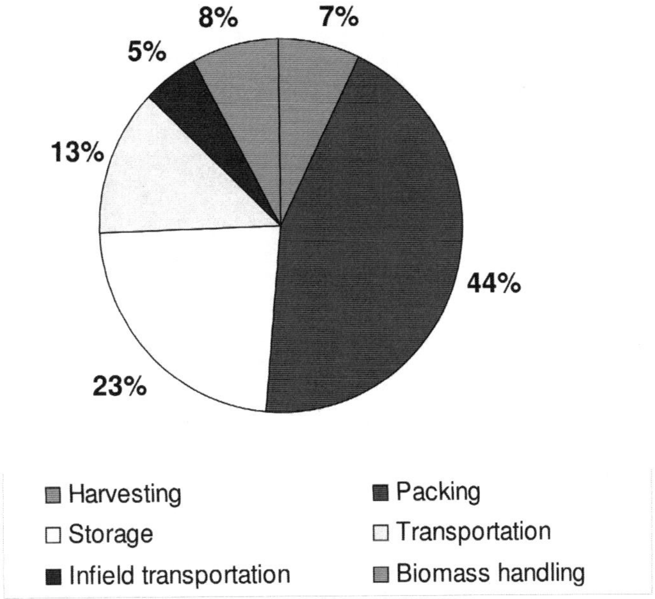

Figure 5: Cost distribution for the Miscanthus production and provision using baling option

5 Conclusions

Biomass feedstock production and provision is an important component of bioenergy system. BioFeed is an optimization model that has been proposed to integrate important feedstock production activities, analyze them as a single system, and determine the optimal system configuration at a regional level. One of the unique features of this model is to incorporate farm level decisions and present an operational blueprint as model solution. The model has been successfully applied for switchgrass production in Illinois. This research built on that work and enhanced the BioFeed model for more comprehensive modeling and analysis. The revised model included biomass processing options such as grinding, chopping and pelletization in addition to baling. It also allowed more detailed modeling of farm biomass handling and infield transportation operation, which included optimization of equipment selection as well as their operation. A numeric scheme was developed to represent the physical form of biomass as it moves through the supply chain, which was then used to ensure the selection of the appropriate equipment for biomass handling and transportation. A super-structure of various possible supply configurations was also developed. This ensured that logically correct sequence of operations were created. The revised model was then applied to the case study of Miscanthus production in southern Illinois. The case study results showed that on-farm biomass grinding was the cheapest packing option, subject to the availability of cheaper on-farm storage options to minimize handling losses. When a sensitivity analysis for technology improvement was carried out, it was observed that increasing the throughput capacities of the packing options, particularly pelletization, led to the biggest cost reduction (11.7% on an average). Based on other simulation studies using BioFeed and not reported here, it was observed that the optimized Miscanthus production cost which optimized the selection of packing option for each farm based on current technology was about 45 \$/Mg, and was dominated by packing costs such as baling or grinding. The optimal solution included use of baling, tub grinding and hammer mill by different farms depending on the farm size and production capacity. Moreover, extending the harvesting window. Overall, the case study and the results emphasize the ability of BioFeed to conduct a comprehensive systems analysis and provide decision support for energy crop production and provision.

Acknowledgment

This work has been funded by the Energy Biosciences Institute through the program titled 'Engineering solutions for biomass feedstock production'.

References

[1] R. D. Perlack, L. L. Wright, A. F. Turhollow, R. L. Graham, B. J. Stokes, and D. C. Erbach. Biomass as feedstock for bioenergy and bioproducts industry: The technical feasibility of a billion-ton annual supply. Technical report, Oak Ridge National Laboratory, 2005.

[2] DOE. Biomass: Multi-year program plan. Technical report, Office of the Biomass Program, Department of Energy, 2008.

[3] Yogendra Shastri, Konstantinos Domdouzis, Ming-Che Hu, L.F. Rodriguez, A. C. Hansen, and K.C. Ting. Systems informatics and analysis of biomass feedstock production for bioenergy: A review. *Biomass and Bioenergy*, Under reivew, 2009.

[4] Yogendra Shastri, Konstantinos Domdouzis, Ming-Che Hu, A. C. Hansen, L.F. Rodriguez, and K.C. Ting. System level analysis of biomass feedstock production for bioenergy sector. In *ASABE Annual Meeting 2009, Paper Number: 095998*, Reno, Nevada, 2009. ASABE.

[5] Konstantinos Domdouzis, L.F. Rodriguez, Yogendra Shastri, Ming-Che Hu, A. C. Hansen, and K.C. Ting. Systems informatics for biomass feedstock production engineering. In *ASABE Annual Meeting 2009, Paper Number: 096702*, Reno, Nevada, 2009. ASABE.

[6] W. Huisman, G Venturi, and J Molenaar. Cost of supply chain for Miscanthus x Giganteus. *Industrial Crops and Products*, 6:353–366, 1997.

[7] S. Sokhansanj, A. Kumar, and A.F. Turhollow. Development and implementation of integrated biomass supply analysis and logistics model (ibsal). *Biomass and Bioenergy*, 30:838–847, 2006.

[8] A. Kumar and S. Sokhansanj. Switchgrass (Panicum vigratum, L.) delivery to a biorefinery using integrated biomass supply analysis and logistics (IBSAL) model. *Bioresource Technology*, 98(5):1033–1044, 2007.

[9] David Styles, Fiona Thorne, and Michael B. Jones. Energy crops in Ireland: An economic comparison of willow and Miscanthus production with conventional farming systems. *Biomass and Bioenergy*, 32(5):407–421, 2008.

[10] Shahab Sokhansanj, Sudhagar Mani, Anthony Turhollow, Amit Kumar, David Bransby, Lee Lynd, and Mark Laser. Large-scale production, harvest and logistics of switchgrass (panicum virgatum l.) current technology and envisioning a mature technology. *Biofuels, Bioproducts & Biorefining*, 3:124–141, 2009.

[11] Emily Heaton, Tom Voigt, and Stephen Long. A quantitative review of comparing the yields of two candidate C4 biomass crops. *Biomass and Bioenergy*, 27:21–30, 2004.

[12] I. Lewandowski and A. Kicherer. Combustion quality of biomass: practical relevance and experiments to modify the biomass quality of Miscanthus giganteus. *European Journal of Agronomy*, 6(3-4):163–177, 1997.

[13] M. Yu, A.R. Womac, C. Igathinathane, P.D. Ayers, and M.J. Buschermohle. Switchgrass ultimate stresses at typical biomass conditions available for processing. *Biomass and Bioenergy*, 30(3):214–219, 2006.

[14] P. Venturi, J.K. Gigler, and W. Huisman. Economic and technical comparison between herbaceous (Miscanthus x Giganteus) and woody energy crops (Salix Viminalis). *Renewable Energy*, 16:1023–1026, 1999.

[15] J.M.O. Scurlock. Miscanthus: A review of European experience with a novel energy crop. Technical report, Oak Ridge National Laboratory, 1999.

[16] M.B. Jones and M. Walsh. *Miscanthus for Energy and Fibre*. London: James & James (Science Publishers) Ltd., 2001.

[17] W. Huisman. Optimising harvesting and storage systems for energy crops in the netherlands. In *Proceedings of the International Conference on Crop Harvesting and Processing*, 2003.

[18] Edwards M.W. Smeets, Iris M. Lewandowski, and Andre P.C. Faaji. The economical and environmental performance of miscanthus and switchgrass production and supply chains in a european setting. *Renewable and Sustainable Energy Reviews*, 13:1230–1245, 2009.

[19] ASABE. Agricultural machinery management. Technical Report ASAE EP496.3 FEB2006, American Society of Agricultural and Biological Engineers, February 2006.

[20] ASABE. Agricultural machinery management data. Technical Report ASAE D497.5 FEB2006, American Society of Agricultural and Biological Engineers, February 2006.

[21] Energy Information Administration. Updated state-level greenhouse gas emission factors for electricity generation. Technical report, Office of Integrated Analysis and Forecasting, Energy Information Administration, U.S. Department of Energy, 2001.

[22] S. Sokhansanj, A.F. Turhollow, and Erin Wilkerson. Development of the integrated biomass supply analysis and logistics model (IBSAL). Technical Report ORNL/TM-2006/57, Oak Ridge National Laboratory, Oak Ridge, TN, March 2008.

(64e) Methods for Determining Chemical Mechanisms of Biomass Torrefaction

Richard Boardman, Energy Systems & Technology, Idaho National Laboratory, Idaho Falls, ID, Jaya Shankar Tumuluru, Biofuels & Renewable Energy Technology, Idaho National Laboratory, Idaho Falls, ID and Jospeh Smith, Advance Process & Decision Systems, Idaho National Laboratory, Idaho Falls, ID

The main objective of torrefaction is to produce a heat-treated solid, char-like material. Torrefaction (or slow to mild pyrolysis) serves to improve the properties of biomass in relation to thermochemical processing techniques for energy generation; for example, combustion and co-combustion with coal. The topic has gathered interest in the past two decades but further understanding is required for optimization of the process thus enhancing economic efficiency, which is crucial to the success of the commercial operations. In particular, there is a noticeable gap in current literature regarding kinetic mechanisms and rate data. Limited kinetic data and reaction models have been recently published for woody materials and recently reed canary grass, wheat straw, and willow. A quantitative understanding of torrefaction for many varieties of biomass may be attained by conducting appropriate torrefaction tests; particularly softwoods and herbaceous biomass that contain a significant fraction of hemicelluloses as glucomannan as opposed to hardwoods and wheat straw which contains relatively high amounts of hemicelluloses as xylan. Organic matter in biomass is comprised of lignins, hemicelluloses, and cellulose. While the ratio of these constituents varies depending on the type plant species and growing conditions, it has been postulated that the thermal-chemical behavior of these constituents in the temperature range of torrefaction (i.e., 200-300°C) is primarily a function of heating rate, peak temperature, time at peak temperature, and gas temporal conditions. Torrefaction appears to be a two-stage process where hemicelluloses decomposition and devolatilization precedes cellulose breakdown and dehydration into smaller chains. Molecular cross-linking has also been observed, which may be a detriment in biochemical conversion of torrefied feedstock. A method for elucidating the mechanisms of hydrogen and function group extraction and chemical cross-linking of the three major constituents has been developed. This requires control of thermal energy deposition and real-time measurement of species evolution, followed by post torrefaction sample chemical and physical analysis. A heated fluid-sand bed is used to achieve consistent heating conditions of closely uniform particles sizes. Species devolatilization is monitored using a set of analytical instrument and laboratory methods for volatile and semi-volatile species determination. Mass evolution rates are correlated to obtain devolatilization rates that can then be applied to single particles or particle ensembles. NMR is useful for determining cross linking that occurs within the samples.

An early set of data, rate correlations, and physical/chemical properties of torrefied corn stover are presented in this paper to illustrate application of the testing and characterization methods.

See more of this Session: Sustainable Biomass Feedstock Production and Supply for the Emerging Biorefinery Industry

(64f) Optimal Design of Biomass Processing Networks for Biofuel Production Under Uncertainty

Jinkyung Kim, Matthew J. Realff and Jay H. Lee, School of Chemical & Biomolecular Engineering, Georgia Institute of Technology, Atlanta, GA

Bio-fuel represents a promising candidate for renewable energy. It presents the biomass industry with exciting opportunities as well as significant challenges. One of the challenges for the emerging industry concerns a very high level of uncertainty around feedstock costs and supply amounts. The uncertainty makes the assessment of the economics and investment decisions very difficult. This study focuses on optimal design of biomass supply chain networks for bio-fuel production under significant uncertainties in biomass feedstock availabilities and their acquisition costs. Biomass yields are extremely sensitive to factors like the weather and supply costs are also affected by emergence of new industries competing for the same types of resources. Designing a processing network robust to uncertainties in price and availability are critical to assess and reduce the risk associated with investment in the bio-fuel industry. It is desirable to account for all possible future scenarios in designing and analyzing such a network. The supply chain network we study covers the Southeastern region of the United States and includes biomass supply locations and amounts, candidate sites and capacities for two kinds of fuel conversion processing, and the logistics of transportation from the locations of forestry resources to the conversion sites and then to the final markets. A two stage stochastic program is formulated and solved given the data for possible scenarios and their probabilities. The first stage decisions (called "here-and-now decisions") are the capital investment decisions including the locations and capacities of the processing plants, and the second stage decisions (called "recourse decisions") are the material flows from biomass sites to processing sites and eventually to the final markets. The MILP solves for the optimal numbers, locations, and sizes of the conversion processing plants, as well as the amounts of biomass, intermediate products, and final products for each scenario in order to satisfy to final market demands, with the objective of maximizing the expectation of the overall profit. The robustness of the nominal design (for a single nominal scenario) vs. the robust design (for multiple scenarios) is analyzed under various scenarios considered.

See more of this Session: Sustainable Biomass Feedstock Production and Supply for the Emerging Biorefinery Industry
See more of this Group/Topical: Sustainable Engineering Forum

(73a) Chemical Engineering Outside the Pipe: Industrial Ecology and Sustainability

Roland Clift, School of Engineering, University of Surrey, Guildford, United Kingdom

The 2010 Danckwerts lecture will be concerned with explaining the emerging field of Industrial Ecology (IE) and why it represents a new and important area for application of the skills of the chemical engineer. That guide to popular understanding, Wikipedia, defines Industrial Ecology (IE) as [1] "a 'systems-based, multidisciplinary discourse that seeks to understand emergent behaviour of complex integrated human/natural systems' [2]. The field approaches issues of sustainability by examining problems from multiple perspectives, usually involving aspects of sociology, the environment, economy and technology. The name comes from the idea we should use the analogy of natural systems as an aid in understanding how to design sustainable industrial systems". Wikipedia also says that [1] "Chemical engineering is the branch of engineering that deals with the application of physical sciences.... and life sciences.... with mathematics to the process of converting raw materials or chemicals into more useful or valuable forms". While surprisingly sound, these definitions do not bring out the close relationship between industrial ecology and chemical engineering. Put simply, chemical engineering principles can be applied to managing material and energy flows and transformations in an economy or an industrial sector or a company - i.e. to industrial ecology - as much as to flows and transformations in pipes and vessels [3]. The common ground is shown by two of the analytical tools which are central to IE: Life Cycle Assessment (LCA) and Material Flow Analysis (MFA). LCA is used to identify and, where possible, quantify the full resource uses and environmental impacts associated with supplying a product or service; "a product is followed from its 'cradle' where raw materials are extracted from natural resources through production and use to its 'grave', disposal" [4]. The all-important Inventory phase of a LCA is essentially a material and energy balance over the whole supply chain. LCA is the tool used to measure the greenhouse gas emissions (sometimes called the "carbon footprint") of a product or company or plant. MFA is another application of material balances: "a method of analyzing the flows of material in a well-defined system" [1], which may be a national or regional economy or an industrial sector or a defined area such as a city. While the concepts behind these basic IE tools are not new to chemical engineers, the purpose may be. The Wikipedia definition highlights that IE aspires to address "whole systems" which may be complex and chaotic. This idea is illustrated by Figure 1, a very broad-brush view of human activities which illustrates some of the ideas in industrial ecology including that of a "circular economy", one of the ideals of IE, in which materials are re-used many times on their passage through the economy. Material is only lost from the economy when it is dispersed; therefore undispersed "waste" is seen as a resource which can be reused. Realising such an economic system goes beyond re-use and recycling; it involves designing and planning for multiple uses and finding "industrial symbioses" in which a waste or low-value product from one industry can be a valuable input to another - which is the heart of the "ecology" metaphor. The result can be an industrial network which looks much like the kind of process system familiar in chemical engineering. One of the differences from a process system is that the viability and success of such a network depend not just on the characteristics of the flows but on the relationships between the different actors in the system. This is one of the reasons why IE examines "problems from multiple perspectives, usually involving aspects of sociology, the environment, economy" as well as technology. Multiple perspectives imply multiple objectives so that single-objective optimisation is rarely appropriate: decision structuring and deliberative processes are also central to the IE agenda [5,6]. Figure 1. The Human Economy [5]: E – emissions. The Wikipedia definition also emphasises that IE is concerned with "issues of sustainability". Among the myriads of definitions of sustainability, one of the most succinct is due to Jackson [7]: "Sustainability is the art of living well, within the ecological limits of a finite planet". "Ecological limits" are at the heart of the problem of sustainability: indefinite growth in consumption is as impossible as "business as usual" in the global economy. Furthermore, "living well" is to be interpreted in an ethical sense, not merely equated with material consumption or physical comfort. Since the first attempts to articulate the meaning of sustainable development [8], the importance of equity has been recognised: one person's quality of life must not be at the expense of the well-being of someone else, living now or at some future date. Therefore a further discipline must be added to the mix: ethics (which is a branch of philosophy, not religious studies). This is one of the reasons behind the increasing interests in ethics shown by many national engineering institutions. Applying ethical principles to globalised supply chains raises some "interesting" questions... References 1. en.wikipedia.org (accessed July 19, 2010) 2. Allenby, B. (2006) "The ontologies of industrial ecology". Progress in Industrial Ecology 3 (1/2): 28-40 3. Clift, R. (2007) "Time to leave the process sandpit". The Chemical Engineer July: 21-2. 4. Baumann, H. & Tillman, A.-M. (2004) The Hitch-hiker's Guide to

LCA. Lund: Studentlitteratur. 5. Mitchell, C.A., Carew, A.L. & Clift, R. (2004) "The role of the professional engineer and scientist in sustainable development", Chapter 2 in Sustainable Development in Practice (ed. A.Azapagic, S.Perdan & R.Clift). Chichester: John Wiley & Sons. 6. Elghali, L. , Clift, R. , Begg, K. G. & McLaren, S. (2008). "Decision support methodology for complex contexts". Engineering Sustainability, 161(1), 7-22. 7. Jackson, T. (2010) "Keeping out the giraffes", p. 20 in Long Horizons (ed. A.Tickell). London: British Council. 8. World Commission on Environment and Development (1987) Our Common Future. Oxford: Oxford University Press. Figure 1. The Human Economy [5]: E – emissions.

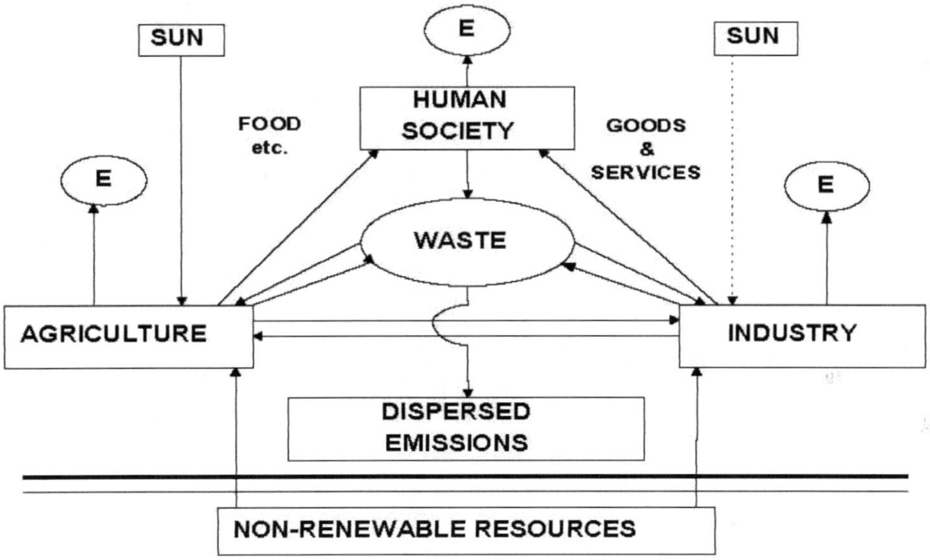

See more of this Session: Danckwerts Lecture
See more of this Group/Topical: Liaison Functions

Thermochemical Conversion of Biomass to Biofuels – Process Demonstration Unit Test Results

Ravi Chandran[1], Dave Newport[1], Kym Arcuri[2], Sean Whitney[1], Daniel Leo[1] and Shawn Freitas[1]
[1]ThermoChem Recovery International (TRI), Baltimore, MD USA
[2]Emerging Fuels Technology (EFT), Tulsa, OK USA

Abstract

TRI has constructed and commissioned a Process Demonstration Unit (PDU) at Durham NC. This integrated gasification and gas to liquids biorefinery plant is designed to process 4 dry tons per day of biomass, generate, clean up, compress and condition syngas, and catalytically convert it to Fischer-Tropsch (F-T) liquids. The gasification system uses TRI's proprietary indirectly heated gasification technology. The synthesis reactor is an Emerging Fuels Technology (EFT) F-T reactor with their proprietary cobalt catalyst to produce fuels and waxes. The biomass feed system and gasifier have operated for 5,400 h, the primary gas cleanup has logged 2,600 h and the overall system with F-T has operated in the integrated mode for 1,300 h, all on woody biomass. The trials have demonstrated clean syngas production with stable H_2 to CO molar ratio (~1.9), 60+% CO conversion in a single stage of F-T and the generation of high quality, sulfur-free, 100% renewable fuels i.e. pure white wax and water clear, medium fraction hydrocarbon liquids. System configuration details are provided and test results on performance, products and emissions are presented.

Introduction

Advances in biomass production and conversion support a future with great potential for sustainability, energy security and climate stability. There are many biological and chemical platforms to convert biomass into valuable products, however most of these technologies can only convert a narrow range of biomass types into a limited number of products. In contrast, gasification is a thermochemical platform that can convert almost all of any carbon-based biomass into a wide range of downstream value-added products. Gasification is an enabling technology that generates a synthesis gas ("syngas") composed of hydrogen and carbon monoxide, supports a wide range of value-added processes and products, and offers the potential to transform regional economies into sustainable clean energy systems.

TRI offers a proprietary gasification technology for biomass that has some unique attributes: production of a medium calorific value syngas, ability to dial in the H_2 to CO molar ratio to match the synthesis reaction step for biofuels and/or biochemical production, feedstock flexibility, generation of the highest syngas flow for synthesis reaction due to the indirect heating method, and opportunity for superior thermal and process integration.

In order to generate commercially scalable data and gain valuable operating experience with different biomass feedstocks, TRI has constructed and commissioned a 4 dry ton per day biomass to Fischer-Tropsch liquids process demonstration unit (PDU) at the Southern Research Institute's Carbon to Liquids Development Center at Durham, NC.

Trials have been carried out in this unit in direct support of two major Department of Energy (DOE) funded commercial biorefinery projects: NewPage's Project Independence and Flambeau River Biofuels' Park Falls project, both located in Wisconsin. The biorefinery at NewPage will process 500 dry tons per day of woody biomass and produce wax and diesel fuels, replace fossil fuel in the pulp mill lime kiln with F-T tail gas, and supply steam and hot water to the mill to displace coal. The Flambeau biorefinery will process 1,000 dry tons of woody biomass to produce wax and diesel fuels and generate green electricity and steam for the mill; this will facilitate the shut down of two natural gas fired boilers.

PDU Description

Figure 1 shows a photograph of the 1 MW thermal throughput PDU test facility. The PDU is made up of the following sub-systems: biomass storage and feed, biomass gasification (steam reformer and carbon trim cell), primary gas cleanup, secondary gas cleanup and conditioning (syngas compression and polishing) and F-T synthesis. The PDU is designed to be integrated to any number of alternate downstream catalytic or fermentation processes to produce fuels and chemicals from syngas.

Figure 1. TRI Process Demonstration Unit

The proprietary biomass feed system is designed to handle a wide range of biomass feedstocks. The feeder utilizes a 3 stage hydraulic piston arrangement to develop a solid biomass plug of defined length and pressure to form the seal between the pressurized reformer and the atmospheric biomass day bin. As the biomass plug forms the primary seal, one of the three pistons is always in a closed position, which prevents a plug blow-out and provides additional safety to syngas leakage. In addition to the 3 stage plug feeder, the biomass feed system includes a 6.5 m^3 day bin, transport screws, weighing screw and high speed biomass injection auger. The biomass day bin, which is on load cells, is batch filled from standard sacks using an overhead house crane. The biomass feed rate is controlled by adjusting cycle frequency, and the weigh screw ensures that the same mass of biomass is fed in each cycle, irrespective of biomass density. For the pilot plant using a single feeder, an intermediate dosing bin is used to smooth out the biomass feed stream to the reformer.

The TRI gasification system utilizes an indirectly heated medium temperature, low pressure fluidized bed steam reformer primary stage and a higher temperature, low-pressure fluidized bed gasifier second stage to generate syngas; here this has been sized to process four oven dried tons per day of biomass feed, with a moisture content of 10% or less. Scaling parameters were employed in designing this system to allow the direct application of the PDU data to commercial-scale units. The endothermic energy required for the steam reforming reaction is provided indirectly through electrical heaters banks. In the case of commercial systems, proprietary pulse heater modules will provide this heat input. The bed material is engineered alumina and is fluidized with superheated steam using a steam grid installed in the bottom of the reformer vessel. A cyclone is suspended inside the reformer to capture fine bed material and larger char particles from the syngas, and return these to the bed through a dipleg. A set of lock hopper valves is provided for discharge of bed material from the reformer vessel during maintenance shut downs. The vessel is fitted with level, density, temperature and pressure measurement instrumentation. A separate small Carbon Trim Cell (CTC) is installed beside the steam reformer to gasify the fine char leaving the reformer. A proprietary solids collection system is installed in the syngas discharge from the reformer to collect ash/char elutriated from the reformer and route it to the CTC. Within the CTC, partial oxidation with oxygen is utilized to convert the residual carbon and to manipulate the H_2: CO ratio to the level required for the downstream gas to liquids process. The CTC is a refractory lined round vessel and fitted with a sparger style fluidization grid. A set of lock hopper valves is provided for discharge of bed material during maintenance shutdowns. Another cyclone is installed in the syngas discharge from the CTC and collects fine biomass ash elutriated from the bed. The cyclone dipleg transfers the ash into a steel drum via a set of lock hoppers. The CTC is fitted with level, density, temperature and pressure measurement instrumentation.

Heat exchangers are used to superheat the fluidization steam and pre-heat the nitrogen and plant air for the gasification process. A majority of the combined syngas from the reformer and carbon trim cell is routed to a Regenerative Thermal Oxidizer (RTO) for destruction.

A slipstream from the combined reformer/CTC syngas is routed through a Heat Recovery Steam Generator (HRSG) where it is cooled. Following the HRSG is a venturi scrubber with a high ΔP throat to capture any entrained particulates, and a gas cooler to condense residual steam and further cool the syngas. As the gas is cooled in the venturi scrubber and gas cooler, tars generated in the gasification process will condense. The tars are captured using a proprietary solvent and recovered using a decanter and then reinjected into the CTC and converted to syngas. The syngas from the gas cooler is then routed to a H_2S scrubber, using caustic soda to absorb trace quantities of H_2S.

After primary gas cleaning, a multi-stage compressor raises the clean syngas pressure up to the level necessary for the desired Fischer-Tropsch product selectivity and catalyst performance. The pressurized syngas is cooled and then passes through a packed ammonia scrubber using water to eliminate HCN and NH_3. The syngas is then heated to F-T temperature and is routed through fixed guard beds to remove traces of COS, H_2S and any HCl passing through the primary gas clean up and ammonia scrubber.

The clean and conditioned syngas is then fed into the Fischer-Tropsch reactor, where H_2 and CO react over a catalyst to form primarily straight-chain hydrocarbons of various chain lengths, generating light, medium and heavy fraction Fischer-Tropsch products. The F-T reaction is exothermic, and heat is removed in the form of steam generated in the reactor cooling jacket. The steam jacket is designed to represent the commercial system with respect to water and steam rates. The F-T reactor is a fixed-bed three tube system where the tube length and gas path conditions are identical to those in commercial scale reactors and scale-up simply involves adding additional tubes. Hence, the PDU results are directly applicable to full-scale commercial systems. The stream leaving the F-T reactor contains heavy waxy hydrocarbons in liquid form, medium and light fraction hydrocarbons in gaseous form, water byproduct in gaseous form and unreacted syngas (tail gas). A hot separator at the exit of the reactor is maintained slightly below the reactor temperature to collect the heavy F-T fraction. The vapor fraction goes through a cooler to condense the water and lighter hydrocarbons fractions, which are further separated in flash tanks. The tail gas, containing unconverted H_2, CO and light fraction hydrocarbons in the range of C1-C5 is vented to the RTO for destruction.

Control is achieved through a commercially available Distributed Control System (DCS) which is connected to a Historian data acquisition system. All sensor data are stored in the Historian for record and analysis. Continuous and discrete analytical data from Gas Chromatography (GC), Fourier Transform Infrared Spectroscopy (FTIR), UV-Vis Spectroscopy, oxygen sensor, isokinetic particulate sampling probes, Draeger tubes, Gastec detector tubes and Thermal Desorption Tubes augment the real-time operating data and provide a comprehensive view of the dynamic process chemistries.

Test Objectives

The primary objective of these PDU test runs were to establish and verify the performance of the integrated TRI-EFT system in converting NewPage Corporation and Flambeau River Biofuels LLC biomass feedstocks to liquids. This was accomplished through parametric testing and continuous operation of the plant over many weeks, measuring the parameters described in the test plan. The specific objectives were to confirm the following:

- syngas composition and yield
- maintainability of syngas composition at H_2 to CO molar ratio of about 1.9
- tar fraction or yield and its approximate composition
- endothermic heat of reaction
- speciation of sulfur, chlorine, and nitrogen in the syngas and the split between the gas and solid phases
- syngas contaminant (tars, particulate, HCl, H_2S, NH_3, COS, HCN) levels upstream of the syngas compressor and evaluate syngas cleanliness for gas compression
- tars, particulate, HCl, NH_3, and H_2S capture removal efficiencies
- effluent (wastewater from gas cleanup and F-T) generation rate and its characteristics

- syngas cleanliness upstream of the F-T reactor and compare it with the targets for contaminant (tars, particulate, HCl, H_2S, NH_3, COS, HCN) levels
- CO conversion rate in the Fischer-Tropsch reactor
- F-T liquids yield and the fraction of heavy and medium hydrocarbons
- catalyst activity and selectivity
- catalyst deactivation rate
- subsystem mass and energy balances

Test Conditions and Results

All of the biomass provided for the trial activities was procured from the same general wood basket area envisioned to supply the commercial plants, and from which long-established biomass suppliers already successfully procure in a fully sustainable manner over millions of green tons of material annually. Several hundred tons of material were procured, prepared (sized and dried) and transported to the Durham test facility in Super Sacks averaging in weight from about 250 – 325 kg each. The biomass nominally was sized to 12 mm cross section by 12 mm to 37 mm long and dried to about 10% moisture by weight. The bulk density was on the order of 165 kg/m^3 or 10.5 lb/ft^3.

Multiple test campaigns were conducted at different operating modes – gasifier; gasifier/primary gas cleanup; and gasifier/primary and secondary gas cleanup/F-T or integrated system. Parametric and continuous tests were carried out with biomass feed rates ranging from 60 to 180 kg/h. During the integrated mode trial period, biomass was fed into the gasifier at rates of between 145 and 160 kg/h, or approximately 4 tons per day. In one campaign, the average elemental composition of the dry biomass as determined by ultimate analysis was: Hydrogen 45.53%, Carbon 52.18%, Nitrogen 1.09%, Sulfur 0.01%, Oxygen 40.48% and Ash 0.71%. The average HHV was 19.85 MJ/kg dry (8,532 Btu/lb dry).

The PDU was run with N_2 co-fluidization to properly simulate both fluidization dynamics and steam-to-carbon ratio in the commercial system; besides, the instrumentation purges in the PDU constituted a high proportion of the syngas flow rate. Consequently, the PDU syngas ended up with a relatively high proportion of N_2. Sample syngas composition produced during one of the integrated trials corresponded on average to: Hydrogen 28.8%, Carbon Monoxide 15.2%, Carbon Dioxide 17.2%, Methane 7%, Nitrogen 30.4%, Ethylene 0.9% and Ethane 0.5%. But for the N_2 dilution, the syngas composition is representative of the gasification process and therefore provides high-quality, scalable data and meets specifications for conversion into liquid hydrocarbon products. In full-scale commercial systems which will be hundreds of times larger than the pilot plant, nitrogen will be1.5-2% by volume of the syngas. With this adjustment, the commercial syngas composition is estimated to correspond to: Hydrogen 40.6%, Carbon Monoxide 21.4%, Carbon Dioxide 24.3%, Methane 9.9%, Nitrogen 1.8%, Ethylene 1.3% and Ethane 0.7%. The calorific value or HHV of this syngas is approximately 12.5 MJ/m^3 dry (335 BTU/dscf). The hydrocarbon content is slightly higher than anticipated and additional process modifications are being implemented to further improve the H_2 and CO content. Across all the test campaigns, the syngas produced had a H_2 to CO molar ratio in the 2.8 to 1.7 range, which shows the ability of the system to target H_2 to CO ratio across a broad spectrum. During the integrated mode trial, the H_2 to CO molar ratio remained stable and averaged 1.9 in a narrow range.

The overall carbon conversion efficiency based upon syngas composition and solids catch analysis varied somewhat across all the trials and approached a high of 98%. Measured and model calculated endothermic heat of reaction for steam reforming exhibited good agreement. Component mass balance closure was also satisfactory (generally within +/-10%).

Tar samples were collected as per the International Energy Agency (IEA) Tar Sampling Protocol (1) for sampling and analysis so that the results are comparable to published reports and literature. Tars characterizations were analyzed by GC/MS. Tars yield were determined by gravimetric analysis. Gasifier tar is defined by the International Energy Agency (IEA) as the generic term for all organic compounds present in the gasification product gas excluding gaseous hydrocarbons (C1 – C6) with a molecular weight less than or equal to benzene, benzene is not considered a tar. Literature supports further definition and suggests that tars be categorized as primary or condensable with single ring aromatics being defined as primary tars and poly-aromatic hydrocarbons (PAH) being defined as condensable tars (2). This distinction is necessary because primary tars often have lower boiling points and remain gaseous, while condensable tars have higher boiling points and can form viscous liquids that may have detrimental effects on downstream processes. The tar produced by this system is approximately 82% primary and 18% condensable. This corresponds to approximately 970 ppm or 3.9 g/m^3 of primary tars and approximately 150 ppm or 0.85 g/m^3 of condensable tars. When correlated with the biomass feed rates, the total tar production is equivalent to around 0.013 kg tar per kg dry biomass. The actual tar levels produced by this system are among the lower levels reported in the literature for similar systems (2-4).

The primary contaminants in the syngas exiting the gasifier were: Carbonyl Sulfide (COS), Cyanide (HCN), Hydrochloric Acid (HCl), Ammonia (NH_3), Naphthalene, Benzene, Toluene, Hydrogen Sulfide (H_2S) and particulates. Due to the novelty of the process, several of the analytical methods had to be refined to minimize interference and enhance accuracy and repeatability of measurements. The high moisture and ammonia content of the pre-cleaned syngas made measurement of H_2S extremely difficult. This is due to the causticity of the ammonia and the dynamic solubility of H_2S and ammonia in water, a relationship that does not affect COS. More robust measurement and analysis methods including OSHA 1008 method were found and implemented to determine H_2S concentration.

NH_3, HCN, HCl, H_2S and COS concentrations in the syngas upstream of the F-T reactor were all below detection limits i.e. < 100 ppb. FTIR measurements indicated naphthalene concentration to be less than 1 ppm downstream of the secondary gas cleanup system.

The wastewater produced did not contain hazardous compounds. Extremely small amounts of solid waste are produced, and such non-hazardous waste is made up primarily of ash as well as some carbon. The Chemical Oxygen Demand (COD) was measured at between 280 and 500 mg/L for all high volume wastewater streams. The Total Suspended Solids (TSS) ranged from 1 to 58 mg/L for all wastewater streams.

The TRI PDU has been operated in integrated mode utilizing 3 separate batches of Fischer Tropsch catalyst prepared by EFT. The F-T system has operated at 60+% CO conversion with the PDU quality N_2-diluted syngas and all of the results from the PDU are in line with the results obtained in laboratory scale reactors at EFT (with ultra clean syngas) and with model predictions. Extended life tests are being conducted in EFT's laboratory scale test unit on both TRI PDU Fischer Tropsch catalyst batches 2 and 3. The life test provides a comparison between catalyst performance using bottled (ultra clean) and biomass derived syngas. The laboratory performance testing also illustrates the longer term transients associated with catalyst activity, methane selectivity and deactivation rates. The design productivity (circa 170-190) has been exceeded through the 2000+ hour run and negligible temperature increases (<3 C) have been required. The deactivation rate is initially very high and decreases to a low <0.2%/day level in approximately 50 days. The rate of deactivation combined with the catalyst activity leads to a catalyst life in excess of 1 year. Similar results have been obtained with the Batch 3 catalyst. This catalyst (preferred formulation) has improved performance with respect to the line out period (20 days), methane selectivity and catalyst activity. The EFT laboratory received a series of wax samples generated during the operation of the TRI/EFT PDU. The samples were individually weighed and categorized by visual inspection based on whiteness of the wax. All of the samples appearing as "white" were composited and then the composite was analyzed by high temperature GC to determine the alpha value and the general hydrocarbon distribution of the material. The alpha value of the wax was 0.92. The medium fraction liquids exhibited properties consistent with ASTM D-975 specifications; for example, the flash point was 64 C and distillation temperature was 296 C.

The F-T system operated satisfactorily with biomass-derived syngas from 100% woody biomass feedstock and successfully produced F-T liquids over an extended period (see Figure 2).

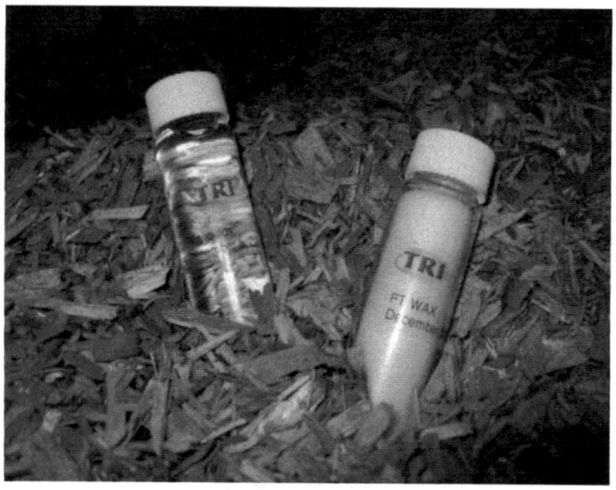

Figure 2. Sample Fischer-Tropsch liquids (wax and diesel) from biomass

Conclusions

TRI has completed ~1,300 hours of fully integrated woody biomass trials at its Durham PDU. The PDU performance over that period has been very encouraging and supportive of the commercial system design and achieved the following milestones:

- Stable gasifier operation from 40 to 120% of rated or design capacity
- Achieved and maintained the target H_2 to CO molar ratio of 1.8 – 2.0
- Very low tar formation (condensable or multi-ring aromatic tars <150 ppm and total tars of ~1.3 by weight of dry biomass)

- Good primary and secondary gas cleanup performance (inorganic contaminants all less than <100 ppb at F-T inlet)
- Good single stage CO conversion in F-T (60+ %) with PDU quality N_2-diluted syngas
- Good quality wax and medium hydrocarbon liquids

More diverse feedstock trials including energy crops and refuse derived fuel (RDF) are planned.

References

1. http://www.tarweb.net/
2. Milne, T. A., Abatzoglou, N. and Evans, R. J. (1998), "Biomass Gasifier 'Tars': Their Nature, Formation and Conversion," NREL Technical Report (NREL/TP-570-25357).
3. Ciferno, J.P. and Marano, J.J. (2002),"Benchmarking Biomass Gasification Technologies for Fuels, Chemicals and Hydrogen Production," E^2S report prepared for US DOE National Energy Technology Laboratory.
4. Kiel, J.H.A. et al. (2004), "Primary measures to reduce tar formation in fluidized-bed biomass gasifiers," Final report SDE project P1999-012, ECN-C-04-014.

Acknowledgement

The authors are thankful to the US DOE National renewable Energy Lab., NewPage Corporation and Flambeau River Biofuels for partial funding of these trials.

(89b) Improvements in Steam Pretreatment of Lignocellulosic Biomass, for Bioethanol Production, Using New/Combination of Catalysts

Sanam Monavari, Mats Galbe and Guido Zacchi, Chemical Engineering, Lund University, Lund, Sweden

Reducing the production costs of ethanol from lignocellulosic material is essential in enabling its commercialization. A process based on enzymatic hydrolysis and fermentation is today regarded as the most promising alternative in converting the biomass to ethanol with high yields and low production cost. However, to reach high yields and high ethanol concentration, cellulosic biomass must be pretreated. This is a crucial step to increase the enzymatic digestibility of biomass and has a key role in the ethanol production cost. Dilute acid pretreatment, using SO2, has been shown to be a leading pretreatment process that is currently under commercial development for conversion of forest and agricultural residues to ethanol. Increased recovery of hemicellulose sugars, improved enzymatic hydrolysis of the solid residue and higher ethanol yields are some certain advantages of the dilute-acid pretreatments. However, handling high acidity of the pretreated material, difficulty in recovery/ degradation of the catalyst, corrosion of the reaction vessels, and formation of the inhibitory compounds to fermentation organisms are some of the drawbacks of the dilute acid processes. Utilization of an organic-acid as the impregnating catalyst instead, would results in a waste stream that is biodegradable and could be used for production of biogas by anaerobic digestion. The latter could also be used to improve the storage of biomass, like ensiling. This study deals with the influence of utilization of organic-acids, as impregnating and preservative catalyst, on the sugar yield from the pretreatment.

See more of this Session: Developments in Biobased Alternative Fuels I
See more of this Group/Topical: Sustainable Engineering Forum

(89c) Rare Earth Metal Triflate Co-Catalyzed Low Severity Dilute Acid Pretreatment of Lignocellulosic Biomass

John Degenstein[1], Melvin P. Tucker[2] and Yun Ji[1], (1)Chemical Engineering, University of North Dakota, Grand Forks, ND, (2)National Bioenergy Center, National Renewable Energy Laboratory, Golden, CO

Dilute acid pretreatment solubilizes significant amounts of hemicellulose; however costs associated with pretreatment are high. Reducing these costs by lowering the severity of pretreatment can make the process more economical, however decreased yields result. Increasing yields in low severity dilute-acid pretreatments is desirable and have been shown by incorporating Lewis acids. We will compare low severity dilute-acid pretreatments incorporating rare earth metal triflate (trifluoromethanesulfonate) Lewis acid co-catalysts with controls. A 300mL steam jacketed reactor vessel with fast heating and cooling times was used for pretreatments. The kinetics of reactions will be determined for lanthanide series triflate co-catalyst pretreatments and compared to controls. In this study, both rare earth metal triflate co-catalyzed dilute acid pretreatments and dilute acid pretreatment controls will be performed at temperatures between 150 oC and 190oC. Rare earth metal triflates (i.e. La(OTf)3) at loadings between 0.5 mM to 10mM will be incorporated with low concentrations of sulfuric acid prior to pretreatment. The yields in enzymatic hydrolysis following co-catalyst pretreatment will be compared with controls.

See more of this Session: Developments in Biobased Alternative Fuels I
See more of this Group/Topical: Sustainable Engineering Forum

(89d) Low Cost Medium for Starch Particle Fermentation

Gordon A. Hill[1], Devin Bear[1], Nancy Bawa[1], Catherine Niu[1] and William J. Roesler[2], (1)Chemical Engineering, University of Saskatchewan, Saskatoon, SK, Canada, (2)Biochemistry, University of Saskatchewan, Saskatoon, SK, Canada

The direct conversion of starch particles to ethanol was developed using barley α-amylase to produce glucose and *Saccharomyces cerevisiae* as the fermentation microorganism. This technology removes the need to cook starch and thus reduces energy costs for the bioethanol industry, so is now used commercially in the USA. However, one serious drawback is the requirement to include high quantities of expensive enzymes to break down the insoluble starch particles. A low cost medium has been developed in this work and shown to produce superior yields of ethanol, 0.51 g ethanol / g starch, compared to expensive medium, 0.46 g ethanol / g starch. The new medium reduces the simultaneous hydrolysis and fermentation medium costs by 80% compared to expensive medium and requires no addition of costly enzymes. In continuous bioreactor runs using the inexpensive medium, steady state conversions of starch to ethanol were achieved at dilution rates up to 0.25 h^{-1} with complete utilization of the starch and high yields of ethanol.

See more of this Session: Developments in Biobased Alternative Fuels I
See more of this Group/Topical: Sustainable Engineering Forum

Production of Carboxylic acids from Acidogenic Fermentation of Algefiber® (Sea Weed Sludge) using a Mixed Culture of Marine Microorganisms

*Sampath A. Karunarathne, M. Clayton Wheeler and G. Peter van Walsum**

Department of Chemical Biological Engineering, Forest Bioproducts Research Institute

University of Maine, Orono, Maine 04496

Abstract

Seaweed (Macro algae) has an increasing interest as a biomass source for biofuel production. In this study, carboxylic acids were produced from seaweed derived biomass using the MixAlco™ process. Algefiber®, derived from red seaweed (*Eucheuma spinosum*) from seaweed processing (FMC Biopolymer), was used as the biomass source. This alkaline pretreated waste biomass was anaerobically digested using a mixed culture of marine-derived microorganisms. The fermentation was carried out at 35^{0}C (mesophilic) and 55^{0}C (thermophilic) in bench scale batch fermenters. The acidogenic digestion and suppression of methanogenesis were achieved by adding small doses of iodoform. Calcium carbonate and ammonium bicarbonate were used as buffers to maintain the pH of the fermentation. The produced carboxylic acids ranged from 2 to 7 carbons. The total carboxylic acid production was found to be higher at lower temperature. Acetic acid was the prominent acid at both temperatures but the proportion of acetic acid in the mixture was greater at thermophilic temperature. Produced carboxylic acids, neutralized with above buffers to form carboxylate salts, can be chemically upgraded to higher alcohols liquid biofuel.

Keywords: Acidogenic, Biofuels, Carboxylic acid, Mixed Culture

Introduction

Conversion of biomass to energy through anaerobic digestion is a natural process that has been used for decades to produce energy. Even though anaerobic digestion focused on methane production in most cases, MixAlco™ process (Hotlzapple et al., 1999) produces carboxylic acids through anaerobic digestion of any biodegradable material (e.g., municipal solid waste, industrial biosludge, sewage sludge, manure, agricultural residues, and energy crops).

* *Corresponding author*: pvanwalsum@umche.maine.edu

Alkaline pretreatment is the first step of MixAlco™ process and it increases the digestibility of biomass (Aiello-Mazzarri and Holtzapple, 2006). Carboxylic acids are intermediate products of anaerobic digestion process which are produced during acidogenic phase (Agbogbo and Holtzapple, 2006). The produced carboxylic acids are neutralized by calcium carbonate to form carboxylate salts, which can be chemically upgraded to mixed alcohols or valuable chemicals. Use of buffers also facilitates a favorable pH for microbial growth. Suppression of methanogenesis is achieved by adding small doses of iodoform to prevent production of methane. Mixed culture fermentation in MixAlco™ is a non-sterile low cost process which does not require enzymes or aseptic conditions (Agbogbo and Holtzapple, 2006). A salt tolerant, mixed culture of microorganisms collected from marine sediments is used as the inoculum for fermentation.

Algefiber® is a waste biomass from seaweed processing (FMC Biopolymer, Rockland, ME). Different types of red seaweeds are processed to extract carrageenan, leaving algefiber® as a waste product which has some organic materials and minerals. Currently it is deposited in landfills and some portion is used as an organic fertilizer and soil amendment in agriculture. However, use of this seaweed-derived biomass has a great potential for biofuel production for two reasons. First, the seaweed processing involves alkaline treatment, therefore the algefiber® can be readily used for fermentation excluding alkaline pretreatment. Secondly, it provides an opportunity to produce bio fuels without interfering with forests or terrestrial food production. This study is focused on investigating the potential of algefiber® to produce carboxylic acids through mixed-culture acidogenic fermentation as a platform for value added chemicals and biofuels.

Materials and Method

Anaerobic batch fermentation was carried out in roller bottles. The roller bottle was a 1L Nalgene centrifuge bottle fitted with an 11 size rubber stopper with 18 mm glass tube inserted through it. The glass tube was capped with a rubber septum and sealed with an aluminum crimp seal. Two pieces of 0.25" stainless steel rods were inserted in to holes of the stopper to ensure the mixing of solids and also to serve as handles for pulling the stopper out of the bottle. The roller bottles were placed in a Wheaton compact roller system located in a temperature (35 ^{0}C or 55 ^{0}C) controlled incubator and were rotated horizontally at 1 rpm.

Algefiber® derived from red seaweed *Eucheuma spinosum* was used as the substrate for fermentation. Approximately 400 mL of 100 g/L substrate broth was inoculated with marine sediments collected from Rockland beach, ME. $CaCO_3$ and NH_4HCO_3 were added as neutralizing agent to control the pH between 6 and 7. Methane inhibition was achieved by adding 1- 4 mg/L of iodoform (CHI_3) from iodoform solution containing 20 g CHI_3/L ethanol.

Total solids (TS) of the algefiber® was determined by drying at 105 ^{0}C and volatile solid (VS) content was determined by ashing at 575 ^{0}C according to NREL Laboratory Analytical Procedures. Carbohydrate content was measured by High Performance Liquid Chromatography (Shimadzu) with Refractive Index Detector and

UV detector, using both the BioRad Aminex HPX-87H column according to NREL Laboratory Analytical Procedure. Carboxylic acids were measured using a Gas Chromatograph (Shimadzu model GC-2010) with Stabilwax DA column and FID detector with helium as a carrier gas. Liquid samples were prepared by mixing 1 mL of the supernatant from the centrifuged liquid sample with 1 mL of an internal standard (1.162 g/L of 10-mM 4-methyl-n-valeric acid) and 1 mL of 3M H_3PO_4. Matreya® volatile acid mix solution (Catalogue no. 1075) was used as the external standard.

Results and Discussion

Algefiber® used in this experiment had a 0.35 g of volatile solid/1 g total solid, but its ash content was significantly high (65% from total solid). This high mineral content could be due to alkaline treatment and extraction of organic fraction during processing. Despite of this low organic content of substrate, acidogenic digestion shows interesting results.

Total carboxylic acid production during the acidogenic fermentation is given in Figure 1. Fermentation at lower temperature gave a higher carboxylic acid concentration than higher temperature for both $CaCO_3$ and NH_4HCO_3 buffered systems. Use of two buffers had a considerable effect on carboxylic acid production. NH_4HCO_3 buffer gave a comparatively higher acid production than $CaCO_3$ (figure 1.).

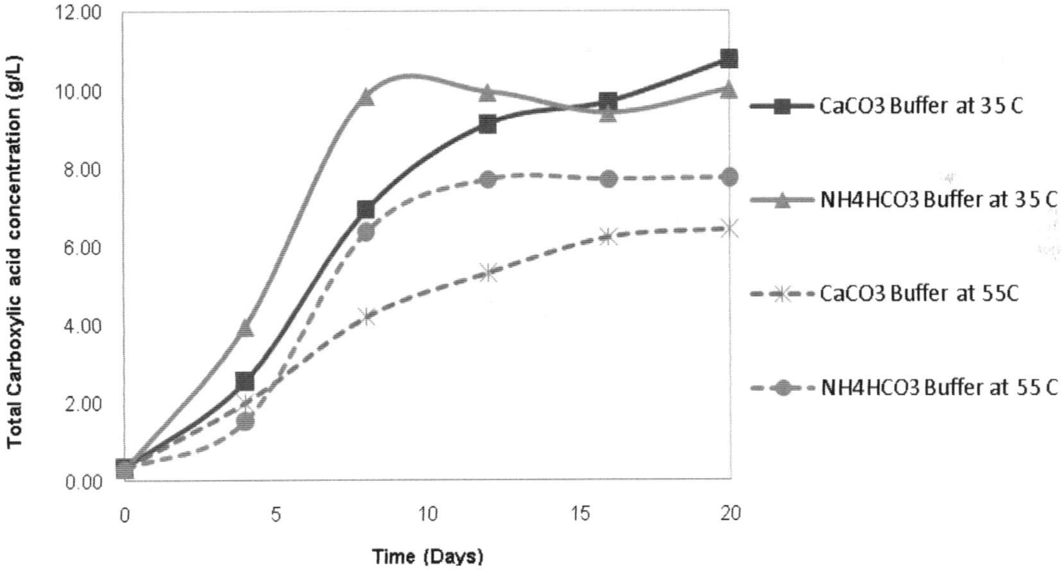

Figure 1. Total carboxylic acid concentration of 100 g/L algefiber® fermented at 35 ^{0}C and 55 ^{0}C using $CaCO_3$ and HN_4HCO_3 buffers

Produced carboxylic acids ranged from acetic (C2) to heptanoic (C7) and acetic acid is the prominent acid in all conditions. Figure 2 shows the effect of temperature on carboxylic acid composition in final carboxylic acid mixture. Percentage of acetic acid is higher at high temperature even though the total carboxylic acid concentration is lower.

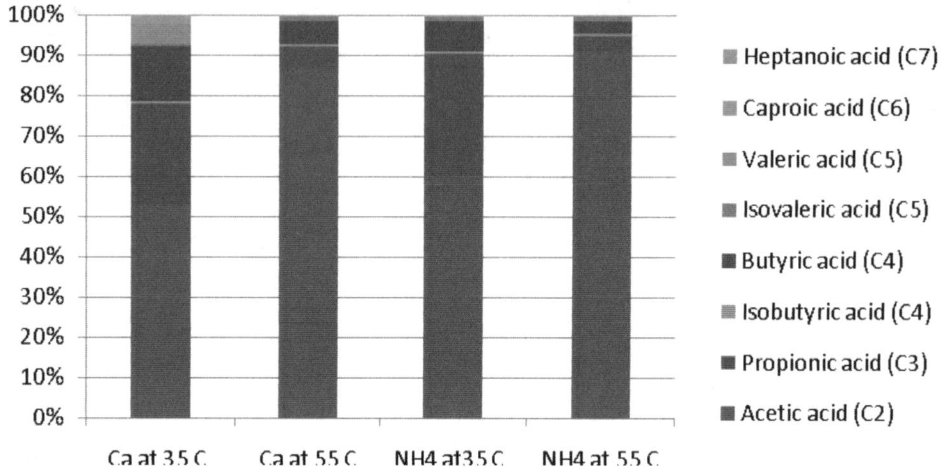

Figure 2. Composition of carboxylic acid produced at 35 ^{0}C and 55 ^{0}C using $CaCO_3$ and HN_4HCO_3 buffers

Conclusion

Production of C2 to C7 carboxylic acids from acidogenic digestion demonstrates the ability of algefiber as organic biomass source for MixAlco™ process. Acetic acid is the prominent acid produced in the acidogenic fermentation. Lower temperature favors higher percentage of longer chain acids and produce higher concentration of carboxylic acid than higher temperature.

Acknowledgements

The authors would like to acknowledge support from FMC Biopolymer Rockland, ME and U.S. Department of Energy (grant #: DE-FG36-08GO18165).

References

Agbogbo FK, Holtzapple MT. (2006). Fermentation of Rice Straw/Chicken Manure to Carboxylic Acids Using a Mixed Culture of Marine Mesophilic Micoorganisms. *Applied Biochemistry and Biotechnology* 129-132:997-1014.

Aiello-Mazzarri C., Agbogbo, F.K., & Holtzapple, M.T., 2006. Conversion of Municipal Solid Waste to Carboxylic Acids Using A Mixed Culture of Mesophilic Microorganisms. *Bioresource Technoogy.* 97, 47-56.

Holtzapple, M.T., Davison, R.R., Ross M.K., Aldrett-Lee, S., Nagwani, M., Lee, C.M., Lee, C., Adelson, S., Kaar, W., Gaskin, D., Shirage, H., Chang, N.S., Chang, V.S., and Loescher, M.E., (1999) Biomass Conversion to Mixed Alcohol Fuels Using the MixAlco Process. *Applied Biochemistry and Biotechnology.* 77–79, 609-631.

(89f) Wet Biomass to Gasoline: An Integrated Hybrid Process of Anaerobic Digestion and Bromine Mediated Biogas Conversion to Hydrocarbon Fuel

Eric W. McFarland[1], Jeffrey H. Sherman[2], Daniel J. Auerbach[2], Sagar B. Gadewar[3] and Vivek Julka[3], (1)Chemical Engineering, University of California, Santa Barbara, Santa Barbara, CA, (2)GRT Inc., Santa Barbara, CA, (3)GRT, Inc., Santa Barbara, CA

Anaerobic digesters convert a variety of biomass feeds into biogas containing methane and carbon dioxide. It is costly to upgrade the product gas for pipeline applications and combustion is relatively inefficient. Using bromine to activate the methane in the presence of carbon dioxide allows subsequent chemical conversion of the methyl bromide without the need of significant processing of the biogas. The overall methane conversion occurs in a three basic steps, i) bromination under bromine limiting concentrations, ii) reaction of methylbromide to form hydrocarbon products, and iii) recovery of the bromine with a catalyst/metal oxide solid reactant. The process does not utilize a synthesis gas intermediate and is tolerant to common methane contaminants including the high concentrations of carbon dioxide present in biomethane. Process optimization and reaction engineering allows us to use relatively non-selective thermal bromination at modest temperatures ~ 400 C for methane activation with conversions in excess of 75%. Methyl-Br has many reactivity features in common with Methyl-OH and thus the wide variety of products available with methanol feeds (e.g. MTO, MTP, MTG) can also be made with methyl bromide and the hydrocarbon products include aromatics suitable for bio-jet fuel. The high molecular weight of the alkylbromides provides a chemical handle for ease of separation. Use of adiabatic reactors and other process options allow minimization of the requirements for costly unit operations with expensive metallurgy though proper choice of the materials of construction is critical. A potential feature to reduce the process cost is the use of a solid-reactant for recovery of the bromine that minimizes bromine or hydrobromic acid inventories and facilitates efficient recovery of the halogen. Bromine has fundamental advantages over chlorine and advanced reactor designs utilizing moving or switched bed zone reactors offer further process cost reduction options. Preliminary engineering-economic models and pilot plant operation show the process to have particular promise as a commercial route to production of hydrocarbon transportation fuels from biomethane derived from byproducts of existing commercial processes and marginal land crops.

See more of this Session: Developments in Biobased Alternative Fuels I
See more of this Group/Topical: Sustainable Engineering Forum

(117a) Opportunities at the Energy-Water Nexus: Leveraging Water Treatment Technologies for Efficiency, Economic Prosperity and Sustainability In the 21st Century

David Klanecky, Dow Water & Process Solutions, Edina, MN

Water treatment is an essential part of creating sustainable potable water supplies and enabling industry and economic development around the world, and demand for water is increasing every year. By 2030, power and industrial needs will constitute 22 percent of global water demand, and forty percent of this growth will be driven by China alone. Technologies such as reverse osmosis, nanofiltration and ultrafiltration are allowing engineers to maximize the quality and quantity of available clean water while helping lower associated capital and operational costs. Desalination, contaminant removal and water reuse applications are continuously becoming more energy-efficient through lower feed pressures, greater active membrane areas and smaller system footprints. In addition to saving energy, water treatment and processing technologies are helping to generate cleaner, more reliable forms of energy through biofuels processing and nuclear applications. Innovations in research and development are quickly evolving water treatment options and helping to change the way we approach water and energy – today and into the future.

See more of this Session: Plenary: Energy and Water Sustainability for a Smart Planet
See more of this Group/Topical: Liaison Functions

(117b) Water and Energy: Integrated Challenges, Integrated Solutions

Jared Ciferno, U.S. Department of Energy, National Energy Technology Laboratory, Pittsburgh, PA

Water, once considered a nearly inexhaustible resource, is becoming increasingly limited, and water requirements for electricity production must compete with other demands, such as agriculture and sanitation. The Department of Energy's (DOE) Fossil Energy Program is conducting a considerable R&D effort directed at the nexus between energy and freshwater resources. Much of this work is being conducted through DOE's National Energy Technology Laboratory. Research areas address improved freshwater management for thermoelectric power plants; reducing water impacts associated with carbon capture and sequestration; and reducing water impacts associated with natural gas and oil production.

Thermoelectric power plants (including coal, oil, natural gas, and nuclear), generate about 90% of the electricity in the United States, and require large quantities of cooling water, a resource that is limited in parts of the nation and throughout the world. Although commercially available cooling technology options can reduce or mitigate water consumption, they result in added cost, increased complexity, and reduce the amount of power available from the plant. DOE is developing new technologies that will reduce the cost and complexity of these systems under its Existing Plants R&D program.

Efforts to capture 90 percent of carbon emissions by using current near-commercial carbon capture and storage (CCS) technologies on coal-based power plants could more than double the amount of water consumed per unit of electricity generated. A key objective of DOE R&D activities is to mitigate the potential impact of CO_2 capture on water resources. This is being addressed under two key inter-related components of its R&D Program – (1) the development of advanced CO_2 capture technologies that require less freshwater and (2) the development of advanced water management technologies that reduce freshwater usage throughout the power plant.

DOE is also addressing water-related issues associated with natural gas and oil production such as produced water and its effects on the environment, treatment of process waters, and the availability of water in arid lands. Produced water is the largest waste stream generated by the oil and natural gas industry. DOE's Oil and Natural Gas Program strategies related to water resources include: enabling sustainable oil/gas water management options that support regional and basin-oriented oil and gas development strategies; developing technologies to improve environmental performance and reduce costs; developing scientific and technical information and decision making tools for risk-based regulation and compliance; and increasing the understanding of environmental benefits and risks.

See more of this Session: Plenary: Energy and Water Sustainability for a Smart Planet
See more of this Group/Topical: Liaison Functions

(117c) Energy, Water, Sustainability for a Smarter Planet

Spike Narayan, IBM, San Jose, CA

Over the last few years, our eyes have been opened to global climate change, and to the environmental and geopolitical issues surrounding sustainability. As an example, by 2015, more than half of the regions in the world will experience some level of stress in water supply and management. Similar issues are surfacing for energy as well. IBM's smarter planet initiative aims to address important sustainability issues by developing solutions that give us the ability to manage our resources through effective sensing, data gathering and analytics to allow predictive real time action. I shall discuss the research activities at IBM to support a smarter planet and describe technologies for intelligent systems, smart energy, water, and healthcare.

See more of this Session: Plenary: Energy and Water Sustainability for a Smart Planet
See more of this Group/Topical: Liaison Functions

(118a) Project Management in a Matrix Organization

Frank van Lier, The Lubrizol Corporation, Wickliffe, OH

Successful project management is a critical aspect for organizational success. It becomes even more challenging when the project manager and team members are part of a matrix organization. The ability to gain buy-in on goals and objectives, access the needed people and resources, and carry out projects requires deft people skills. Benefits can include higher quality outcomes and increased implementation speeds. This presentation will cover one industry veteran's perspective on how to accomplish project objectives in a highly matrixed organization.

See more of this Session: Project Management: What's New? What Works?
See more of this Group/Topical: Management Division

(118b) Environmental Law 101 Module

Chris A. Paul, Mcafee & Taft, Tulsa, OK

This tutorial, Environmental Law 101 Module, provides a summary of legal issues for engineering professionals and organizations most relevant for new project managers. An extended abstract will be submitted prior to the conference.

See more of this Session: Project Management: What's New? What Works?
See more of this Group/Topical: Management Division

Planning, Managing, and Best Practices for Project Procurement

Dr. Eldon R. Larsen, Marshall University Graduate College, South Charleston, WV

Paul Jason Dean, Kentucky Transportation, Ashland, KY

Abstract

Project procurement is all about acquiring materials, services, equipment, supplies, and anything else outside of the project team that is needed to complete a project satisfactorily. Procurement of goods and services can be one of the costliest portions of a project's budget. Procurement can also be a significant source of project risk and schedule delay. Yet, it is often underestimated in importance to the project. In this paper we discuss various aspects of project procurement, including contract management, and present some best practices.

Procurement Planning

Before a project team starts into actual project procurement, it is important to plan how procurement will take place, what procedures will be followed, what documentation will be needed, what levels of approval and signatures will be required, and how all of this affects the rest of the project planning, risks, resources, and so on. Additionally, the project team needs to decide what will be purchased, versus what the team will provide. Procurement involves both buyers and sellers. In fact, the project team could be a buyer, a seller, or both.

Often, procurement planning will include a determination of potential sellers and requirements for those sellers. Some organizations have lists of pre-qualified and/or preferred sellers to whom they specifically go for goods and services. Other organizations are required to openly solicit bids from anyone interested in providing the needed good or service. Thus, it is important to understand what policies will govern any given project in your organization.

When planning project procurements, there are many factors to consider:

- Project or product scope constraints related to performance, timing, and cost
- Health, safety, and environmental requirements
- Security concerns
- Intellectual property concerns
- Legal issues surrounding licenses, permits, regulations, safety, and employment
- Partnership arrangements or teaming agreements
- Organizational procurement policies and procedures
- Previous experience with sellers
- Lessons learned regarding procurements for previous projects
- Insurance needs
- Roles and responsibilities
- Past and present information about potential sellers

- Unique requirements for the project
- Previously established agreements and contracts
- Make-or-buy decisions
- How to manage multiple sellers, where applicable
- Important procurement lead times needed in the project plan
- Needed procurement completion dates
- The potential for needed insurance contracts and performance bonds
- Documentation that will be required of contractors by the buyer before, during, and at completion of contract work
- Seller selection criteria, including making these public when required
- Qualified sellers lists giving the names of potential sellers who have already been identified as qualified to provide a specific good or service

It is also important to look at whether it is better to make or to buy some goods needed for the project, and whether to supply internally, or to buy from outside, some needed services. Sometimes, rather than a make-or-buy decision, a project manager could also be faced with a *rent-or-buy* decision, similar to a make-or-buy decision, but where something is potentially rented rather than purchased.

Sometimes procurement planning will involve determining whether the buyer wants to use a *sole source* or a *single source*. Though these terms seem very similar, there is an important technical difference. A *sole source* is used when a specific seller is the only one who can provide exactly what the buyer needs, perhaps due to intellectual property protection like patents. In contrast, a *single source* might be used because of preferences of the buyer due to factors like negotiated long-term special pricing or special qualifications or performance history and a desire to maintain long-term relationships between the buyer and seller.

The result of procurement planning is a plan that guides how the procurement process will be managed. This plan can range from somewhat informal to very detailed and very formalized, depending upon the size and the needs of the project and according to organizational policies. Once procurement planning has occurred, it is important to go back an integrate this planning into the main plan for the project and into the work breakdown structure, network diagram, and Gantt chart as needed.

It is important in the procurement planning to clearly define roles and responsibilities for people involved in the procurement process. In larger projects it is highly recommended that an owner's technical representative be appointed with clearly defined responsibilities in monitoring project progress, quality, billing, and so forth as deemed appropriate for the project—many inexperienced project managers regret later on that they did not pay enough attention to this role.

Also, an important product of procurement planning should be a decision as to what kind of contracts, if any, will be used and how bidding on those contracts is to take place. Also, if independent estimates are to be obtained for the contract work, the procurement planning process should determine how this is to be obtained.

Because procurement often involves obtaining or providing project-needed goods and services through contracts and purchase orders, there is a need for project managers to understand contract administration and various purchasing terms and procedures.

Contracts

Contracts are legal, mutually binding agreements between buyers and sellers. When a buyer enters into a contract, he is legally bound to provide the monies agreed-upon for the sellers' products. Because contracts are legally binding, disagreements can be subject to resolution in the courts, if needed. Sometimes contracts can also take the form of *agreements, understandings,* or *purchase orders*.

Sometimes project managers decide to enter into contracts with sellers in order to share project risks with a contractor, when the project manager does not want to absorb the full risk, or when the project team is not adequately qualified or able to provide a needed good or service.

It is important to write project contracts so that they meet the specific needs of the project. Even if projects seem to be very similar, each project is still unique in some way. Thus, each contract can vary even for similar projects.

Who writes the contract—the buyer or the seller? In practice it could be either, depending upon the situation. However, if someone outside of the project team is going to write the contract, it is strongly advisable for the project manager to draft what he or she thinks should be in the contract, before seeing what the other party writes. This will help the project manager (1) to think about what is really needed and (2) not to become so focused only on what the other party writes that he forgets to include other important aspects that should be included in the contract. The party who writes the contract often will write it to be in his favor, so caution is important. Take time to get the contract right.

Contracts need to be approved and signed within an organization's policies, so the project manager needs to become familiar with those policies. Additionally, it is good practice to become very familiar with one's purchasing department and contract administrators in organizations where those will be called upon for help.

There are many types of contracts, some beyond the scope of this book. The more common, standard contract types are discussed below. Keep in mind that not all contract types are appropriate for all organizations. Organizational policies could greatly restrict what contract options are allowable. For example, in many governmental organizations there are laws that specify what types of contracts may be used.

Fixed-Price or Lump-Sum Contracts

Fixed-price or lump-sum contracts are those in which the total set price is agreed upon for a clearly defined good or service. The buyer must clearly define what good or service is desired. The seller is legally bound to provide the good or service for the fixed price. If there are mutually agreed-upon changes to the project scope, then the price can be adjusted accordingly. The simplest form of a fixed-price contract is a purchase order for a product to be provided by a certain date for a certain price. However, there are several variations of fixed-price contracts, and some of them can include incentives to meet specific project objectives. A few of these variations follow.

Firm Fixed Price Contracts (FFP). FFP contracts are one of the most commonly used project contracts. People like these contracts because the price is clearly agreed upon and understood. Any increased costs will be completely born by the seller, unless there is a mutually agreed-upon scope change, in which case the resulting price changes are born by the buyer.

Fixed Price Incentive Fee Contracts (FPIF). FPIF contracts are similar to FFP contracts, *except* that FPIF contracts allow for performance variations with incentives to meet pre-defined performance measures of success. These measures of success can and should be related to cost, schedule, and project and product performance. The final price is not known until the project is completed. There can be pre-defined price ceilings, above which the seller bears any additional cost and must still complete the contract obligations.

Fixed Price with Economic Price Adjustment (FP-EPA). FP-EPA contracts are often used when the contract spans a long period of time, like several years, which can affect the base cost of raw materials and labor costs. For example, over several years the prices for steel, concrete, and lumber can vary greatly, making it extremely difficult for a contractor to accurately estimate long-term costs of building materials. Therefore, FP-EPA contracts provide for pre-defined final price adjustments due to changes in factors like material costs, inflation, or regulatory changes. These contracts should use reliable financial indices and should be used to protect both the buyer and the seller.

Cost Reimbursable Contracts

Cost reimbursable contracts reimburse the seller for his costs, plus a fee to allow the seller to make a profit. The sellers costs often include both direct and indirect or overhead costs, and the indirect costs are usually but not always some percentage of the direct costs. Additionally, there can be incentives to meet specific project objectives. There are several common variations in cost reimbursable contracts, some of which follow.

Cost Plus Fixed Fee Contracts (CPFF), also called Cost Plus Percentage of Cost Contracts (CPPC). CPFF and CPPC contracts are used when the buyer and seller agree that the seller is to be reimbursed for allowable costs, which are not fixed in advance, plus the seller is to receive a fee based upon a percentage of the estimated cost when the contract is initially signed, which might not be the final actual cost. The fee is paid at the end of the job and does not normally depend on other factors like contractor performance. If the project scope changes through an approved change control process, then the fee could change.

Cost Plus Incentive Fee Contracts (CPIF). In some ways the CPIF contract is similar to the CPFF contract, except that for the CPIF contract the contractor or seller receives a fee called an incentive bonus, which bonus is based upon meeting certain specified performance criteria in the contract. The performance criteria could involve cost metrics, schedule metrics, and product and project scope quality and performance metrics. For these contracts, if the seller is able to provide the required product or service for a final actual cost that is less than originally estimated when the contract was signed, then there is often some provision for both the seller and the buy to share in the savings, and thus there is an incentive for the seller to pay careful attention to costs. There could also be monetary incentives for the seller if the contract is completed early, but not always. Every project's contract will be different in what incentives are specified.

Cost Plus Award Fee Contracts (CPAF). CPAF contracts are very similar to CPIF contracts. The buyer reimburses the seller for allowable costs. The seller's fee is based upon meeting agreed-upon subjective measures of performance specified in the contract, and is usually not subject to appeal.

Time and Materials (T & M) Contracts

Time and materials contracts tend to be a kind of hybrid between fixed priced contracts and cost reimbursable contracts. Similar to cost reimbursable contracts, time and materials contracts are open-ended in that the actual allowable costs to be reimbursed could be more or less than the original estimated cost. Sometimes these contracts specify fixed-prices per unit of a good or service that will be reimbursed. Often there are *not-to-exceed* limits on how much will be reimbursed.

Common Procurement Documents

There are a number of important documents frequently involved in procurement activities. Some of these documents have several different names, depending in part on the industry or organization using them. Depending on the needs of the project, the procurement documents should provide information in a manner that provides pertinent information and helps achieve the procurement objectives as easily and effectively as possible. Standardized formats, checklists, clear statements of work, clear expectations, complete legal provisions and constraints are all useful and important. Sometimes public disclosure and advertising are either required or recommended, in order to be fair to all participants. Generally sellers should follow buyers' document guidelines. Below, some of the more common documents are discussed.

Request for Quotation (RFQ). Requests for quotations often are used to ask for small-dollar amounts or off-the-shelf commodities not needing much or any customization. Usually RFQs do not require negotiation, unless it is minimal.

Request for Information (RFI). Requests for information ask for information about a potential seller that would help the buyer to understand that seller's qualifications to do the work needed in a procurement.

Request for Proposal (RFP). Requests for proposals are sometimes called *solicitations* or a *request for tender* (RFT). The seller's response would be a *proposal.* RFPs generally ask for proposals for customized goods or services which are sometimes high-priced or extensive in nature. The seller's proposal should include relatively detailed information about the approach the seller would use to provide the needed good or service. RFPs can result in negotiations with respect to schedule, costs, and performance or scope or the work to be accomplished.

Invitation for Bid (IFB). Invitations for bids or IFB are sometimes called *requests for bids* or RFB. These documents are similar to RFPs, but the IFB terminology is more common for government projects' sealed bidding.

Letter of Intent. A letter of intent is simply a document prepared by the buyer stating the intent of the buyer to purchase a good or service from a specific seller.

Procurement can involve many other documents. The project manager, buyers, and potential sellers should all become familiar with the specific documents and procedures required by the particular buyer's organization.

Bidder Conferences

For large project needs, where multiple potential sellers will be bidding on work to be performed for a buyer, the buy might hold a bidder conference. Bidder conferences might also be called contractor conferences, vendor conferences, or pre-bid conferences. These meetings occur will all the bidders interested in bidding, generally at the same time and location, *prior to* any actual bidding.

These conferences are used for the buyer to present pertinent information to all the interested potential bidders. The information would include clarification on the specific requirements, expectations, and constraints of the work to be provided to the buyer. These meetings provide an opportunity for bidders to ask clarifying questions and to better understand the buyer's needs. *All* interested bidders should be given exactly the same information and opportunities to obtain specifications and clarifying information.

It is highly recommended that the buyer provide each interested bidder with a complete and detailed record of the bidder conference, including answers to questions, specifications and drawings, handouts, presentation slides, and so forth. The buyer might even consider providing each interested bidder with a videotape of the meeting. Most important is that each bidder is treated equally.

It is also recommended that the buyer provide the bidders with a clear understanding of the evaluation criteria and procedures that will be used to select the final bidder which will provide the needed work to the seller.

Independent Estimates

For potentially large, expensive contracts, buyers often asked a qualified independent agent or organization to make an independent estimate of what the contract work *should cost.* Sometimes this independent estimate is done internally by the buyer. The independent estimate is used as a kind of check on the bidders' proposals. Significant differences between the independent estimate and bidders' proposals might indicate that there is a misunderstanding of what the buyer wants, and there was not sufficient information given to the bidders. If a long period of time has elapsed between the time of the independent estimate and the bidders' proposals, inflation or other economic factors might be affecting the discrepancies.

Contract Negotiations

For many contracts, especially those where the work is extensive and expensive, once a specific seller has been selected by the buyer, they will enter into contract negotiations. The objective is to reach mutual agreement on all the details of the contract between them: what will be provided, when, the price, payment arrangements, how conflicts and disputes will be resolved, roles and responsibilities, authorities, terms and laws, rights to intellectual property, and criteria and procedures for contract closure. Additionally, and very importantly, the contract needs to make clear the procedures to follow for any changes to contract scope, schedule, or costs.

For small contracts, there might not be any contract negation at all. However, it is advisable for the project manager to check with legal counsel or procurement experts to help determine when negotiations are needed and how they are to take place.

Contract Closeout

Contract closeout is done to formally signify that the work of the contract has been accepted by the buyer and that the terms of the contract have been achieved. This procedure verifies that all contract deliverables have been delivered in an acceptable manner. Additionally, all contract documents are finalized, recorded, and archived for future reference. Contracts can also be closed out early if for some reason it is necessary to end to contract before the contract work has been completed. This obviously might result in a contract dispute and a negotiated settlement of the contract, the procedure for which should have been detailed in the contract itself.

Green Project Procurement

Green Procurement is a new and emerging trend that focuses on the environmental sustainability of a product. The focus on green products is to reduce carbon dioxide emissions, increase water efficiency, increase energy savings, improve indoor air quality and achieve an overall smaller environmental footprint.

Many government entities and companies are enacting green procurement activities with a view of minimizing their overall environmental impact, and with the ultimate goal of specifying environmental preference for a product that truly has credibility as being green. A claim that an environmental product is green is simply not justification to mandate the product into a project. A green product has to be as good as or better than the traditional alternative. When specifying products many organizations and developers initially analyze the upfront or initial cost of acquiring a product. A new way of thinking must be utilized where the overall life cycle cost and the environmental footprint of the product is considered. The planning phase of project development offers the best opportunity to investigate and specify green products. This is where alternatives can be identified and where the environmental technical standards are outlined, along with performance and functional requirements.

The U.S. Green Building Council has a program called LEED (Leadership in Energy and Environmental Design) Green Building Rating System and accelerates the global adoption of sustainable green building and development practices through the creation and implementation of universally understood and accepted tools and performance criteria.

E-procurement

E-procurement is the utilization of the internet to facilitate purchase and sale of supplies, work and services through the Internet. E-procurement is a tool that allows a user to reach many more suppliers/contractors that a project sponsor would not be able to reach in traditional publications. Government, private and trade association web sites allow the buyers and sellers of goods and services to view all available opportunities for bid invitations. E-procurement is heavily utilized in large supply chain industries or just in time manufacturing facilities.

Many agencies, suppliers and contractors utilize their own e-procurement web pages or that of an industry trade group to promote services, proposal, and opportunities in a manner that allows them to reach more users. The federal government, states, and many large companies have web pages that outline procurement opportunities, policies, procedures, rules, regulations, and frequently asked questions tabs which outline material relevant to procurement in their organizations. This is all in an effort to disseminate information to their respective audiences.

Procurement Best Practices

Before completing this chapter on procurement, let's look at some of the *best practices* in the procurement field. The following best-practices summary is based mainly on work by William Gindlesperger (2009) and by Mark Smith (2003).

- Establish common procurement definitions, and the organization's standardized contract clauses that can be adapted as needed by specific project needs
- Create and deploy an integrated contract management system with a complete record and audit trail; where possible make it web-based
- Identify needs for new contracts and sources of alternative supply quickly
- Identify risks and potential exposure, due to contracts
- Create a pre-qualified supplier pool in which all suppliers are uniformly and objectively evaluated—know the suppliers in the pool and maintain relationships with fewer, yet higher performing suppliers
- Create purchase order templates pre-populated with approved suppliers and supplier profile data
- Maintain, update and utilize software packages (internal and external) that outline average unit bid prices broken down by quantity, county, state, region and the United States, to allow for more accurate pricing and competitive bidding
- Create thorough bid specifications with full details before an invitation for bid
- Create a fair bidding environment and match specifications with capable suppliers
- Streamline procurement procedures and create efficiencies to save time and money; facilitate access to approved suppliers
- Create mutual acceptance of the process that every supplier understands
- Call for, and award bids in a timely manner
- Teach suppliers that past *performance* is important to receive contract opportunities, and then *price* is the essential to win those contracts
- Maintain and nurture positive relationships, with contractors, trade organizations, government entities, utility companies, suppliers, etc, as a means to reach out and ask questions and utilize their expertise when a problem or question arises
- Be active and participate in partnering – make it a plan for success, as this allows for solutions and answers to problems in a timely manner and may resolve many issues where future litigation may arise
- Adhere to and practice a code of ethics
- Establish transparency by sending bid results to all suppliers that submitted a bid
- Monitor every step of the workflow process—ensure clear communications, attention to details, and a deliverable end product; flag contract expiration dates and confirm contractual milestones
- Unbundle or break down large project into smaller contracts to ensure greater participation of smaller contractors

- Have an owner's technical representative systematically ensure that work performed is per the contract—establish clear accountability, and correct problems before they grow
- Complete the project with accurate invoicing

Summary

Project procurement can be a very important and costly part of a project, and a very important part of project management. The upfront time invested in planning for procurements is extremely important in order to facilitate a procurement process that flows as smoothly and effectively as possible. Project managers should seek to establish procurement relationships between the buyer and sellers where both sides are winners; this is in the interest of everyone involved.

References

Gindlesperger, Willaim, *Supply Chain Management Review*, July 8, 2009.

Smith, Mark. *Supply & Demand-Chain Executive*, October 28, 2003.

About the Authors

Dr. Eldon R. Larsen is and AIChE Fellow and is a professor of engineering at the Marshall University Graduate College in South Charleston, WV. He serves as a Director in the AIChE Management Division board. He has taught project management courses in industrial and academic settings for over 16 years.

Paul Jason Dean works for the state of Kentucky in Transportation and manages projects constantly as part of his daily work.

(128a) Plenary Presentation- Lisa Jackson (Invited)

Lisa Jackson, EPA (Invited speaker), Washington, DC

.

See more of this Session: MAC Chemical Engineering Forum
See more of this Group/Topical: Liaison Functions

(128b) The Pipeline for Future Engineers and Its Impact on US Competitiveness

Emmanuel A. Dada, ChemProcess Technologies, LLC, League City, TX, Rosemarie D. Wesson, Division of Chemical, Bioengineering, Environmental and Transport Systems, National Science Foundation, Arlington, VA, Christine Grant, Chemical and Biomolecular Engineering, North Carolina State University, Raleigh, NC, Derrick K. Rollins Sr., Chemical and Biological Engineering, Iowa State University, Ame, IA and Gilda A. Barabino, Biomedical Engineering, Georgia Institute of Technology, Atlanta, GA

.

See more of this Session: MAC Chemical Engineering Forum
See more of this Group/Topical: Liaison Functions

(128c) Questions and Answer Panel Discussion

Emmanuel A. Dada, ChemProcess Technologies, LLC, League City, TX, Rosemarie D. Wesson, Division of Chemical, Bioengineering, Environmental and Transport Systems, National Science Foundation, Arlington, VA, Christine Grant, Chemical and Biomolecular Engineering, North Carolina State University, Raleigh, NC, Derrick K. Rollins Sr., Chemical and Biological Engineering, Iowa State University, Ame, IA and Gilda A. Barabino, Biomedical Engineering, Georgia Institute of Technology, Atlanta, GA

Panel discussion consisting of Award recipients.

See more of this Session: MAC Chemical Engineering Forum
See more of this Group/Topical: Liaison Functions

(142a) BioDiesel Production From Algae - Obstacles and Challenges

Heike Fruhwirth[1], Clemens Borkenstein[1], Josef Knoblechner[1], Robert Raudner[1], Edgar Ahn[1] and Matthäus Siebenhofer[2], (1)R&D, BDI - BioDiesel International AG, Grambach/Graz, Austria, (2) Department of Chemical Engineering and Environmental Technology, Graz University of Technology, Graz, Austria

The transport sector accounts for more than 30% of the total energy consumption in the European Community (EU). It is 98% dependent on fossil fuels with the crude oil feedstock being largely imported. The transport sector is considered to be the main reason for the EU failing to meet the Kyoto targets [1]. Bio fuels provide the best option to replace a significant share of fossil fuel, Europe has defined ambitious targets for the development of bio fuels. The aim is to improve European domestic energy security, ameliorate the overall CO_2 balance and sustain European competitiveness.

Traditionally, vegetable oils – especially rapeseed oil – were used as feedstock material for the European biodiesel production. These feedstocks are limited in quantity and their use for biofuel production has been heavily debated recently ("food vs. fuel"). Alternative types of feedstock – like used cooking oil (UCO), animal fat or non-food vegetable oils increasingly gain share in the raw material spectrum. Innovative technologies are needed to produce high-quality biodiesel in an energy efficient way, from a wider range of feedstock resources. One prospective feedstock for BioDiesel production is photosynthetic microalgae. These organisms use sunlight as energy source to fix CO_2 and form oxygen and valuable biomass. The potential of algal biomass is being increasingly discussed as a promising source of biofuels which does not compete with food crops. Algae have been grown successfully on a small scale for years in the pharmaceutical industry, for health care products and more recently also for specialized aquaculture, a lot of effort is done to achieve a cost-effective cultivation on large scale for biofuel purposes. Following an extensive "waste to fuel" philosophy great efforts are taken to extend this approach into algae technology. In the present algae R&D project a new comprehensive process to use microalgae as a new source of raw materials for BioDiesel production is being developed. Microalgae are cultivated on a laboratory scale as well as a pilot plant scalewith a focus on the design of photo-bioreactors and algae harvesting facilities as well as process modeling for scale up purposes. Although proven technologies exist for esterification of difficult raw materials, the conversion of algae oil is a challenging procedure. Technologies as well as analytical standards therefore had to be modified to meet the special requirements of algal oil. The goals in the development of an algae conversion technology of 'BioDiesel from Algae' project are: • Development of analytical methods with respect to information on compatibility with engines for different biodiesel products/qualities • Improvement of algae oil pre-processing techniques • Improvement of phase separation technology • Integration of the up- and down-stream processing steps • Improvement of the economic balance per unit biomass • Compliance with the European BioDiesel Standard EN14214 Results of the present project status will be presented and compared with state of the art BioDiesel production.

[1] DG Research, EUR 22066, "Biofuels in the EU, a vision for 2030 and beyond", Final report of the Biofuels Research Advisory Council, 2006.

See more of this Session: Developments in Biobased Alternative Fuels II
See more of this Group/Topical: Sustainable Engineering Forum

(142b) Novel Photobioreactor for Biodiesel and Electricity

Gordon A. Hill[1], Pranabendu Mitra[1], Divya Sasi[1] and Erin E. Powell[2], (1)Chemical Engineering, University of Saskatchewan, Saskatoon, SK, Canada, (2)Bioprocessing Unit, Saskatchewan Research Council, Saskatoon, SK, Canada

The development of new technologies to protect the environment and provide renewable fuel is of growing interest in North America. High greenhouse gas discharges, primarily due to carbon dioxide formed by burning fossil fuels, and dwindling supplies of petroleum are two motivating factors that make these technology developments imperative. In this work, we report on the design and utilization of a novel photobioreactor that can enhance the growth of algae such as *Chlorella vulgaris*. We achieved an increase in productivity for producing biodiesel by capturing carbon dioxide faster than in traditional bioreactors or open ponds, and we also efficiently produce electricity when utilizing this photobioreactor as a microbial cathode. By coupling the photobioreactor to an existing bioethanol fermentor, a complete microbial fuel cell is achieved that acts as a living battery, and a net sink of carbon dioxide occurs due to the complete utilization of carbon dioxide by the algae. Biodiesel and electricity production rates as well as the economics of the incorporation of photobioreactors into an existing starch-based bioethanol facility will be presented.

See more of this Session: Developments in Biobased Alternative Fuels II
See more of this Group/Topical: Sustainable Engineering Forum

Micro Chemical Processing Technology for Production of Biodiesel Fuel

Tricia Thomas[1], Robert Dacus[1], Jennifer Lewis[1], Rob Mebane[1], Jim Hiestand[1], Rob Bailey[2], Mary Lowe[2], Frank Jones[1]*

*1-University of Tennessee at Chattanooga, 2-Loyola University Maryland, *Author to whom correspondence should be addressed*

Biodiesel is widely recognized as an environmentally friendly alternative to petroleum-based fuels. Although biodiesel has become more readily available in recent years, the processes currently used in commercial production of biodiesel are relatively inefficient. Improvements in these processes are required in order to lower costs of biodiesel relative to petroleum diesel and drive the growth of a robust biodiesel industry.

Traditional commercial biodiesel production processes rely on homogeneous catalysis of the transesterification reaction of vegetable oil with an alcohol, typically methanol. The use of a fixed solid catalyst instead of a liquid catalyst is recognized as a first step in improving the efficiency of biodiesel production. The use of the solid catalyst eliminates the production of soap, an undesirable by-product that results from using a liquid basic catalyst in conjunction with an oil feedstock that contains free fatty acids. Thus, solid catalysts reduce downstream processing and enable the use of lower-quality feedstock oils which may be fairly high in free fatty acids. Suitable metal catalysts are being sought by researchers around the globe; success has been demonstrated using various metal oxides including tin, titania, zirconia, and alumina based oxides [1, 2, 3].

A second step in improving biodiesel production efficiency involves reducing processing time. Many current commercial biodiesel production processes are batch processes with fairly long reaction times on the order of hours. One approach to address this issue is the use of the microscale which is especially efficient in reducing processing times. In traditional fixed-catalyst systems, reactant fluids are brought into contact with solid catalysts where reaction occurs. The rate-limiting step is frequently the mass transfer of the reactants to (or products away from) the catalyst. In a microscale system, the reactants flow into catalyst-coated channels having cross dimensions of 100 microns or less. The narrow cross dimension brings the molecules into close proximity to the catalyst, and reaction occurs at the channel wall. The use of the microscale can thus significantly reduce reaction time.

The goals of the authors are to (1) identify heterogeneous catalysts for biodiesel production that are suitable for use in microfluidic systems, and (2) use microfluidic systems to reduce processing times and costs associated with biodiesel production from a variety of feedstocks. For this work, computer simulations were used to design microchannels for use in the production of fatty acid methyl ester (FAME) biodiesel from oil feedstocks and methanol. The microchannels were modeled as being completely coated with a fixed solid metal oxide catalyst. The kinetic properties of the metal oxide catalyst were approximated from bench-top scale batch studies conducted at temperatures ranging from 60° C to 180°C. The best microchannel designs have been fabricated and are currently being evaluated.

Catalyst Evaluation

Fatty acid methyl esters (FAME), or biodiesel fuels, can be produced in the presence of an appropriate catalyst by the reaction between methanol and a triglyceride to produce FAME and glycerol (Figure 1). Alternatively, free fatty acids can be reacted with methanol in the presence of an appropriate catalyst to produce FAME and water (Figure 2). Increased temperatures are frequently employed to speed the rate of reaction

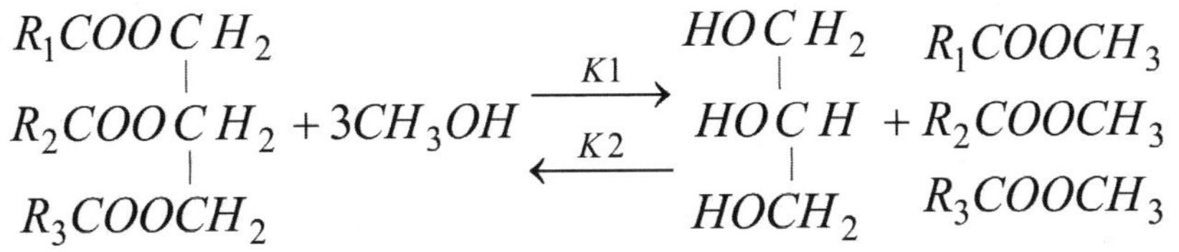

Triglyceride *Methanol* *Glycerol* *Methyl Esters*

Figure 1. Transesterification reaction of oil with methanol produces glycerol and biodiesel fuel (methyl esters)

$$RCOOH + CH_3OH \underset{K2}{\overset{K1}{\rightleftarrows}} H_2O + RCOOCH_3$$

Free fatty acid *Methanol* *Water* *Methyl Ester*

Figure 2. Reaction of free fatty acid with methanol produces water and biodiesel fuel.

An ideal catalyst should be inexpensive yet produce high levels of conversion of a variety of feedstocks, preferably at relatively low temperature and pressure; for this work an added property is that the catalyst should adhere to silicon surfaces. Nickel and nickel oxide [Ni(II)O] were evaluated with multiple feedstocks. Two types of batch reactors were utilized. Experiments performed at atmospheric pressure under reflux conditions (~64.5°C, the boiling point of methanol) were conducted in stirred glass vessels. These experiments used a large excess of methanol since a significant portion of the methanol was present in the vapor state. Methanol vapor was refluxed back into the reactor from a condenser located above. Experiments performed at elevated temperatures (120°C and 180°C) were conducted in closed stainless steel vials with no stirring. At these elevated temperatures the reactor pressurizes to some value just above the vapor pressure of methanol: 7 bar at 120°C or 27 bar at 180°C. The lack of stirring in these stainless steel reactors negatively impacts the reaction rate because oil and alcohol are reluctant to mix.

Nickel oxide nanopowder (Sigma-Aldrich, nominally 10 nm in diameter) and nickel nanopowder (Sigma-Aldrich, 10-20 nm) catalysts were evaluated for the production of biodiesel from methanol and soybean oil (SBO). Chemical analyses were performed with a 400MHz NMR spectrometer. Results are shown in Figures 3 and 4. Figure 3 demonstrates that the lack of stirring in the stainless steel vials could be overcome by increasing temperature. Increased temperature can also compensate for the use of an inferior catalyst, as shown in the comparison between nickel and nickel oxide catalysts in Figure 4. For the conditions and catalysts evaluated, attaining reasonably high levels of conversion required 2 to 7 hours.

Additional experiments have addressed the use of the nickel oxide catalyst with alternative feedstocks including corn oil, olive oil, algal oil, and tall oil. The use of tall oil, a by-product from kraft pulping of softwoods, was of particular interest since tall oil is actually a mixture of free fatty acids (primarily oleic) rather than a triglyceride. Nickel oxide was an effective catalyst for converting all of these feedstocks, and was most efficient when used with the tall oil – a marked contrast to common homogenous base catalysts. Conversions after two hours are shown in Figure 5. No soap by-product was created with any of the feedstocks, in contrast to traditional homogenous catalysts.

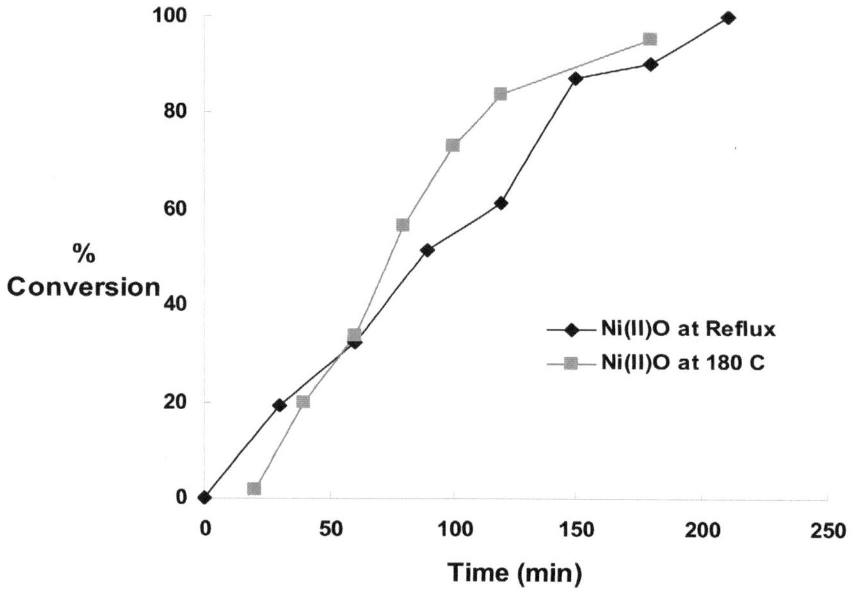

Figure 3. Conversions of SBO into fatty acid methyl esters (FAME) at both reflux temperature (64.5°C) and 180°C in the presence of Nickel (II) Oxide.

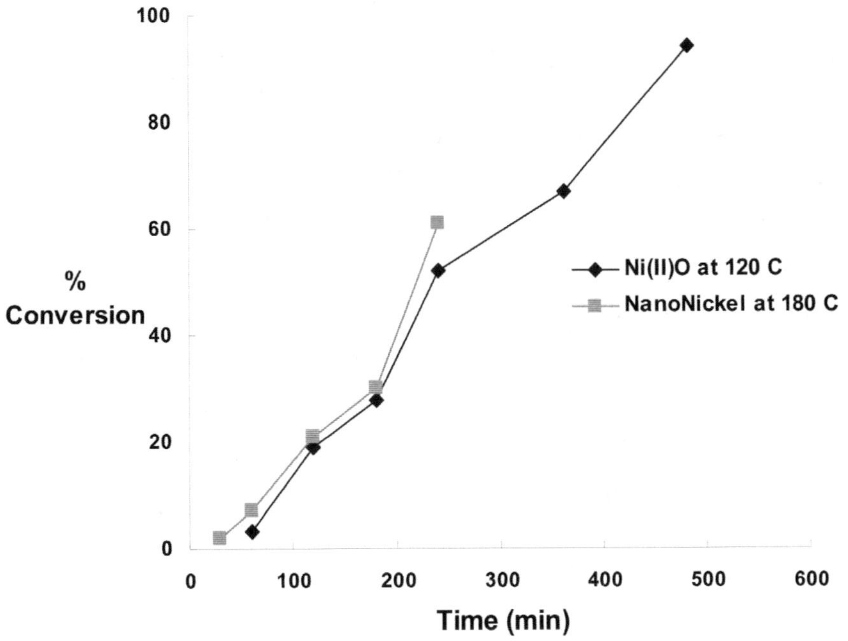

Figure 4. Conversions of SBO into FAME using NiO at 120°C and Ni at 180°C.

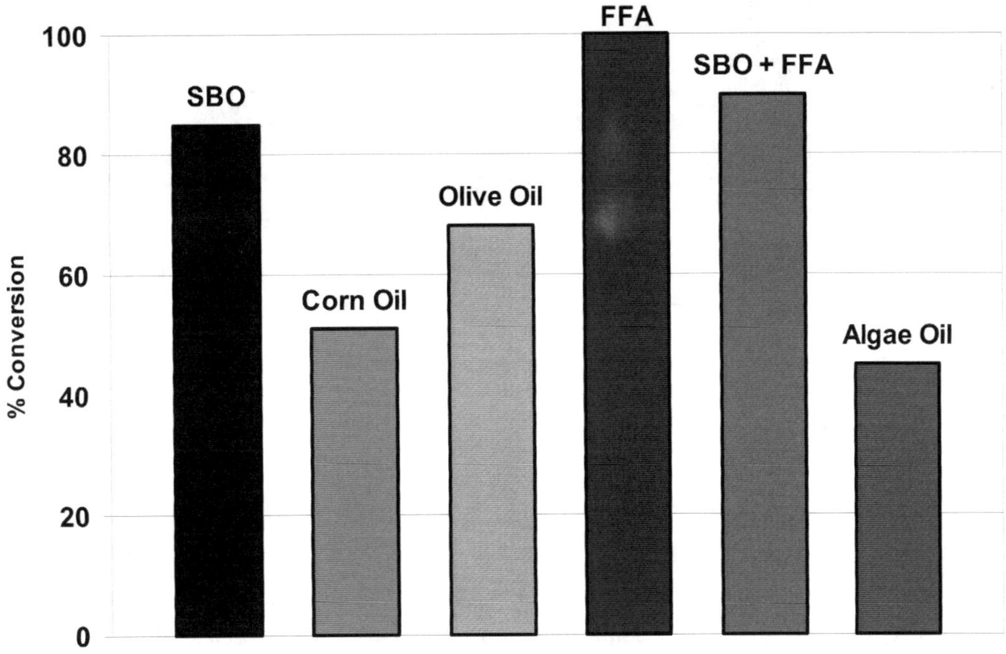

Figure 5. Comparison of the conversion to FAME from various feedstocks in stainless steel vials in the presence of Ni(II)O at 180°C with a residence time of 2 hours.

Computational Studies

Commercial software (CFD-ACE+ from ESI Group) combining computational fluid dynamics, heat transfer, and heterogeneous reaction kinetics was used to design the microchannel reactors. Previous work using CFD-ACE+ to simulate flow and chemical reaction in microreactors has demonstrated that simulation results can be successfully benchmarked with experimental data (4, 5, 6). Design considerations included flow rate, pressure drop, catalyst choice, and channel dimensions and packing.

Figure 6 shows images of similar real and virtual microchannels. The image on the left is an scanning electron microscope (SEM) image of a polydimethylsiloxane (PDMS) channel with a repeating 3-triangle packing pattern from a previous project. This configuration is also being used in the present project. The triangular obstructions provide fluid mixing, favorable heat transfer, and additional reactive surface. The image on the right is a partial computational grid used for this project that contains approximately 800,000 cells per 1 centimeter of channel length, a 100 x 100 micrometer cross section. This grid was used to simulate the reaction behavior of SBO/methanol feed to FAME and glycerol.

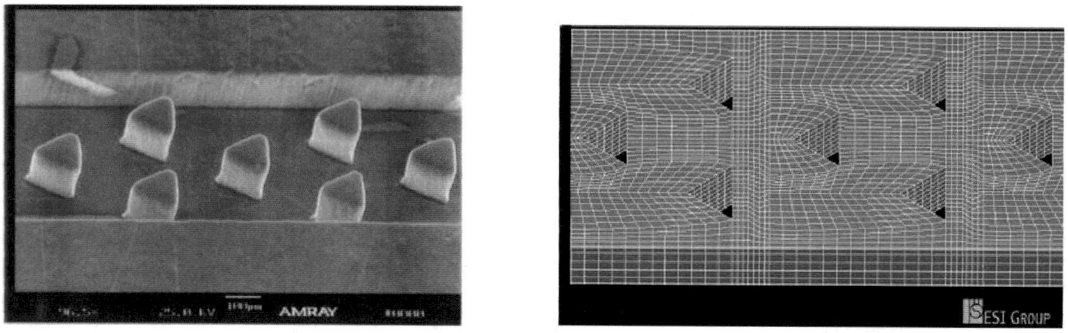

Figure 6. Fabricated (left) and simulated (right) microchannels

Conversion results of a microreactor simulation are shown in Figure 7. This simulation uses a constant temperature of 60°C and an average entry velocity of 0.00167 m/s. The feed has a 12-to-1 molar ratio of methanol to SBO. A first order reaction rate of 0.0002 s^{-1} from our catalyst studies was applied at all solid surfaces. The reaction channel is one centimeter long; however, over 95% of the soy bean oil is converted to biodiesel in the 800 micron entry length shown. Conversion is 99.98% at the end of the channel, which has a residence time of only 6 seconds. Although this prediction contains many assumptions and may be a bit optimistic, it is promising when compared with the hours of processing time required by traditional techniques.

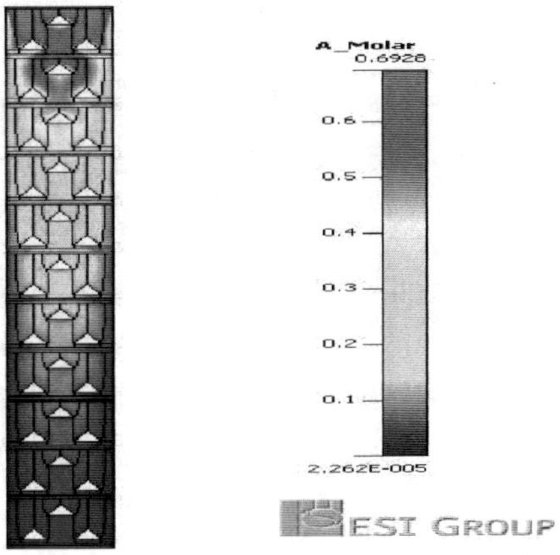

Figure 7. Top view of a simulated single-channel microreactor showing the molar concentration profile for soybean oil ("A") in the first 800 microns of a 1 cm long x 100 μm deep x 100 μm wide microchannel. SBO feed enters at the top and reacts along the channel. The scale on the right shows that high concentrations of SBO are shown in fuchsia and low concentrations are shown in blue. Complete conversion (>99.98%) is achieved in a 1 cm channel with a residence time of 6 seconds.

Some practical design constraints concerning microreactor fabrication and operation need to be considered. Since the point of using microreactors is that the small scale reduces reaction time, the effect of varying cross section was examined (Figure 8). Reducing channel cross section from 100μm x 100μm to 50μm x 50 μm reduces residence time from 6 seconds to less than 1 second. It also reduces channel length required for complete (99.9+ %) conversion from about 10 mm to less than 1 mm. However, the pressure drop increases from about 4 atmospheres to 19 atmospheres, too high for safe operation.

Broader and shallower channels are found to have lower pressure drops. A cross section of 50μm deep x 500μm wide has a pressure drop of about 1 atm per centimeter of channel length. A broader shallower channel is also somewhat easier to fabricate and coat evenly with catalyst. This design has been fabricated and is presently being tested.

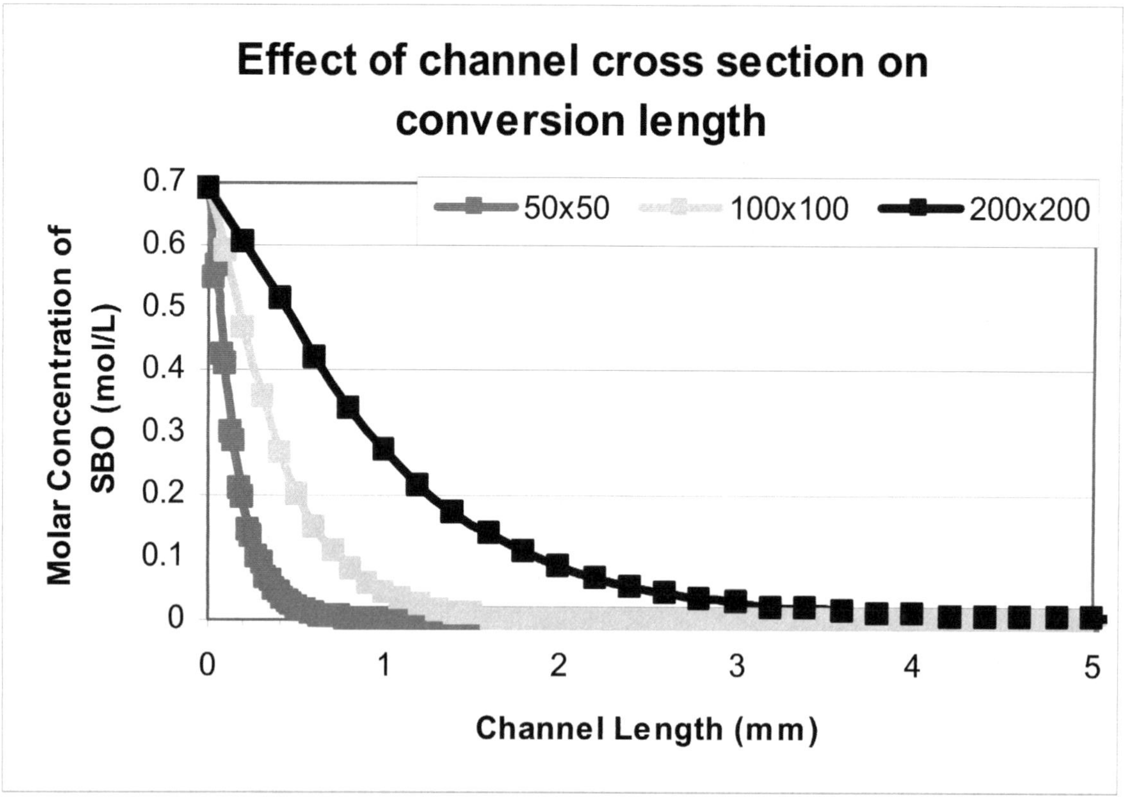

Figure 8. As channel cross sections get smaller, conversion occurs in significantly shorter channel lengths. However, pressure drops can rise to precarious levels. For 50 µm x 50µm, the pressure drop is about 19 atm. For 200 µm x 200 µm the pressure drop is only about 1.5 atm.

Conclusions

Bench-top batch studies at temperatures ranging from 60°C to 180°C indicate that nickel oxide is an appropriate catalyst for the production of biodiesel (FAME) from methanol and a variety of oils with conversion approaching 100% in two hours for tall oil at 180°C. Simulations demonstrate the feasibility of using microchannels coated with nickel oxide catalyst to achieve high levels of conversion of soybean oil at moderate temperatures and pressures. Simulations show that residence time is reduced from hours in traditional industrial systems to only seconds. Microreactors have been fabricated according to the preferred design (50µm deep x 500µm wide), and the performance characteristics of these microreactors are currently being evaluated.

In practice, several hundred microreactors would be stacked and plumbed together to achieve the desired production rate, with scale-up being linear. The resulting processing plant would be approximately the size of a computer tower. Target production rate for one "plant" is 500-1000 L of fuel per week. One advantage of the small scale production would be the ability to optimize individual plants for a given feedstock. This capability would promote the use of diverse oil feedstocks, a necessity given the limited amount of any single oil source and the impact of oil type on process optimization [7, 8, 9].

Works Cited

1. Lotero, E., J.G. Goodwin, D. A. Bruce, K. Suwannakarn, Y. Lie, D.E. Lopez. "The Catalysis of Biodiesel Synthesis," *Catalysis*, **19**, 41-83 (2006)
2. McNeff,C. V., L. C. McNeff, B. Yan, D. T. Nowlan, M. Rasmussen, A. E. Gyberg, B. J. Krohn, R. L. Fedie, T. R. Hoye, "A continuous catalytic system for biodiesel production," *Applied Catalysis A: General*, **343**, 39-48 (2008)
3. Sivasamy, A., K. Y. Cheah, P. Fornasiero, F. Kemausuor, S. Zinoviev, S. Miertus, "Catalytic Applications in the Production of Biodiesel from Vegetable Oils," *ChemSusChem*, **2**, 278-300, (2009)
4. Bailey, Rob; Frank Jones, Ben Fisher, and Bill Elmore, "Immobilized Enzymatic Microbioreactor Design Enhancement using Computational Simulation." *Applied Biochemistry and Biotechnology*, Vols. 121-124, 639-652 (2005)
5. Jones, Frank; Scott Forrest, Jim Palmer, Zonghuan Lu, John Elmore, and Bill Elmore, "Immobilized Enzyme Studies in a Micro-Scale Bioreactor." *Applied Biochemistry and Biotechnology*, Vols. 113-116, 261-272 (2004)
6. Jones, Francis; Zonghuan Lu, Bill B. Elmore, "The Development of a Novel Micro-Scale System as an Immobilized Enzyme Bioreactor," *Applied Biochemistry and Biotechnology*, Vols. 98-100, 627-640 (2002)
7. Chisti, Y., "Biodiesel from microalgae," *Biotechnology Advances*, **25**, 294-306 (2007)
8. Groom, M. J., E. M. Gray, P. A. Townsend, "Biofuels and Biodiversity: Principles for Creating Better Policies for Biofuel Production," *Conservation Biology*, **22**, 602-609 (2008)
9. Moser, Bryan R., "Biodiesel production, properties, and feedstocks," *In Vitro Cell. Dev. Biol. – Plant* **45**, 229-266 (2009)

(142d) Sulfur Level Changes in Brown Grease Conversions with Sulfuric Acid and Heterogeneous Zirconia-Supported Metaloxides Catalysts

Manhoe Kim[1], Craig DiMaggio[2], Shuli Yan[3], Steven O. Salley[4] and K. Y. Simon Ng[4], (1)National Biofuel Energy Lab, Next Nery, Detroit, MI, (2)National Biofuel Energy Laboratory, Detroit, MI, (3) Wayne State University, Detroit, MI, (4)Chemical Engineering and Materials Science, Wayne State University, Detroit, MI

Brown grease recovered from waste water is less expensive lipids and would make biodiesel much more competitive with petroleum-based fuels. Brown grease has been converted with homogeneous sulfuric acid and metaloxides (ZnO2, ZnO2-La2O3, ZnO2-Al2O3 and ZnO2-NiO) supported ZrO2 catalysts in batch reactors. The supported catalysts were prepared via impregnation and sol-gel method. The metaloxides loading on the supports were analyzed with SEM-EDS. Sulfur and metal concentrations in both reactant and product were measured with an inductively coupled plasma spectrometry. Total acid numbers of the brown grease and products were determined according to ASTM D 664. Composition of glycerides in brown grease was measured with a HPLC. The yields of fatty acid methyl esters in the samples were quantified using a GC-MS. Amounts of free fatty acids and macromolecules (bp>240°C) were 87.5% and 15±5%, respectively. Maximum FAME yield was detected in the range of 78~85%. Sulfur content in brown grease was very high as 640 ppm. The sulfur levels in the converted FAME phases decreased to 300~330 ppm by the methanolysis. About a half of the sulfur in the brown grease removed by methanolysis can be regarded as ester-bonded sulfates. The other half of the sulfur can be regarded as sulfonates, sulfones and reduced sulfurs. These sulfurs were difficult to be removed by methanolysis.

See more of this Session: Developments in Biobased Alternative Fuels II
See more of this Group/Topical: Sustainable Engineering Forum

(142e) Plantwide Design and Control of Biodiesel Production Processes Via Two-Step Syntheses or by Simultaneous Esterification/ Transesterification

Jian-Kai Cheng[1], Yin-Heng Shen[1], Yong-Tang Jhuang[2], Chuan-Chen Chao[1], Jeffrey D. Ward[1], I-Lung Chien[2] and Cheng-Ching Yu[1], (1)Department of Chemical Engineering, National Taiwan University, Taipei, Taiwan, (2)Department of Chemical Engineering, National Taiwan University of Science and Technology, Taipei, Taiwan

Biodiesel made from vegetable oil, animal fat or waste food oil is an alternative diesel fuel. It has attracted increasing interest because of decreasing petroleum reserves and concerns about environmental protection. At present, there are several ways to produce biodiesel, e.g., pyrolysis, micro-emulsification and transesterification [1-2]. Among of the options for biodiesel production, transesterification with methanol as the other reactant is the most common and profitable method [1-2]. Transesterification can be alkali-, acid- or enzyme-catalyzed [3], where alkali-catalyzed transesterification is most often used commercially because of its faster reaction. Recently, however, a main challenge for biodiesel production is the cost of feedstock and its impact on food price. Lower-cost alternative feedstocks, such as waste cooking oil or yellow grease, are becoming more attractive even though they usually contain high free fatty acid (FFA) content which causes problems with soap formation in the presence of alkaline catalyst. If a feedstock with FFA is considered, because of the restriction of FFA content in the alkali-catalyzed process, two-step process with esterification to use up FFA followed by transesterification is required. An additional cost for this esterification pretreatment should be included in the alkali-catalyzed process. Therefore, the acid-catalyzed process with simultaneous esterification and transesterification reaction may be more attractive. In this work, we investigate the design and control of biodiesel production processes with oil feedstock containing FFA. Two processes are considered: one is alkali-catalyzed (sodium hydroxide as catalyst) and the other is acid-catalyzed (oleophilic acid catalyst). The economics of the two processes are evaluated, and the effect of FFA content in the oil feed is also explored. To simplify the analysis of kinetics and thermodynamics, soybean oil is regards as pure triolein, a triglyceride in which all three fatty acid chains are oleic acid. Parameters of the kinetics for the sodium hydroxide catalyst are taken from Noureddini et al [4], and those for oleophilic acid catalyst are regressed from data given by Lien et al [5].

For the transesterification reactor system, because the reactor effluent exhibits phase separation (two-liquid phase), a reaction/separation system is devised to improve the conversion. The system has a decanter after each reactor and with an internally recycled glycerol phase rich in methanol which can dramatically reduce the feed ratio of methanol to triglyceride. Because there is a strong interaction between the reactor cost and the separation cost (recycle cost) in the biodiesel process, we design the reactor system and separation system simultaneously. We start with an alkali-catalyzed plant-wide process without FFA content in the feedstock to demonstrate the benefit of this reaction system. Comparison is made between the conventional (without recycling glycerol phase) and the proposed (including recycling glycerol phase) biodiesel production processes. The results show that a 16% reduction in the total annual cost (TAC) can be achieved using simultaneous reaction with internally recycled glycerol phase. This also corresponds to a 22% reduction in the energy cost.

Next, we look at the effect of FFA content on the economic comparison between the alkali-catalyzed (two-step process) and the acid-catalyzed process (simultaneous esterification and transesterification reaction), in which a reaction system with internally recycled glycerol phase is used. In this case, when the FFA content increases, it can be expected that the acid-catalyzed process will be more economically competitive compared to the alkali-catalyzed process.

Finally, plantwide operability of the entire biodiesel production plant is evaluated. The results indicate that the proposed process can handle 20% production rate changes with reasonably good dynamic performance.

References:

1. Ma, F. R.; Hanna, M. A., Biodiesel production: a review. Bioresour Technol 1999, 70, (1), 1-15.

2. Sharma, Y. C., et al., Advancements in development and characterization of biodiesel: A review. Fuel 2008, 87, (12), 2355-2373.

3. Fjerbaek, L., et al., A Review of the Curr ent State of Biodiesel Production Using Enzymatic Transesterification. Bioresour Technol 2009, 102, (5), 1298-1315.

4. Noureddini, H.; Zhu, D., Kinetics of transesterification of soybean oil. J Am Oil Chem Soc 1997, 74, (11), 1457-1463.

5. Lien, Y. S., et al., Biodiesel Synthesis by Simultaneous Esterification and Transesterification Using Oleophilic Acid Catalyst. Ind Eng Chem Res 2010, 49, (5), 2118-2121.

See more of this Session: Developments in Biobased Alternative Fuels II
See more of this Group/Topical: Sustainable Engineering Forum

(142f) Production of Biodiesel Using Dimethyl Carbonate as the Methylating Agent: A Glycerol-Free Biofuel

Michael Miguez[1], Tracy J. Benson[2] and Samir Budhathoki[2], (1)Dan F. Smith Department of Chemical Engineering, Lamar University, Beaumont, TX, (2)Chemical Engineering, Lamar University, Beaumont, TX

Biodiesel is currently produced to supplement the supply of petroleum diesel, and much research is performed to find more efficient means of production. Biodiesel offers some desirable properties over petroleum-based diesel, being both environmentally friendly and lubrication. Currently, biodiesel is produced by reacting triglycerides, chiefly oils from row crops, with methanol by means of different catalysts. The desired product is a mixture of fatty acid methyl esters (FAME's), but a glycerol byproduct is also generated. The glycerol has to be separated in downstream processes, which requires more equipment and process unit operations. Biodiesel can also be produced by reacting triglycerides with dimethyl carbonate (DMC). The DMC reaction generates a more desirable product than the methanol reaction, as it does not generate any products requiring downstream separations. This new type of reaction will be analyzed to determine which conditions work best to generate biodiesel. The DMC reaction is initially carried out at various temperatures with various catalysts in batch reactors to generate biodiesel. The specific products from reacting triglycerides with DMC in a 1:1 mole ratio are 2 moles of FAME's and 1 mole of a glycerol derivative which has one fatty acid group and a carbonate group attached. For simplification, it is referred to as a glycerol carbonate fatty acid ester. A GC-MS is used to analyze the products to determine the components generated. Once enough data is available to determine specific reaction conditions, a larger scale reactor will be designed to produce enough biodiesel to test for ASTM standards and eventually test in a diesel-powered pressure washer. Initial results have indicated that the DMC reaction has a higher transition state than the methanol process. However, modeling efforts using Aspen 7 have revealed an overall lower energy demand for the complete reaction/downstream process. Complete reaction and modeling results will be presented.

See more of this Session: Developments in Biobased Alternative Fuels II
See more of this Group/Topical: Sustainable Engineering Forum

(175a) A Survey of Separation Technologies for Drinking Water Treatment in Developing Country Settings, with a Focus On Membrane Systems

Jeff Albert[1], Ranjiv Khush[1], Brad Lang[1], Joel Segre[2], Nicky Khaki[1] and Veena Srinivasan[3], (1)The Aquaya Institute, San Francisco, CA, (2)McKinsey and Company, San Francisco, CA, (3)School of Earth Sciences, Stanford University, Stanford, CA

A wide array of separation technologies are in use to treat drinking water in less-industrialized economies. These technologies are deployed in centralized, semi-centralized, and household applications. This presentation will provide an overview of those applications as well as a more detailed discussion of membrane-based systems.

Membrane technologies have come into increasingly widespread use in water and wastewater treatment in industrialized countries. The conventional wisdom is that they represent high-end state-of-the-art methods whose costs remain prohibitively high for application in developing country markets.

We have spent nearly two years researching the mature drinking water refill industry in Southeast Asia and the emerging industries in South Asia and East Africa. We have met with entrepreneurs operating existing water treatment and refill businesses, government regulators, local financial institutions, and water treatment equipment suppliers to understand the current state of the sector and evaluate its future prospects. We have also been working with local entrepreneurs to develop new water treatment and vending enterprises.

Membrane systems are already in widespread use in the Asian water refill sector, primarily in the form of reverse osmosis incorporated into multibarrier designs employing pretreatment (in addition to a separate optional disinfection stage). We estimate that at least 5,000 Filipino water refill businesses employ RO as a core element of their treatment process, where supply chains for membranes and other system components are well-developed. Other membrane designs have recently begun to be deployed as standalone systems, most notably hollow-fiber ultrafiltration.

The convergence of increasing affordability of membrane-based treatment systems with increasing capacity to purchase, operate, and maintain these systems suggests that the less industrialized economies will represent a massive emerging market for membrane technologies in drinking water treatment in the future.

See more of this Session: Separations Technologies in Water Supply in Developing Communities
See more of this Group/Topical: Separations Division

(175b) Sustainable Mitigation of Global Arsenic Crisis: Challenges and Progress

Arup SenGupta, Lehigh University, Bethlehem, PA

In Bangladesh, Vietnam, eastern part of India, Cambodia, Thailand, Argentina and several other countries, drinking water drawn from underground sources has been responsible for widespread arsenic poisoning affecting millions of people. Although the genesis of arsenic contamination is yet to be fully understood, natural geochemical weathering of subsurface soil is the sole contributor of dissolved arsenic in groundwater. To this end, the collaborative work between Lehigh University in Pennsylvania, USA and Bengal Engineering and Science University (BESU) in West Bengal, India and Institute of Technology in Cambodia (ITC) has been directed toward providing arsenic-safe water in remote villages in affected areas. During the last fifteen years, two hundred low-cost, easy-to-operate arsenic removal units have been installed at the existing wells to insure a supply of safe drinking water. Currently, about two hundred thousand people (200,000) drink arsenic-safe water in this region.

The two most significant scientific advancements that have emerged from this ongoing project are the development of: the first reusable polymer based arsenic-selective adsorbent; and the inexpensive methodology for safe containment of arsenic-laden sludge in rural environment.

The lecture will emphasize the underlying role of chemical science and chemical engineering principles in providing sustainable solutions to a global water-related crisis.

See more of this Session: Separations Technologies in Water Supply in Developing Communities
See more of this Group/Topical: Separations Division

(175c) Community Water Supply: Technology, Economics and Community Capacity

John Wind, Chevron Energy & Technology Company, Richmond, CA and Paola de Cecco, San Jose State University

The sustainability of community water supply projects in "developing" countries is critical for increasing long-term access to clean drinking water. Important sustainability parameters include water supply & purification technology, project development, financing and local capacity building, and O&M programs. Community involvement in the design, project implementation and long-term operations is critical to water supply project success. We will demonstrate the importance of these factors in the water and health project in Ngelenge, Tanzania, a partnership between Engineers Without Borders-USA and the Ngelenge Development Association, a local NGO. The role of separations technology (e.g. flocculation, filtration) and groundwater well development will be described in detail, to show the role of chemical engineering skills in international community development projects.

See more of this Session: Separations Technologies in Water Supply in Developing Communities
See more of this Group/Topical: Separations Division

A novel technology to remove arsenic from drinking water for Bangladesh tubewells

Ashok Gadgil [1,2], Susan Addy [2], Case Van Genuchten [2]
1. Lawrence Berkeley National Laboratory, Berkeley, CA
2. Civil and Env. Engineering, University of California, Berkeley, CA

Abstract

Bangladesh and neighboring areas face large health threats from drinking arsenic-contaminated ground water. Arsenic levels in Bangladesh ground water are typically several hundred μg/L (compared to WHO recommendation of 10 μg/L for the MCL). About 50 million people drink such water, with hundreds of thousands already showing serious adverse health effects in what is described as the largest mass poisoning in history. The challenge is to develop a method for arsenic remediation that is (1) technically effective for removing arsenic down to 10 μg/L in the presence of other competing ions in the water, (2) affordable to most of the local population, (3) robust and easy to operate and maintain, and (4) does not require use of other toxic or hazardous chemicals. We describe a novel method that aims to meet these goals. ElectroChemical Arsenic Removal (ECAR) uses a small DC current and ordinary steel electrodes to produce a specific type of iron rust in the arsenic-contaminated ground water that binds to the arsenic and can be removed by filtration. We describe results using synthetic groundwater prepared in the laboratory, and also preliminary results from real Bangladesh groundwaters. We describe the design of a small ECAR reactor to treat 100 L of water at time, for a technical trial in West Bengal (India). Lastly, we show results from Extended X-ray Absorption Fine Structure (EXAFS) analysis that suggests the structure of the iron precipitate and the dominant mode of arsenic surface complexation.

1 Introduction

Arsenic contamination in groundwater has been discovered in Argentina, Chile, Mexico, China, Hungary, Vietnam, Cambodia, West Bengal (India), and Bangladesh [1, 2]. In most areas, surface waters are severely microbially contaminated, leaving no viable alternative drinking water source. In Bangladesh alone, 57 million people are exposed to arsenic levels of up to 3200 μg/L [3], well in excess of the maximum contaminant level (MCL) recommended by the World Health Organization of 10 μg/L [4]. A recent decade-long cohort study published in *The Lancet* found that 1 in 5 (21.3%) of all deaths in Bangladesh are attributable to arsenic in drinking water [5].

Arsenic exposure leads to dermatologic, neurologic, vascular, and carcinogenic effects [6]. It is known to cause skin, lung, bladder, urinary tract, and kidney cancer [7, 8]. Exposed children also suffer from a demonstrable decrease in intellectual function indicated by IQ [9, 10]. Increased health care and loss of productivity and income cost the average household as much as $84 per year [11] - a crippling burden on yearly income for those making US $1-2 per day.

Over twenty years after arsenic contamination was discovered in groundwater [12] no effective and affordable arsenic treatment technology has been implemented on a large scale [13-15]. Arsenic remediation units (ATUs) have failed to reach even 1% of the population at risk in Bangladesh [14]. The same is true for many safe water alternative methods such as pond sand filters, rainwater harvesting, shrouded shallow dug wells, and piped water systems [14]. Lack of proper maintenance is consistently cited as a reason for ATU failure [13-17]. When ATUs are available and maintained, high upfront or ongoing costs prevent poor users from gaining access. The challenge is to develop a method for arsenic remediation that is (1) technically effective for removing arsenic down to 10 μg/L in the presence of other competing ions in the water, (2) affordable to most of the local population, (3) robust and easy to operate and maintain, and (4) does not require use of other toxic or hazardous chemicals.

ElectroChemical Arsenic Remediation (ECAR) is a form of Electrocoagulation (EC) that has been developed to meet the needs of an appropriate community scale implementation scheme that is financially viable, locally affordable, and offers long-term sustainable safe water access in rural areas [18, 19]. In ECAR, electrolytic oxidation of a sacrificial iron anode produces Hydrous Ferric Oxide (HFO; also called Fe(III) precipitates) in arsenic-contaminated water. Arsenic forms complexes with HFO, which then aggregate to form a floc that can be separated from water [20]. As(III) oxidation to As(V) occurs during the ECAR process [21, 22] either through electrolytic action at the electrode [21] or via highly reactive radical species produced by the oxidation of Fe(II) by dissolved oxygen [23, 24]. ECAR is promising due to many advantages over chemical coagulation - including pH buffering ability, ease of operation, amenability to automation, low maintenance, low sludge production, small system size, and the benefit of side reactions like electro-oxidation and electro-flotation [20, 21]. In ECAR, adsorbent media with a high capacity is generated during treatment, with no need for media regeneration or hazardous chemicals.

In this paper, we describe ECAR treatment results using arsenic-contaminated synthetic groundwater prepared in the laboratory, and also preliminary results from real groundwater in Bangladesh and Cambodia. We describe the design of a small ECAR reactor to treat 100 L of water at time, for a planned technical trial in West Bengal (India). Lastly, we show results from Extended X-ray Absorption Fine Structure (EXAFS) analysis that suggest the structure of the Iron precipitate and the dominant mode of Arsenic surface complexation.

2 ECAR Performance in Synthetic Groundwater

Arsenic removal is known to be highly sensitive to groundwater composition, specifically the presence of phosphate, silicate, natural iron and, to a lesser extent, carbonate, and calcium [24-26]. Each of these ions is present in contaminated Bangladesh groundwater [3]. Co-occurring Ions affect arsenic removal by competing with arsenic for sorption sites on HFO [24, 27], or by affecting the structure of HFO formed from Fe(II) oxidation [28, 29]. To understand ECAR arsenic removal performance in a relevant groundwater composition, batch tests were performed in Synthetic Bangladesh Groundwater (SBGW; Table 1) designed to mimic tubewell-water parameters in Bangladesh measured by BGS [3].

Table 1. Composition of Synthetic Bangladesh Groundwater (SBGW) for a representative initial total arsenic concentration of 600 μg/L .

Ion	Na^+ (mM)	Ca^{2+} (mM)	Mg^{2+} (mM)	Cl^- (mM)	HCO_3^- (mM)	SiO_3^{2-} (mM)	SO_4^{2-} (μM)	PO_4^{3-} (μM)	As(III) (μM)	As(V) (μM)	As(tot) μM	pH
SBGW	6.0	1.5	0.33	3.5	4.5	0.70	84	42	4.0	4.0	8.0	7.0

Batch experiments were performed in an electrochemical cell using an iron anode and copper electrode in arsenic-spiked SBGW, containing both As(III) and As(V). Contaminated water was treated in 3-liter batches using a galvanostatic current. Treated solution was stirred for an additional 60 minutes with no electrodes. All tests were duplicated and the results averaged.

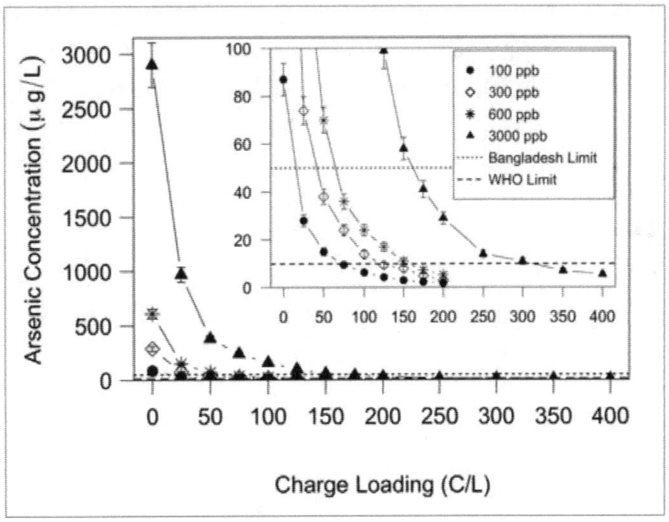

Figure 1. Aqueous arsenic concentration as a function of charge loading in Synthetic Bangladesh Groundwater (SBGW) for initial arsenic concentrations of 100 – 3000 μg/L (1:1 As(V):As(III)). Dotted and dashed lines indicate the legal Bangladesh limit (50 μg/L) and the WHO MCL (10 μg/L) respectively. Detail near the WHO MCL is shown in the inset.

Aqueous arsenic concentrations as a function of charge loading (i.e. the total charged passed through the electrochemical cell during electrolysis) for initial arsenic concentrations of 100 – 3000 μg/L are shown in Figure 1. In every case, ECAR reduces total arsenic in SBGW to well below the WHO MCL of 10 μg/L. The effect of initial concentration is to increase the charge loading required to reach the WHO MCL (see inset, Figure 1). Charge loading is directly related to the concentration of dissolved iron in solution by Faraday's law [30, 31], and thus can be thought of as a proxy for the HFO adsorbent dosage. An increase in the initial concentration is expected to require an increase in HFO adsorbent dosage to reach the WHO MCL.

It should be noted that numerous complicated processes are occurring simultaneously during ECAR, including the electrochemical dissolution of the electrode, Fe(II) oxidation and HFO hydrolysis, surface adsorption, and coagulation [20, 32]. ECAR operating parameters, such as the current density (current per active electrode area), charge loading, operating current, and post-electrolysis mixing time all have the potential to affect arsenic removal in subtle and sometimes complicated ways. It is beyond the scope of this paper to discuss or test the effect of each, though many have been explored to some extent in our lab [18].

3 Solid/Solution Separation

For robust and locally affordable operation in the field, a low cost method capable of separating arsenic-laden solids from water in a reasonable amount of time is needed. Low cost settling and decantation was compared to rapid but higher cost filtration using 0.1 μm filters after ECAR treatment in SBGW (Figure 2). Almost two days of settling was required for the former method to show solid separation comparable to filtering, or for the supernatant in the settling solution to show acceptable (< 10 μg/L) arsenic levels. Variations in the required settling time were observed with different ECAR operating conditions, but > 1 day was always required. Significantly faster settling was observed when arsenic-spiked 0.1M NaCl solution was used in place of SBGW. Some ions found in groundwater are known to constrain precipitating HFO to small oligomers [29, 33, 34] which can be colloidally stable. X-ray Absorption Spectroscopy Spectra (discussed below) supports this hypothesis.

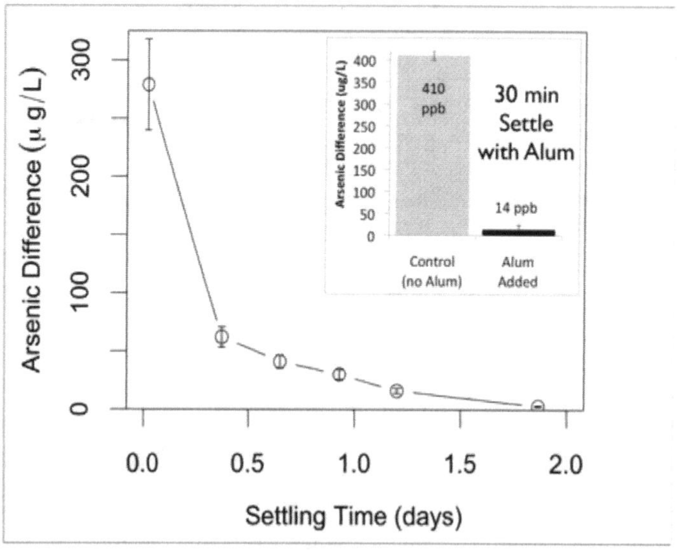

Figure 2. The difference in supernatant aqueous arsenic concentration between filtered (0.1 μm filters) and unfiltered samples while settling after ECAR treatment. The Inset shows data after 30 min of settling with and without 100 mg/L alum.

Addition of alum [$Al_2(SO_4)_3$ $18(H_2O)$] solution as a coagulant significantly decreases the settling time of ECAR HFO particles in SBGW. Adding 100 mg/L alum immediately after electrolysis in combination with a 2 min rapid stir and a 25 min slow stir shows separation comparable to high cost filtration (see inset, Figure 2) after only 30 min of settling (57 min total with stir time). Subsequent tests showed that 25 mg/L alum is sufficient to achieve < 10 μg/L arsenic in the supernatant within 4 hours of settling (including stir time). Wholesale alum sells for about $0.08 - 0.24/kg [35], adding < 1% to the total operating costs.

4 ECAR Performance in Real Groundwater

4.1 Bangladesh

Six groundwater samples were obtained from arsenic contaminated tubewells in rural villages of Jhikargachha, Abhaynagar, and Sonargaon Upazilas in Bangladesh. One liter of water from each well was collected after approximately 5 minutes of pumping (to avoid bacterial contamination and oxygenated water in the well-head) and stored in tightly capped polyethylene bottles filled to the brim (for full sample collection and treatment procedures, see [18]).

Figure 3 shows the initial (immediately before treatment) and final (post-treatment) aqueous arsenic concentrations for Bangladesh groundwater samples treated with ECAR. In every case, ECAR is able to reduce initial arsenic concentrations of up to 510 μg/L to less than the WHO MCL of 10 μg/L in real groundwater matrices. Samples BGS-1 to -4 were treated in the beaker scale setup described above. Samples BGW-5 and BGW-6, were treated in a bench-scale continuous flow prototype (described in [18]). One sample (BGW-7) was stored for 12 days and then filtered (no ECAR treatment) to determine the removal effect of naturally occurring iron precipitates in the tubewell water. Consistent with other passive sedimentation tests [24, 36], only a fraction of the arsenic was removed, and the final arsenic concentration (144 μg/L) was well above both the WHO MCL and the legal Bangladesh limit (50 μg/L).

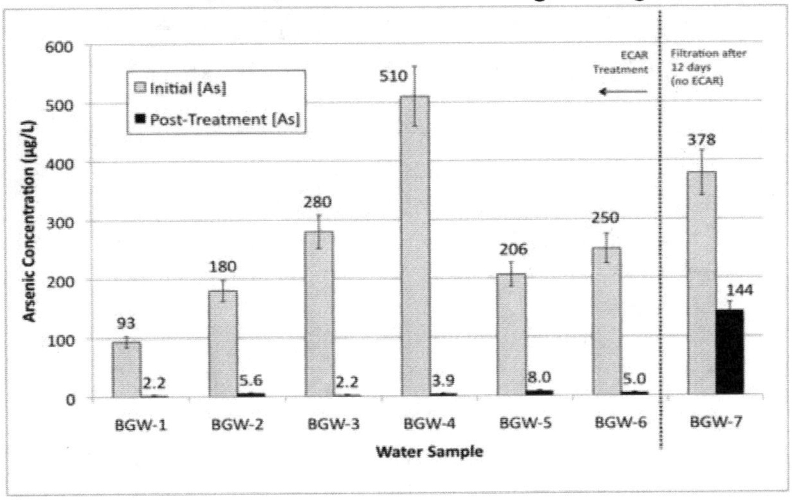

Figure 3. Arsenic concentration in samples collected from six tubewells in Bangladesh before (initial [As]) and after (Post-Treatment [As]) ECAR treatment. The rightmost bars show initial and post-treatment arsenic concentrations using passive sedimentation only (no ECAR treatment) after 12 days of sample storage.

4.2 Cambodia

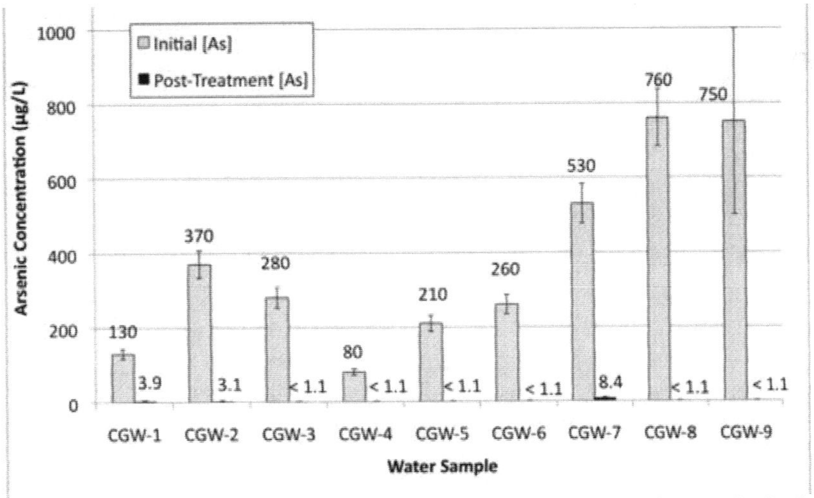

Figure 4. Arsenic concentration in samples collected from nine tubewells in Cambodia before (initial [As]) and after (Post-Treatment [As]) ECAR treatment.

Up to 1 million are at risk due to arsenic in drinking water in the Mekong Delta region of Cambodia and South Vietnam [2], primarily from aquifers with a different overall water composition than Bangladesh [25]. Nine arsenic-contaminated tube wells were chosen from the three communes Preak Russei, Dei Edth, and Preak Aeng in the Mekong Delta region of

Kandal Province, Cambodia (initial arsenic concentrations 80 – 750 μg/L). Arsenic concentrations before and after treatment with ECAR (using the bench-scale continuous flow prototype) are shown in Figure 4. In every case, ECAR-treated water contains less than 10 μg/L arsenic. In 6 of the 9 cases, final arsenic concentrations were below the detection limit of ICP-MS (< 1.1 μg/L).

5 Development and Performance of 100L Batch Prototype

A small 100L batch scale ECAR prototype has been designed, built, and tested in the lab for removing arsenic from SBGW. The prototype (Figure 5) comprises a cylindrical tank for dosing and mixing (Fig. 5a), connected to a sedimentation tank (Fig. 5b) for coagulant addition and solid/solution separation. The electrode assembly consists of 10 parallel mild steel plates (5 anode and 5 cathode) with alternate plates connected in series. The configuration allows for easy reversal of current, allowing each plate to alternate between anode and cathode to prevent extensive rust build up and passivation. A DC motor attached to a small impellor pushes water under the base plate and up between the electrode plates, allowing for efficient and even mixing between the plates.

a) b)

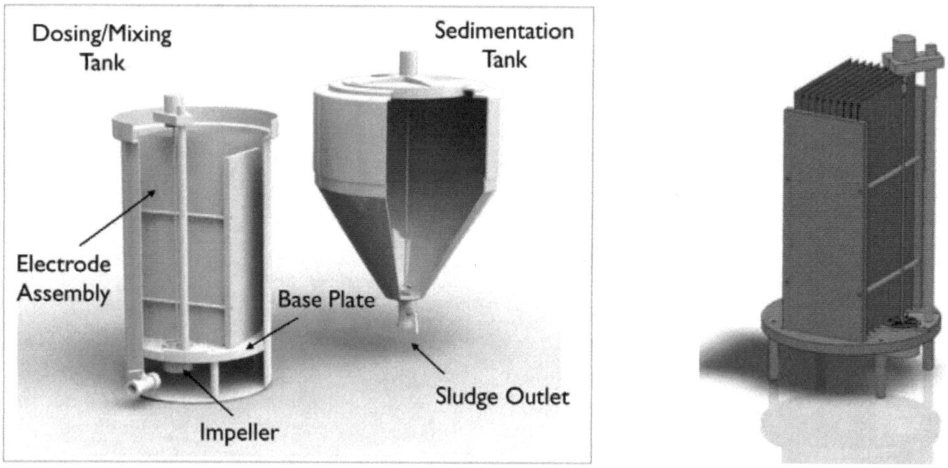

Figure 5. Cutaway rendering of (a) the full 100 L batch prototype (b) the electrode assembly and base plate only.

The prototype has successfully and repeatedly remediated SBGW with up to 3000 μg/L initial arsenic (with equal parts of As(III) and As(V)) to less than the WHO MCL in the lab, showing equal or better performance (in terms of required charge loading) than the bench-scale electrochemical cell. With the addition of a small amount of alum (25 mg/L), settling occurs within 4 hours, leaving a clear supernatant solution with < 10 μg/L arsenic. Sludge is easily accessed and collected through a valve at the cusp of the sedimentation tank. The amount of sludge collected per 100 L batch (including alum) is 10 – 20g for initial arsenic concentrations of 600 – 3000 μg/L. The prototype will be field tested in the highly arsenic-contaminated Murshidabad district of West Bengal, India, beginning in November of 2010.

6 Preliminary Results from XAS

X-ray absorption spectroscopy (XAS) is a powerful tool that allows for characterization of element-specific surface reactions on a molecular scale [37]. Fe K-Edge XAS spectra of arsenic-laden ECAR precipitates were collected to determine the predominate iron oxide structure generated *in-situ* in SBGW, and to determine if any change in structure occurs when precipitates are generated at different current density, J (current per unit active electrode area). As K-Edge XAS spectra were collected to determine the predominant surface complex responsible for arsenic immobilization during the ECAR process.

Samples of arsenic-laden ECAR precipitates were collected after treating SBGW (initial arsenic concentration 600 μg/L) at current densities 0.02, 1.1, and 5.0 mA/cm^2. Solids were extracted using 0.1 μm filters and stored as a wet paste in a deoxygenated environment at 1°C. Fe K-Edge XAS spectra were collected at beamline 10.3.2 of the Advanced Light Source (ALS) of Lawrence Berkeley National Laboratory (Berkeley, CA) at room temperature in transmission mode and calibrated using an Fe Foil. AsK-Edge XAS spectra were collected at beamline 11-2 of the Stanford Synchrotron Radiation Laboratory (SSRL; San Jose, CA) at room temperature in fluorescence and transmission mode and calibrated using an As Foil. Spectra were processed and analyzed using SixPack Software [38]. Full details of processing and analysis will be described elsewhere, but generally follow the procedures of [39]. A reference spectra for synthetic Geothite was provided by Dr. Brandy Toner; processing and analysis are similar to those described in [39].

Fourier transforms were performed over the k-range 3-11 Å^{-1} and 4-13.5 Å^{-1} for Fe and As respectively. By applying the Fourier transform to the $\chi(k)k^3$ data (where $\chi(k)$ is the Extended X-ray Absorption Fine Structure, or EXAFS, spectra), it is possible to extract individual frequencies that contribute to the overall data and convert them to peaks resembling a radial distribution function (RDF). These peaks are related to the interatomic distance between the absorbing atom and its nearest neighbors but vary by a *phase shift,* ΔR, resulting in a peak shift of approximately -0.3 to -0.5 Å depending on the type of interacting atoms.

6.1.1 Iron Shell-by-shell fits

The Fourier transformed Fe data for ECAR precipitates prepared at current densities J = 0.02, 1.1, and 5.0 mA/cm^2 are compared to that of Goethite in Figure 6a. The first peak represents the first shell (Fe-O) bonds within the FeO_6 octahedra, and is shared between the ECAR samples and Goethite. The best theoretical curve fit to ECAR samples (overlaid in Figure 6a) indicates two distinct Fe-O distances at 1.93 and 2.04 Å, suggesting a small, but significant distortion of the FeO_6 octahedra. These distances are in agreement with EXAFS measurements of FeO_6 octahedra in amorphous and crystalline iron phases [33, 34, 39].

The second shell of iron (oxyhydr)oxides is typically due to Fe-Fe pairs and contains more meaningful structural information. Here there is a clear difference between the ECAR samples and Goethite (Figure 6a). In the polyhedral approach, the polymerization of individual FeO_6 octahedra into crystalline iron phases occurs by corner-, edge-, and face- sharing octahedral linkages [40, 41]. These different topologies have characteristic Fe-Fe interatomic distances (indicated by a vertical line in Figure 6a) that can be distinguished in the second shell peak. Theoretical curve fitting showed that, although multiple Fe-Fe paths are seen in the fit of goethite, only one Fe-Fe path at approximately 3.06 Å was required for the best fit of EC precipitates. The spatial resolution, given by $\pi/2\Delta k$ where Δk is the k-range of the data used in transform [42, 43], is 0.2 Å. This is sufficient to resolve second-shell contributions from paths at 3.23 and 3.43 Å. Forcing additional Fe-Fe paths with these distances into the fit yielded physically impossible $N_{Fe\text{-}Fe}$ (coordination number) values and excessive (and often negative) values of σ^2 (mean squared displacement parameter). This suggests that FeO_6 octahedra joined by corner-sharing linkages were not present in detectable amounts. Further confirmation is provided by the lack of a characteristic beat-node in the back-transformation of the second shell peak for ECAR precipitates (data not shown).

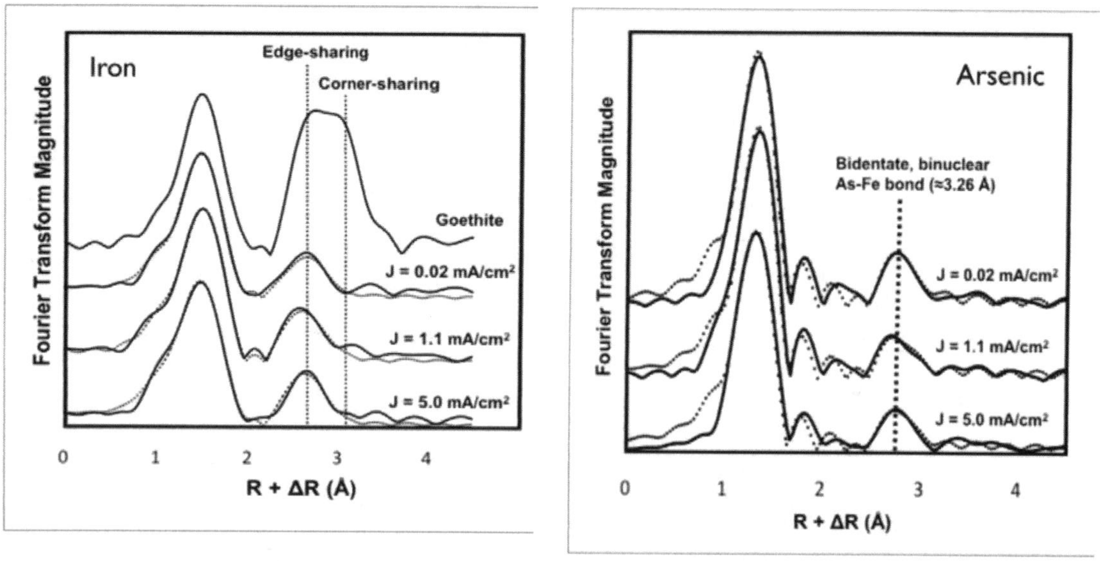

Figure 6. (a) Fe- and (b) As-K-Edge Fourier transform magnitude uncorrected for phase shift, for Goethite (Fe only) and ECAR particulates generated at current densities J = 0.02, 1.1, and 5.0 mA/cm^2 (data = solid, shell-by-shell fits = dotted).

The absence of any corner-sharing octahedra can be explained by the ions present in SBGW. Amorphous iron phases similar to EC precipitates exhibit strong affinities for sorption of SiO_3^{2-}, PO_4^{3-}, and AsO_4^{3-} [44], all present in SBGW (Table 1). These oxyanions sorb strongly onto the growth site where FeO_6 octahedra would form double corner-sharing linkages [33, 34, 45, 46] a phenomenon known as surface poisoning. The result of surface poisoning is a 2-dimensional cluster of edge-sharing FeO_6 octahedra with SiO_3^{2-}, PO_4^{3-}, and AsO_4^{3-} bound by corner sharing with adjacent FeO_6 octahedra. Nanoparticulate iron oxides have very high surface area to volume ratios, which leads to enhanced contaminant sorption per adsorbent mass. However, the bound oxyanions also help generate colloidally stable FeO_6 clusters that are difficult to separate from solution. This is likely the cause of the long (2 day) settling times observed for ECAR particles (Figure 2).

6.1.2 Arsenic Shell-by-Shell Fits

First shell As-O fits (Figure 6b) for all ECAR samples are consistent with previous studies of tetrahedrally coordinated As(V) sorbed to iron (oxyhydr)oxides [47, 48], supporting As(III) oxidation to As(V) during the EC process. As(III) oxidation is also supported by the XANES (X-ray Absorption Near Edge Spectra) structure (data not shown).

Similar to Fe analysis, the second peak of the transformed As data is the most important to properly characterize the structure of bound As. If the surface complex is inner-sphere, three possible geometries exist: (1) bidentate, mononuclear (^2E) complexes that share edges between AsO_4 and FeO_6 polyhedra (As-Fe distance ~ 2.8-2.9 Å), (2) AsO_4 tetrahedra bound in bidentate, binuclear complexes (^2C) that bridge oxygen atoms of adjacent FeO_6 octahedra (As-Fe distance ~ 3.2-3.3 Å), and (3) monodentate, mononuclear (^1C) complexes that share single corners between AsO_4 and FeO_6 polyhedra (As-Fe distance ~ 3.5-3.6 Å) [49]. If As is present as an outer-sphere complex, there would be a contribution to the signal above 4 Å, which does not appear in our data. The results of theoretical curve fitting suggest that the ^2C configuration is the primary geometry seen in the data. The data did not support the presence of a large fraction of the ^2E and ^1C geometries, and did not support the formation of AsO_4 polymers.

Due to the nanoscale nature of the EC precipitates suggested by shell-by-shell fits of the Fe k-edge peaks, the reduced contribution of the ^2E geometry is somewhat surprising. Short-ranged polymers consisting of only small clusters of edge-sharing FeO_6 octahedra should exhibit maximized edge-sharing sites [50]. The lack of support for this geometry may be due to the higher surface free energy of the ^2E geometry suggested by [51].

6.1.3 Effect of Current Density

The Fourier transformed spectra for Fe and As (Figure 6) shows little to no change across current densities $J = 0.02 - 5.0$ mA/cm^2, indicating that both the predominate iron oxide structure of precipitates and predominant arsenic surface complex are likely the same within this range. Shell-by-shell fits confirm that best fit parameters are consistent between samples for both Fe and As. This suggests that it is not possible to tune the iron oxide structure for improved As-adsorption by adjusting the current density within the tested range. However, it also suggests that the arsenic adsorption capacity of EC precipitates will be similar across current densities, allowing adjustment of the current density parameter to focus on other goals, such as short treatment-time or low power consumption.

7 Conclusions

Electrochemical arsenic remediation (ECAR) has been successfully tested in synthetic Bangladesh water and found capable of mitigating initial arsenic (including As(III)) from up to 3000 μg/L to below the WHO MCL of 10 μg/L. Separation by low-cost sedimentation can occur in <4 hours with the addition of 25 mg/L alum. ECAR treatment has been demonstrated to successfully mitigate real Bangladesh groundwater (initial arsenic 93 – 510 μg/L) and real Cambodia groundwater ((initial arsenic 80 – 760 μg/L). In all cases, ECAR was able to reduce arsenic levels to below the WHO MCL of 10 μg/L. A small-scale batch 100 L prototype has been fabricated and successfully tested in the lab using synthetic Bangladesh groundwater. It will be field tested in West Bengal, India in November 2010.

Shell-by-shell fits to Fe K-Edge XAS spectra showed only edge linkages and did not support corner linkages, likely caused by SiO_3^{2-}, PO_4^{3-}, and AsO_4^{3-} ions sorbing to corner linkage sites. This constrains HFO generated in ECAR to stable colloidal particles, explaining the long observed settling time when no coagulants are added. As K-Edge spectra fits suggest arsenic is bound to HFO by strong bidentate, binuclear complexes. Changing the current density of ECAR operation within a range of $J = 0.02$ to 5.0 mA/cm^2 caused no change in the generated HFO structure or arsenic complexes formed.

Acknowledgements

We gratefully acknowledge support for this work by The Richard C. Blum Center for Developing Economies, a USEPA P3 Phase II award, The Sustainable Products and Solutions Program at Haas School of Business at UC Berkeley, an award from the UC Berkeley Bears

Breaking Boundaries Contest, and Jane Lewis Fellowship support to one of the authors (CVG). We are also thankful to Iqbal and Kamal Quadir, the CE290 Spring 2007 BEAR team, the many current and past student-members of BAAG, and particularly to Mathew Marcus, Sirine Fakra, John Bargar, Jackie Pena, Brandy Toner, Jonathan Slack, and Howdy Goudey.

References

1. Petrusevski, B., et al., *Arsenic in Drinking Water*, 2007, IRC International Water and Sanitation Centre, Thematic Overview Paper 17: Netherlands.
2. Berg, M., et al., *Magnitude of arsenic pollution in the Mekong and Red River Deltas--Cambodia and Vietnam.* Science of the Total Environment, 2007. **372**(2-3): p. 413-425.
3. BGS, *Arsenic Contamination of Groundwater in Bangladesh*, D.G. Kinniburgh and P.L. Smedley, Editors. 2001, WC/00/19, British Geological Survey: Keyworth, U.K.
4. WHO, *Guidlines for Drinking-Water Quality*, 1993, The World Health Organization: Geneva, Switzerland.
5. Argos, M., et al., *Arsenic exposure from drinking water, and all-cause and chronic-disease mortalities in Bangladesh (HEALS): a prospective cohort study.* The Lancet, 2010. **10**: p. 60481-60483.
6. Chowdhury, U.K., et al., *Groundwater Arsenic Contamination in Bangladesh and West Bengal, India.* Environmental Health Perspectives, 2000. **108**(5): p. 393-397.
7. Tchounwou, P., A. Patlolla, and J. Centeno, *Carcinogenic and systemic health effects associated with arsenic exposure - A critical review.* Toxicologic Pathology, 2003. **31**(6): p. 575-588.
8. Abernathy, C., D. Thomas, and R. Calderon, *Health effects and risk assessment of arsenic.* Journal of Nutrition, 2003. **133**(5, Suppl. 1): p. 1536S-1538S.
9. Wasserman, G.A., et al., *Water Arsenic Exposure and Children's Intellectual Function in Araihazar, Bangladesh.* Environmental Health Perspectives, 2004. **112**(13): p. 1329-1333.
10. Wang, S.X., et al., *Arsenic and fluoride exposure in drinking water: Children's IQ and growth in Shanyin county, Shanxi province, China.* Environmental Health Perspectives, 2007. **115**(4): p. 643-647.
11. Roy, J., *Economic benefits of arsenic removal from ground water -- A case study from West Bengal, India.* Science of the Total Environment, 2008. **397**(1-3): p. 1-12.
12. Garai, R., et al., *Chronic arsenic poisoning from tube-well water.* Journal of the Indian Medical Association, 1984. **82**(1): p. 34-35.
13. Kabir, A. and G. Howard, *Sustainability of arsenic mitigation in Bangladesh: Results of a functionality survey.* International Journal of Environmental Health Research, 2007. **17**(3): p. 207-218.
14. Ahmed, M., et al., *Ensuring safe drinking water in Bangladesh.* Science, 2006. **314**(5806): p. 1687-1688.
15. Hossain, M.A., et al., *Ineffectiveness and Poor Reliability of Arsenic Removal Plants in West Bengal, India.* Environmental Science and Technology, 2005. **39**(11): p. 4300-4306.
16. Hoque, B.A., et al., *Demand-based water options for arsenic mitigation: an experience from rural Bangladesh.* Public Health, 2004. **118**(1): p. 70-7.
17. Ahmad, J., et al., *Willingness to Pay for Arsenic-Free, Safe Drinking Water in Bangladesh*, 2003, Water and Sanitation Programme--South Asia, The World Bank: New Delhi, India.

18. Addy, S.E.A., *Electrochemical Arsenic Remediation for Rural Bangladesh, PhD Thesis*, 2008, University of California - Berkeley. Available at: http://repositories.cdlib.org/lbnl/LBNL-1405E/.
19. Addy, S.E.A., et al., *Electrochemical Arsenic Remediation for Rural Bangladesh*, in *Proceedings of the Symposium on Sustainable and Safe Drinking Water in Developing and Developed Countries: Where Science Meets Policy,* 2008: Chapel Hill, NC. Available at: http://www.escholarship.org/uc/item/7bp8p38b.
20. Mollah, M.Y.A., et al., *Fundamentals, present and future perspectives of electrocoagulation.* Journal of Hazardous Materials, 2004. **114**(1-3): p. 199-210.
21. Kumar, P.R., et al., *Removal of arsenic from water by electrocoagulation.* Chemosphere, 2004. **55**(9): p. 1245-1252.
22. Wan, W., et al., *Effects of water chemistry on arsenic removal from drinking water by electrocoagulation.* Water Research, 2010. **in press**.
23. Hug, S.J. and O. Leupin, *Iron-Catalyzed Oxidation of Arsenic(III) by Oxygen and by Hydrogen Peroxide; pH-Dependent Formation of Oxidants in the Fenton Reaction.* Environmental Science and Technology, 2003. **37**(12): p. 2734-2742.
24. Roberts, L.C., et al., *Arsenic Removal with Iron(II) and Iron(III) in Waters with High Silicate and Phosphate Concentrations.* Environmental Science and Technology, 2004. **38**(1): p. 307-315.
25. Hug, S.J., O.X. Leupin, and M. Berg, *Bangladesh and Vietnam: Different groundwater compositions require different approaches to arsenic mitigation.* Environmental Science and Technology, 2008. **42**(17): p. 6318-6323.
26. Guan, X., et al., *Removal of arsenic from water: Effect of calcium ions on As(III) removal in the KMnO4–Fe(II) process.* Water Research, 2009. **43**(20): p. 5119-5128.
27. Meng, X.G., et al., *Combined effects of anions on arsenic removal by iron hydroxides.* Toxicology Letters, 2002. **133**(1): p. 103-111.
28. Ahmad, J., S. Misra, and B. Goldar, *Rural communities' preferences for arsenic mitigation options in Bangladesh.* Journal of water and health, 2006. **4**(4): p. 463.
29. Voegelin, A., et al., *Effect of phosphate, silicate, and Ca on Fe(III)-precipitates formed in aerated Fe(II)- and As(III)-containing water studied by X-ray absorption spectroscopy.* Geochimica Et Cosmochimica Acta, 2010. **74**(1): p. 164-186.
30. Gu, Z., et al., *Estimating Dosing Rates and Energy Consumption for Electrocoagulation Using Iron and Aluminum Electrodes.* Industrial and Engineering Chemistry Research, 2009. **48**(6): p. 3112-3117.
31. Lakshmanan, D., D.A. Clifford, and G. Samanta, *Ferrous and Ferric Ion Generation During Electrocoagulation.* Environmental Science and Technology, 2009. **43**: p. 3853-3859.
32. Holt, P.K., Barton G.W. Mitchell C.A., *The future for electrocoagulation as a localised water treatment technology.* Chemosphere, 2005. **59**(3): p. 355-367.
33. Doelsch, E., et al., *Speciation and crystal chemistry of Fe(III) chloride hydrolyzed in the presence of SiO4 ligands. 2. Characterization of Si-Fe aggregates by FTIR and Si-29 solid-state NMR.* Langmuir, 2001. **17**(5): p. 1399-1405.
34. Rose, J., et al., *Nucleation and growth mechanisms of Fe oxyhydroxide in the presence of PO4 ions. 1. Fe K-edge EXAFS study.* Langmuir, 1996. **12**(26): p. 6701-6707.
35. Jiang, F., et al., *Estimation of costs of phosphorous removal in wastewater treatment facilities: Adaptation of existing facilities* 2005, Georgia Water Planning and Policy Center, Andrew Young School of Policy Studies, Georgia State University, Atlanta.

36. WaterAid, *Rapid Assessment of Household Level Arsenic Removal Technologies. Phase I--Final Draft Report*, 2001, BAMWSP/DFID/WaterAid.
37. Norman, D., *X-ray absorption spectroscopy (EXAFS and XANES) at surfaces* Journal of Physics C-Solid State Physics, 1986. **19**(18): p. 3273-3311.
38. Webb, S.M., *SIXPack: A graphical interface for XAS analysis using IFFEFFIT presented by Samuel Webb.* Physica Scripta, 2003. **T115**.
39. Toner, B.M., et al., *Biogenic iron oxyhydroxide formation at mid-ocean ridge hydrothermal vents: Juan de Fuca Ridge.* Geochimica et Cosmochimica Acta, 2009. **73**(2): p. 388-403.
40. Manceau, A. and J.M. Combes, *Structure of Mn and Fe Oxides and Oxyhydroxides - A Topical Approach by EXAFS.* Physics and Chemistry of Minerals, 1988. **15**(3): p. 283-295.
41. Combes, J.M., et al., *Formation of ferric oxides from aqueous solutions: A polyhedral approach by X-ray absorption spectroscopy: I. Hydrolysis and formation of ferric gels.* Geochimica Et Cosmochimica Acta, 1989. **53**(3): p. 583-594.
42. Bunker, G., *Introduction to EXAFS : a practical guide to X-ray absorption fine structure spectroscopy.* 2010, Cambridge, UK; New York: Cambridge University Press.
43. Ulery, A.L. and L.R. Drees, *Methods of soil analysis. Part 5, Mineralogical methods*2008, Madison, WI: Soil Science Society of America.
44. Leupin, O.X. and S.J. Hug, *Oxidation and removal of arsenic (III) from aerated groundwater by filtration through sand and zero-valent iron.* Water Research, 2005. **39**(9): p. 1729-1740.
45. Waychunas, G.A., et al., *Surface Chemistry of Ferrihydrite. 1. EXAFS Studies of the Geometry of Coprecipitated and Adsorbed Arsenate.* Geochimica Et Cosmochimica Acta, 1993. **57**(10): p. 2251-2269.
46. Rose, J., et al., *Nucleation and Growth Mechanisms of Fe Oxyhydroxide in the Presence of PO4 Ions. 2. P K-Edge EXAFS Study.* Langmuir, 1997. **13**(6): p. 1827-1834.
47. Morin, G., et al., *Extended X-ray absorption fine structure analysis of arsenite and arsenate adsorption on maghemite.* Environmental Science and Technology, 2008. **42**: p. 2361-2366.
48. Manning, B.A., et al., *Arsenic(III) and Arsenic(V) reactions with zerovalent iron corrosion products.* Environmental Science and Technology, 2002. **36**(24): p. 5455-5461.
49. Ona-Nguema, G., et al., *EXAFS analysis of arsenite adsorption onto two-line ferrihydrite, hematite, goethite, and lepidocrocite.* Environmental Science and Technology, 2005. **39**(23): p. 9147-9155.
50. Waychunas, G.A., J.A. Davis, and C.C. Fuller, *Geomety of Sorbed Arsenate on Ferrihydrite and Crystalline FeOOH - Reevaluation of EXAFS results and Topological Factors in Predicting Sorbate Geometry, and Evidence for Monodentate Complexes.* Geochimica Et Cosmochimica Acta, 1995. **59**(17): p. 3655-3661.
51. Sherman, D.M. and S.R. Randall, *Surface complexation of arsenic(V) to iron(III) (hydr)oxides: structural mechanism from ab initio molecular geometries and EXAFS spectroscopy.* Geochimica Et Cosmochimica Acta, 2003. **67**(22): p. 4223-4230.

(183a) Investing In 2011

Larry Grogan, Grogan Advisory Services, Johnson City, TN

Market woes are just the beginning to managing your portfolio and goals. Now, we have new taxes going into effect which will impact your income potential. So, what can you do with your portfolios to manage the new tax environment? We will look at specific investments, as well as, investment strategies to minimize the new tax environment.

See more of this Session: Investing In 2011
See more of this Group/Topical: Liaison Functions

Transesterification of lipids with supercritical alcohols for biofuels production

Esteban A. Brignole
Planta Piloto de Ingeniería Química – Universidad Nacional del Sur – CONICET
8000 Bahía Blanca, Argentina

Since the early studies of Saka and Kusdiana on the noncatalytic transesterification of vegetable oils with supercritical alcohols, a large collection of reaction data for different vegetable oils and lipid-containing materials, with methanol or other alcohols, has been reported in the open literature. The wealth of information gathered in the last decade will be discussed to present general guidelines to carry out these processes under optimum conditions regarding alcohol/fatty oil molar ratio, operating temperature and pressure and contact time. Very promising have been the studies that indicate that neither water nor free fatty acids interfere with the transesterification process. On the contrary, it seems that their presence in the mixture fed to the reactor is beneficial for the transesterification reaction. This property and the absence of a catalyst simplifies the biodiesel purification, produces better quality glycerin, and much less waste than the conventional catalytic processes.

The experimental studies have progressed from closed batch reactors, to windowed reactors, to continuous reactors. The problems associated to reaction system density, phase conditions, thermodynamic modeling, and effect of cosolvents have all been given considerable attention. The conditions that cause degradation of the fatty esters have also been investigated. As an additional application the process has been used in the recovery and conversion to biodiesel of the lipids contained in algae cells.

Preliminary economic evaluations of large scale production processes have been made. These evaluations indicate that this technology has economic potential and is environmentally benign.

(225a) Introduction and Session Objectives

Phillip Westmoreland, Chemical and Biomolecular Engineering, North Carolina State University, Raleigh, NC, Peter T. Cummings, Department of Chemical and Biomolecular Engineering, Vanderbilt University, Nashville, TN and Sharon C. Glotzer, Chemical Engineering, University of Michigan, Ann Arbor, MI

This session will examine how modeling can be used for discovery and innovation, high fidelity prediction and as a substitute for physical experiment. Also reviewed is the state of code development and management.

See more of this Session: Innovation and Insight via High-Fidelity Simulation and Large-Database Studies of Micro, Macro and Multi-Scale Phenomena: Invited Speakers
See more of this Group/Topical: Topical 2: Simulation Based Engineering and Science

(225b) Inventing a New America through Discovery and Innovation in Science, Engineering and Medicine

Peter T. Cummings, Department of Chemical and Biomolecular Engineering, Vanderbilt University, Nashville, TN and Sharon C. Glotzer, Chemical Engineering, University of Michigan, Ann Arbor, MI

Simulation-Based Engineering and Science (SBE&S) involves the use of computer modeling and simulation to solve mathematical formulations of physical models of engineered and natural systems. The impacts of SBE&S can be observed in critical industries, such as pharmaceuticals,medical imaging, and telecommunications. In addition, the infrastructure of the Nation relies on SBE&S as it enables the efficient construction and design of our highways, automobiles and aircraft, buildings and power grids. While these examples are far from exhaustive, it is clear that SBE&S is an integral component of research and development, with far-reaching impacts on our lives and national competitiveness.

See more of this Session: Innovation and Insight via High-Fidelity Simulation and Large-Database Studies of Micro, Macro and Multi-Scale Phenomena: Invited Speakers
See more of this Group/Topical: Topical 2: Simulation Based Engineering and Science

(225c) Integrated, High Fidelity, Multiscale Process Models for the Process Industries

Chau-Chyun Chen, R&D, Aspen Technology, Inc., Burlington, MA

For decades chemical engineers worldwide have applied 1st-principles-based chemical process modeling and simulation technology to capture process understanding and carry out engineering studies in research and development, process design, optimization and control, and planning and scheduling. Integrated, high fidelity, multiscale process models are the key to the ability to model the physical and chemical world and to generate economic value. We examine the process models for carbon capture and present the model in various scales, from molecules to single and multi-phase systems, process units, process plants, sites, enterprise and the environment. Different models capture knowledge of different nature and with different decision variables. Models of finer scale support models of higher scale with information on key decision variables. Innovation to further enhance model fidelity of various scales and integration of the models in one software environment to support collaboration are essential to deliver higher economic value for the process industries.

See more of this Session: Innovation and Insight via High-Fidelity Simulation and Large-Database Studies of Micro, Macro and Multi-Scale Phenomena: Invited Speakers
See more of this Group/Topical: Topical 2: Simulation Based Engineering and Science

(225d) Petascale High-Fidelity Simulation of Turbulent Reactive Flows: Challenges and Opportunities

Jacqueline H. Chen, Combustion Research Facility, Sandia CA, Livermore, CA

The rapid growth in computing power has presented both opportunities and challenges for high-fidelity simulations of turbulent reacting flows. The advent of petascale supercomputers has made it possible to glean fundamental physical insight into fine-grained 'turbulence-chemistry' interactions in canonical laboratory-scale turbulent flames with direct numerical simulations (DNS). The unique benchmark DNS data are also used to develop and validate predictive models used to design future fuel-efficient combustors utilizing alternative fuels for transportation and power generation. Such simulations are costly, requiring several million cpu-hours on a petascale computer, over a billion grid points, and generating 100's of terabytes of data. The turbulent combustion simulation enterprise involves collaborations with computer scientists in the performance monitoring and optimization of the software on petascale architectures, on automating workflow for providing runtime diagnostics, and in-situ feature detection/tracking/ visualization of time-varying multi-scale, multi-variate data. Aspects of these interdisciplinary collaborations will be described along with reactive flow examples that illustrate the science role of DNS. Outstanding challenges with extracting salient information from 100's of terabytes of data, and strategies for mapping DNS solvers to heterogeneous petascale architectures with accelerators will also be discussed.

See more of this Session: Innovation and Insight via High-Fidelity Simulation and Large-Database Studies of Micro, Macro and Multi-Scale Phenomena: Invited Speakers
See more of this Group/Topical: Topical 2: Simulation Based Engineering and Science

(225e) Integrated Computational Materials Engineering: A Transformational Discipline for Improved Competitiveness and National Security

John Allison, Ford Motor Company, Dearborn, MI

A new field of study is evolving within the materials profession called Integrated Computational Materials Engineering (ICME). ICME involves development of unified materials models which integrate information across length and time scales and across knowledge domains and enables integration of analysis of manufacturing, design and materials into a holistic system. It promises to revolutionize the way the engineering community deals with materials and the way the materials community provides input to the engineering and scientific communities. ICME has significant potential for accelerating development of new materials and reducing the cost of development of new products. The talk will draw on the presenter's experience at Ford Motor Company in development and implementation of a successful ICME tool called Virtual Aluminum Castings. This talk will also describe the conclusions and recommendations of a recent National Academies report on ICME.

See more of this Session: Innovation and Insight via High-Fidelity Simulation and Large-Database Studies of Micro, Macro and Multi-Scale Phenomena: Invited Speakers
See more of this Group/Topical: Topical 2: Simulation Based Engineering and Science

(225f) Open Discussion On High Performance Computing in Chemical Engineering

Sharon C. Glotzer, Chemical Engineering, University of Michigan, Ann Arbor, MI, Peter T. Cummings, Department of Chemical and Biomolecular Engineering, Vanderbilt University, Nashville, TN and Phillip Westmoreland, Chemical and Biomolecular Engineering, North Carolina State University, Raleigh, NC

This will be an open discussion with the speakers and attendees to identify key elements of high performance computing for innovation, prediction, substitution for physical experiment. An objective will be define actions and approaches in which Simulation Based Engineering and Science can benefit from cross-application, common infrastructure or collective development.

See more of this Session: Innovation and Insight via High-Fidelity Simulation and Large-Database Studies of Micro, Macro and Multi-Scale Phenomena: Invited Speakers
See more of this Group/Topical: Topical 2: Simulation Based Engineering and Science

(226a) Process Technology Evaluation for Lignocellulosic Bioethanol Production: Plantwide Configurations Using a Dynamic Modeling Approach

Ricardo Morales-Rodríguez[1], Anne S. Meyer[2], Krist V. Gernaey[3] and Gürkan Sin[1], (1)CAPEC, Department of Chemical and Biochemical Engineering, Technical University of Denmark, 2800 Kgs. Lyngby, Denmark, (2)BioEng, Department of Chemical and Biochemical Engineering, Technical University of Denmark, 2800 Kgs. Lyngby, Denmark, (3)PROCESS, Department of Chemical and Biochemical Engineering, Technical University of Denmark (DTU), Kgs. Lyngby, Denmark

Biofuels have the potential to provide a more sustainable transport fuel resource compared to fossil fuels in the future. A number of countries led by USA and EU have established ambitious targets for increasing the share of renewable energy in the transport sector, e.g. USA has set targets for replacement of 20% of total gasoline for road transport by biofuels by 2022.

Some hurdles need to be overcome to allow meeting these targets. One of them is the provision of an improved technical and economical feasibility of lignocellulose based biofuels production technologies (such as bioethanol, biohydrogen, biobutanol, biomethanol, fischer-tropsch diesel, etc). Most of the efforts for developing and evaluating bioethanol production have been based on experimental lab-scale studies and steady state modeling approaches. The experimental studies can be costly and inefficient in terms of time and resources[1]-[5], while steady state models have a number of limitations such as:

a) The analysis of the complete plant involving different operational scenarios (fed-batch, continuous, continuous-recycle) cannot be performed. b) Dynamic process responses to disturbances, e.g. changes in feedstock composition among others, could not be analyzed. c) Control strategies needed to optimize and operate the plant cannot be evaluated. d) Dynamic phenomena such as inhibition of enzymatic activity or growth/ decay of microorganisms involved in the different process units are not accounted for.

In order to overcome the above-mentioned limitations, a dynamic modelling framework (DMF)[6] consisting of two phases, is employed to generate and evaluate several technological configurations based on a conventional process flowsheet[5] for bioethanol production. In the first phase, a collection, analysis and identification of the most promising mathematical models is carried out. This includes unit operations such as pretreatment[7], enzymatic hydrolysis[8], co-fermentation[9], and downstream processes among others. In the second phase of the DMF, the generation of different process configurations is carried out employing diverse operational scenarios such as, fed-batch, continuous, continuous-recycle which are simulated using an object-oriented modeling tool (MatLab/Simulink). The benchmarking criteria used to find the best process configurations include the ethanol productivity and optimal operation of the complete process flowsheet for bioethanol production from lignocellulosic biomass. The dynamic model-based simulation framework can contribute to development of economically feasible bioethanol production strategies and process configurations. These strategies could be improved by generating and evaluating novel process flowsheet configurations and operational recipes. The simulation platform can also be used for operator training in the plant, e.g. to anticipate and predict possible responses of the plant as a consequence of different types of disturbances.

References [1]. America's Energy Future Panel on Alternative Liquid Transportation Fuels; National Academy of Sciences.(2009) Liquid Transportation Fuels from Coal and Biomass: Technological Status, Costs, and Environmental Impacts National Academy of Engineering; National Research Council. [2]. Aden, A., Ruth, M., Ibsen, K., Jechura, J., Neeves, K., Sheehan, J., Wallace, B., Montague, L., Slayton, A., Lukas, J., 2002. National Renewable Energy Laboratory Technical Report. NREL/TP-510-32438. [3]. Larsen, J., Petersen, M.Ø., Thirup, L., Li, H.W., Iversen, F.K., 2008. Chem. Eng. Technol. 31, 765-772. [4]. Cardona, C.A., Sánchez, O.J., 2007. Bioresources Technol. 98, 2471-2457. [5]. Margeot, A., Hahn-Hagerdal, B., Edlund, M. Slade, R., Monot, F., 2009. Curr. Opin. Biotechnol. 20, 371-380. [6]. Sin, G., Meyer, A. S., Gernaey, K. V., 2010. Comput. Chem. Eng. doi:10.1016/j. compchemeng.2010.02.012. [7]. Lavarack B.P., Griffin G.J. Rodman D., 2002. Biomass Bioenergy. 23, 367-380. [8]. Kadam, K.L. Rydholm, E.C. McMillan, J.D., 2004. Biotechnol. Prog. 20, 698-705. [9]. Krishnan M.S., Ho N.W.Y. Tsao G.T., 1999. Appl. Biochem. Biotechnol. 77-79, 373-388.

See more of this Session: Integrated Processes for Biochemical Conversion of Renewable Feedstocks to Fuels and Chemicals I
See more of this Group/Topical: Sustainable Engineering Forum

(226b) Comparison of One and Two Step Methods for Producing Jet Fuel Building Blocks From Cellulosic Biomass: Hemicellulose Release and Levulinic Acid Production From Maple Wood

Taiying Zhang, Bourns College of Engineering---Center for Environmental Research & Technology, University of California, Riverside, Riverside, CA, Jian Shi, Center for Environmental Research and Technology, Bourns College of Engineering, University of California, Riverside, Riverisde, CA and Charles Wyman, College of Engineering-Center for Environmental Research and Technology, University of California, Riverside, Riverside, CA

Xylose and levulinic acid are important building blocks for making jet fuel from cellulosic biomass. Although hemicellulose can be easily released by hot water or dilute acid pretreatment, levulinic acid is traditionally produced by reacting biomass directly with concentrated acid, degrading much of the hemicellulose sugars in very costly equipment. In this study, a two-step process was developed to increase yields of xylose and levulinic acid with less acid by releasing hemicellulose sugars before introducing dilute acid to produce levulinic acid from the resulting cellulose/lignin residue. This integrated process increased levulinic acid yields by up to ~15% compared with direct reaction with acid in a single step. In addition, yields of levulinic acid, xylose, xylose degradation products, total soluble carbon, and total organic carbon were followed throughout the process to compare with the traditional one step method. Furthermore, product yields and distributions were compared for use of sulfuric and hydrochloride acids over a range of acid concentrations, temperatures, and reaction times.

See more of this Session: Integrated Processes for Biochemical Conversion of Renewable Feedstocks to Fuels and Chemicals I
See more of this Group/Topical: Sustainable Engineering Forum

(226c) Sensitivity and Uncertainty Analysis of Updated Corn Stover to Ethanol Process Design and Economic Models

Ling Tao, David Humbird, Andy Aden and Eric C. D. Tan, National Bioenergy Center, National Renewable Energy Laboratory, Golden, CO

The Energy Independence and Security Act of 2007 (EISA 2007)1 mandates 36 billion gallons per year of renewable fuel by 2022, of which 15 billion gallons are traditional or corn-based, and 16 billion are derived from cellulosic biomass. In order for cellulosic biofuel production to ramp in 2012 as dictated by the EISA, it becomes increasingly important to understand its production economics. This information is used to predict the cost-competitiveness of cellulosic biofuels with petroleum-derived fuels and corn ethanol. A number of different conversion technologies exist for the conversion of cellulosic biomass to a liquid fuel. Our (National Renewable Energy Laboratory, NREL) efforts have been focused on both biochemical and thermochemical conversion to ethanol. In biochemical conversion, biocatalysts such as enzymes and microorganisms are used along with heat and chemicals to first convert biomass to an intermediate mixed sugar stream and then to ethanol or other fermentative biofuel. Researchers of NREL have described such processes in technical and economic detail in two previous design reports (Aden et al. 20022; Wooley et al. 19993). We have updated the ongoing process design in the last AIChE meeting, with the most current data and integration efforts at NREL as well as research funded by the US Department of Energy (DOE), and other sources (e.g., industry, academia). In this paper, alternative process conditions and parameters influencing the biofuel production cost are studied using sensitivity and uncertainty analysis. We have focused our studies on three major areas, feedstock compositional variability, yield variability and financial assumptions variability. Different feedstocks (or compositional variation of the same feedstock) can have a significant impact on equipment design, raw material utilization, utilities, and overall process economics. Feedstock compositional variability data was taken from a previous NREL study that evaluated compositions for 508 commercial hybrid corn stover samples collected from 47 sites in eight Corn Belt states after the 2001, 2002, and 2003 harvests (Templeton, et al., 2009)4. Monte Carlo analysis is carried out in order to assess the probability distribution of the production cost in order study potential interactions among parameters studied. The target of this sensitivity analysis is to provide an understanding of the impact of likely process variables and how these might be controlled to a definable degree.

References

(1) EISA. EISA of 2007 Calls for Additional Production of Biofuels. http://www.renewableenergyworld.com/rea/partner/stoel-rives-6442/news/article/2008/01/eisa-of-2007-calls-for-additional-production-of-biofuels-51063. 2007.

(2) Ibsen, K., Jechura J, Neeves, K., Sheehan, J., Wallace, R. Lignocellulosic Biomass to Ethanol Process Design and Economics Utilizing Co-Current Dilute Acid Prehydrolysis and Enzymatic Hydrolysis For Corn Stover. 2002. NREL Report NREL/TP-510-32438, http://www.nrel.gov/docs/fy02osti/32438.pdf

(3) Wooley, R., Ruth, M., Glassner, D., Sheehan, J. Process Design and Costing of Bioethanol Technology: a Tool for Determining the Status and Direction of Research and Development. Biotechnology Progress, 1999, 15(5), 794-803.

(4) Templeton D.W., Sluiter A.D., Hayward T.K. Hames B.R., and Thomas S.R., Assessing Corn Stover Composition and Sources of Variability via NIR, Cellulose, 16(4), 2009.

See more of this Session: Integrated Processes for Biochemical Conversion of Renewable Feedstocks to Fuels and Chemicals I
See more of this Group/Topical: Sustainable Engineering Forum

Two-stage gas stripping recovery process for enhanced butanol production in acetone-butanol-ethanol fermentation by *Clostridium beijerinckii*

Chuang Xue[a,b], **ShangTian Yang**[a,*]

a Department of Chemical and Biomolecular Engineering, Ohio State University, OH 43210, USA, Yang.15@osu.edu

b Department of Bioscience and Bioengineering, Dalian University of Technology, Dalian 116024, China

Abstract

Butanol is an important chemical and a potential biofuel that can be produced from biomass feedstock in acetone-butanol-ethanol (ABE) fermentation by solventogenic *Clostridia*. However, conventional ABE fermentation is limited by the low butanol product titer (<15 g/L), yield (~0.2 g/g glucose), and productivity (~0.3 g/L/h) due to strong butanol toxicity. In this work, a two-stage gas stripping recovery process was developed for online butanol recovery from ABE fermentation by a butanol-tolerant mutant strain of *Clostridium beijerinckii* immobilized in a fibrous bed bioreactor operated under repeated batch mode. In the successive batch fermentations with online gas stripping, total glucose utilization was 474.9 g/L and ABE production and productivity were 172.1 g/L and 0.53 g/L/h, respectively. In contrast, conventional batch fermentation without online gas stripping consumed 81.0 g/L glucose with ABE production and productivity of 25.5 g/L and 0.48 g/L/h, respectively. Overall, online gas stripping increased ABE productivity and yield 10.4% and 12.5%, respectively. Butanol, acetone and ethanol concentrations were 150.5, 36.9 and 8.6 g/L, respectively, in the condensate from the first-stage gas stripping and reached 434.2 g/L, 107.9 g/L and 19.2 g/L in the final product from the two-stage gas stripping. Acetic and butyric acids were not found in the condensate from the two-stage gas stripping process. The high butanol concentration obtained in the two-stage gas stripping process makes it highly energy efficient in recovering and purifying butanol from dilute ABE fermentation broth.

Exploring Three Pathways to Convert a Hemi-Cellulose Rich Pre-Pulping Extract into Alcohols via the MixAlco™ Process

Abigail S. Engelberth

G. Peter van Walsum

M. Clayton Wheeler

University of Maine, Department of Chemical Engineering

Forest Bioproducts Research Institute

Abstract

Conversion of a hemi-cellulose rich pre-pulping extract into long chain alcohols is of great interest to the developing forest product biorefinery. Pulp production requires, in most part, only the cellulose, leaving much of the hemi-cellulose available for extraction from the wood chips with subsequent conversion into petroleum compatible alcohols using the MixAlco™ process. The hemi-cellulose can be extracted without interference to current pulp yield. The by-products produced from the hemi-cellulose stream could assist in alleviating the problems facing the liquid fuel supply and could boost the revenue of the pulp mill. Three possible chemical routes explored for upgrading the hemicelluloses are: CKA: carboxylates → ketones → alcohols , CEA : carboxylates → esters → alcohols, CHEA: carboxylates → acids → esters → alcohols. The three scenarios were modeled using ASPENPlus™ to determine material balance and energy requirements. Sizing and energy demands from ASPENPlus™ were input into CAPCost to determine the bare module and operating expenses. The output from these two programs was compared to determine which pathway would be most economically viable based on: alcohol yield, alcohol energy density, and operating and capital costs.

Introduction

The chemical upgrading of a biodegradable material to alcohols by way of carboxylate salts produced from a mixed culture fermentation is known as the MixAlco™ process [1]. The MixAlco™ process can be applied to a variety feedstocks ranging from municipal solid waste to agricultural residues [1]. This study is a techno-economic analysis of the upgrading of a pre-pulping hemi-cellulose rich stream via the MixAlco™ process. The purpose of this study is to compare the material and energy outputs for three biochemical upgrading scenarios using the carboxylate platform.

The process for all three pathways begins with a mixed-culture acidogenic fermentation to convert the hemi-celluloses into carboxylate salts. Once the salts have been formed, the pathways begin to differ. For the first pathway – the CKA route –the salts are decomposed into ketones and the resulting ketones are hydrogenated into alcohols. The second pathway – the CEA route - esterifies the salts and then subjects the esters to hydrogenolysis for alcohol formation. The third pathway – the CHEA route – first converts to salts into acids and then extracts the acids from the water using liquid-liquid extraction and the recovered acids are then esterified and undergo hydrogenolysis to form alcohols.

A 'near-neutral' hemi-cellulose extraction method has been studied and has little effect on the quality of the pulp produced [2, 3]. The hemi-cellulose rich stream used for this study was based on the work performed by Mao *et al.* (2008) which uses 1000 tonne/day pulp production as a basis for estimations. The contents and flowrate of the stream from the extractor are shown in Table 1. The hemi-cellulose concentration in the stream is 13.2 g/L. The hemicellulose-rich stream exiting the extractor is at 160°C[2].

Table 1: Contents and flowrate of stream from hemicellulose extractor.

Component	Mass flow rate (kg/hr)
Water	120960.4
Salt	843.8
4-O-MGA	1758.3
Hemicellulose	1947.0
Acetate	1466.7
Lignin	961.9

Process Description and Modeling Notes

Each of the three upgrading pathways was modeled using AspenPlus™ Version 7.2. The aim of this project was to create a model in Aspen that would use realistic components to obtain mass and energy balances that could be used to compare the three pathways so as to determine the most advantageous. The reactions were modeled as best case scenarios, meaning that they all were stoichiometrically based and were completed at 100% conversion. The processes were modeled in this way as a stepping off point to determine with which pathway to move forward. The processing operations for the three upgrading pathways have been separated into areas and the following section explains each area and makes notes of parameters and modules used during the simulations.

Components

The components used during simulation were not necessarily the components that would actually appear in the process because the Aspen properties database does have limitations. The components used were model components and were chosen based on available components in the Aspen database that most closely matched their real-world counterparts. The virtual components and their real-world equivalents for all three pathways are listed in Table 2.

Table 2: Realistic compounds used in the MixAlco process and their Aspen equivalent counterparts.

Compound	Actual Formula	Aspen Equivalent Component Name	ASPEN Formula
Water	H_2O	Water	H2O
Glucose	$C_6H_{12}O_6$	Dextrose	C6H12O6
Mannose	$C_6H_{12}O_6$	Dextrose	C6H12O6
Galactose	$C_6H_{12}O_6$	Dextrose	C6H12O6
Xylose	$C_5H_{10}O_5$	D-Xylose	C5H10O5
Arabinose	$C_5H_{10}O_5$	D-Xylose	C5H10O5
Glucose-oligomer		Dilactic-Acid	C6H10O5

Mannose-oligomer		Dilactic Acid	C6H10O5
Galactose-oligomer		Dilactic Acid	C6H10O5
Xylose-oligomer		Glutaric-Acid	C5H8O4
Arabinose-oligomer		Glutaric-Acid	C5H8O4
Lignin	Variable	Guaiacol (solid)	C7H8O2-E1
Sodium Bicarbonate	$NaHCO_3$	Sodium-Bicarbonate	NAHCO3
4-O-MGA (methylgallic acid)	$C_7H_8O_5$	D-Gulonic-Acid-.Gamma-Lactone	C6H10O6-N5
Undissociated Acetate	$C_2H_3O^+$	Sodium-Acetate	C2H3NAO2
Calcium-Oxide	CaO	Calcium Oxide	CAO
Acetic Acid	$C_2H_4O_2$	Acetic-Acid	C2H4O2-1
Propionic Acid	$C_3H_6O_2$	Propionic-Acid	C3H6O2-1
Calcium Acetate	$Ca(CH_3CO_2)_2$	CA(CH3COO)2	CA(AC)2
Calcium Propionate	$Ca(C_2H_5CO_2)_2$	CA(CH3CH2COO)2	CA(PROP)2
Acetone	C_3H_6O	Acetone	C3H6O-1
Pentanone	$C_5H_{10}O$	Diethyl-Ketone	C5H10O-4
Isopropyl-Alcohol	C_3H_8O	Isopropyl-Alcohol	C3H8O-2
Pentanol	$C_5H_{12}O$	3-Pentanol	C5H12O-D4
Carbon Dioxide	CO_2	Carbon-Dioxide	CO2
Carbon Monoxide	CO	Carbon-Monoxide	CO
Hydrogen	H_2	Hydrogen	H2
Char	C	Carbon-Graphite	C
Lignin oil	Variable	1-Butyne	C4H6-1
Calcium carbonate	$CaCO_3$	Calcium-Carbonate-Aragonite	CACO3
Carbonic acid	H_2CO_3	Carbonic-Acid	H2CO3
Sodium Acetate	$NaC_2H_3O_2$	Sodium-Acetate	C2H3NAO2
Sodium-Propionate	$NaC_3H_5O_2$	Sodium-Propionate	C3H5NAO2
Ammonium Acetate	$C_2H_3O_2NH_4$	Ammonium-Acetate	C2H7NO2
Ammonium Propanoate	$C_3H_5O_2NH_4$	Ammonium-Propionate	C3H9NO2
Ammonium Bicarbonate	NH_4HCO_3	Ammonium-Hydrogen-Carbonate	NH4HCO3
Ammonia	NH_3	Ammonia	H3N
Octanol	$C_8H_{18}O$	1-Octanol	C8H18O-1
Octyl Acetate	$C_{10}H_{20}O_2$	N-Octyl-Acetate	C10H20O2-D4
Octyl Propanoate	$C_{11}H_{22}O_2$	Octyl-Propanoate	C11H22O2-N2
Ethanol	C_2H_6O	Ethanol	C2H6O
Propanol	C_3H_8O	1-Propanol	C3H8O-1

Evaporation

The first step in the process is to remove about a quarter of the water from the hemi-cellulose rich stream in order to concentrate the stream for fermentation and to relieve some downstream costs associated with a high-volume stream. Evaporating directly after extraction takes advantage of the energy present in the stream and does not require as much steam for evaporation. If the evaporator were placed further downstream, then more steam would be required to evaporate an equal amount of water. Prior to evaporation a quantity of either calcium carbonate or ammonium carbonate is added in sufficient quantity to neutralize the amount of acetate present in the stream. The carbonate is recycled from a downstream unit. The neutral pH affords any dissociated acetic acid present from extraction to remain in solution so that it is not lost to the vapor during evaporation. In Aspen, evaporation is modeled as a flash tank powered by 50 psig steam (saturated water vapor). The steam use is calculated by a calculator block which varies the steam flow-rate based on the amount of water to be evaporated. The liquid from the evaporator is sent to the fermentation train.

Fermentation

The fermentation scheme for the MixAlco process makes use of a mixed culture of microorganisms to convert the hemicelluloses into carboxylate salts. The mixed culture allows for the hydrolysis, fermentation and salt formation to occur within the same reactor [4]. The salts are formed by the addition of a buffering agent: calcium carbonate for both the CKA and CHEA routes and ammonium carbonate for the CEA route. The fermentation was modeled in Aspen using a series of four stoichiometric reactors. The reactions to take place in each of the vessels can be summarized as: 1) convert oligomers to monomers with the addition of water, 2) convert monomers to acids, 3) convert acids to carbolxylate salts, and then finally 4) convert H_2CO_3 made during monomer conversion into water and carbon dioxide, which is what would occur at the pH in the reactor. The CO_2 made is vented and recovered for further downstream processing. Table 3 better summarizes the reactions occurring during fermentation. The salts formed during fermentation are mostly calcium acetate, but carbon chains up to calcium heptanate have been observed. For this model only calcium acetate and calcium propionate were used since higher chain calcium salts were not present in the Aspen properties database. Fermentation is maintained at 40°F for a residence time of 2 weeks. The reactors were modeled as four separate blocks to ease trouble-shooting if issues were to arise.

Table 3: Summary of reactions that occur during fermentation for all three pathways.

Reactor Number	Reaction Description	Reaction Stoichiometery
1	Oligomers →Monomers	$C_6H_{10}O_5 + H_2O \rightarrow C_6H_{12}O_6$ (C6 Sugars) $C_5H_8O_4 + H_2O \rightarrow C_5H_{10}O_5$ (C5 Sugars)
1	Acetate → Acetic Acid	$C_2H_3O_2Na + H_2O \rightarrow C_2H_4O_2 + NaOH$
2	Monomers → Carboxylic Acids	$5{\cdot}C_6H_{12}O_6 \rightarrow 10.5{\cdot}C_2H_4O_2 + 2.6{\cdot}C_3H_6O_2 + 1.3{\cdot}CO_2 + 1.3{\cdot}H_2O$ $6{\cdot}C_5H_{10}O_5 \rightarrow 10.5{\cdot}C_2H_4O_2 + 2.6{\cdot}C_3H_6O_2 + 1.3{\cdot}CO_2 + 1.3{\cdot}H_2O$ 6·4-O-MGA $\rightarrow 5.4{\cdot}C_2H_4O_2 + 6.3{\cdot}C_3H_6O_2 + 6.2{\cdot}CO_2 + .2{\cdot}H_2O$
3	Carboxylic Acids → Carboxylate Salts (Calcium Salts)	$2{\cdot}C_2H_4O_2 + CaCO_3 \rightarrow Ca(C_2H_3O_2)_2 + H_2CO_3$ $2{\cdot}C_3H_6O_2 + CaCO_3 \rightarrow Ca(C_3H_5O_2)_2 + H_2CO_3$
3	Carboxylic Acids → Carboxylate Salts (Ammonium Salts)	$C_2H_4O_2 + NH_4HCO_3 \rightarrow CH_3CO_2NH_4 + H_2CO_3$ $C_3H_6O_2 + NH_4HCO_3 \rightarrow CH_3CH_2CO_2NH_4 + H_2CO_3$
4	Carbonic acid split	$H_2CO_3 \rightarrow CO_2 + H_2O$

Dewatering – Mechanical Vapor Recompression

The salts produced during fermentation are dried for use in thermal deoxygenation, esterification or acid recovery. The water was removed from each stream using mechanical vapor recompression. This process was modeled in Aspen beginning with the wet carboxylate stream entering a flash tank. The tank is set to an outlet temperature of 96°C with the vapor leaving the tank is sent to a centrifugal compressor in which it is compressed to a pressure of 1.3 times its entering pressure. The vapor stream is then sent to a heat exchanger where it is fully condensed. The heat generated during condensation is used to drive the evaporator. The operational cost for the vapor recompression unit only relies on the electricity required to power the compressor and is an efficient use of energy [5]. The vapor recompression unit removed 97.5% of the water that was initially present. The salts leaving the MVR unit were mostly dry and were then sent downstream to the next unit depending on the pathway. The salts leaving the MVR unit in the CHEA pathway were dried to a lesser extent because more water was required for the liquid-liquid extraction unit. Figure 1 displays a simplified diagram of the first three units that are common in all three upgrading pathways. The units are color coded and labeled to simplify discussion of the cost break down of each pathway.

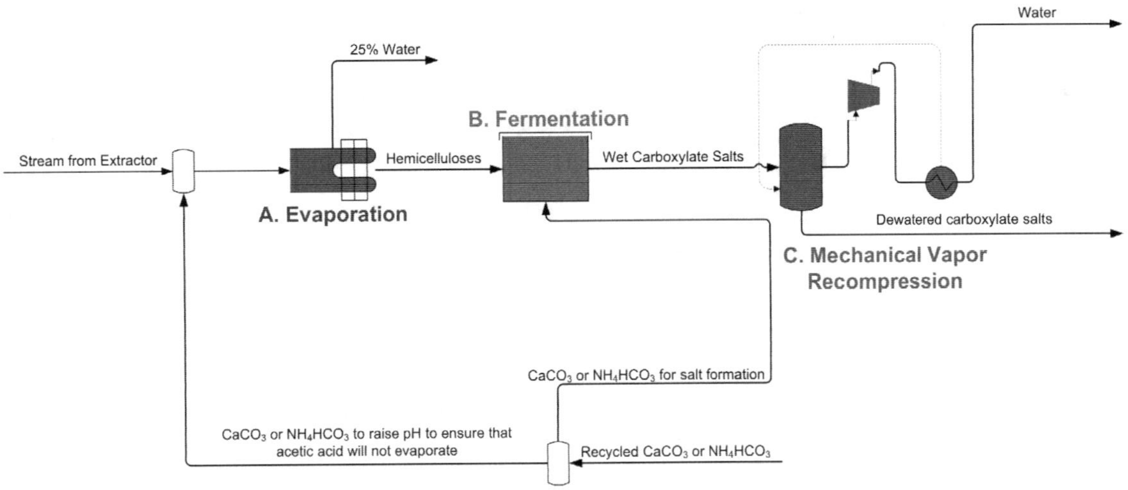

Figure 1: Simplified diagram of the first three units that are common to all three upgrading pathways. The alkaline used in the recycle depends on the pathway. Calcium carbonate is used for the CKA and the CHEA routes while ammonium bicarbonate is used for the CEA route.

Thermal Deoxygenation

The next step in the CKA pathway is to convert the dried carboxylate salts into ketones. This is achieved using a technique termed 'thermal deoxygenation' or TDO. TDO is a process that uses high temperatures to heat the carboxylate salts in order to release bound oxygen to form a ketone. Calcium carboxylate salts decompose to form a ketone and calcium carbonate at or above 430°C with conversion yields as high as 99.5% [1, 6, 7]. Table 4 summarizes the reactions occurring in the TDO reactor – modeled as a stoichiometric reactor with 100% conversion – in the Aspen model. For this step, we have assumed that the lignin present in the stream will decompose into char and a lignin-oil.

Table 4: Summary of reactions that occur during thermal deoxygenation.

ReactionNumber	Stoichiometry
1	$Ca(C_2H_3O_2)_2 \rightarrow C_3H_6O + CaCO_3$
2	$Ca(C_3H_5O_2)_2 \rightarrow C_5H_{10}O + CaCO_3$
3	$C_7H_8O_2$[Lignin] $\rightarrow 2{\cdot}CO + C$[Char) $+ H_2 + C_4H_6$[Lignin-oil]

Hydrogenation

During hydrogenation the ketones are converted into alcohols with the addition of hydrogen. The alcohols produced in for this study are isopropanol and 3-pentanol. Longer chain alcohols are possible, but where not modeled because of the limited variety of carboxylate salts in the Aspen properties database. The hydrogenation reactions were modeled using a stoichiometric reactor with 100% conversion. The reaction temperature was set to 40°C. Table 5 summarizes the reactions that occur during hydrogenation. Figure 2 illustrates the configuration of the processing units that are specific to the CKA route which is subsequent to the dewatering step. This figure is in direct succession from Figure 1 and is also color coded to aid in explanation for the breakdown of costing for each pathway.

Table 5: Reactions that occur during hydrogenation to produce isopropanol and 3-pentanol.

Reaction Number	Stoichiometry
1	$C_3H_6O + H_2 \rightarrow C_3H_8O$
2	$C_5H_{10}O + H_2 \rightarrow C_5H_{10}O$

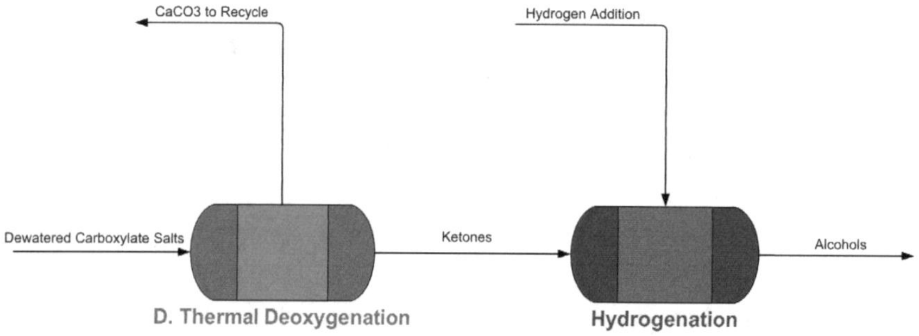

Figure 2: Configuration of units specific to the CKA route starting after the dewatering unit. Inlet stream to the thermal deoxygenation unit is the same as the outlet stream from the mechanical vapor recompression unit in Figure 1.

Esterification

The first step in the CEA route after dewatering is the conversion of ammonium carboxylate salts into esters. This is also a step used during the CHEA route which instead formed esters from the addition of octanol to the carboxylic acids. Both routes modeled the esterification using a stoichiometric reactor set to 80°C with 100% conversion. The salts or acids are esterified with octanol to make long chain octyl esters. The octanol is recovered and recycled during the process. The octyl esters leaving the reactor are sent to a distillation tower to remove the water formed during esterification from the octyl esters. The distillation tower was modeled as a rigorous distillation unit and the column diameter, height and number of trays were calculated for use in costing the process. The reactions that were used for the CEA and CHEA route are shown in Table 6.

Table 6: Esterification reactions for both the CEA and the CHEA pathways.

Reaction Number	Stoichiometry
	CEA Pathway
1	$C_8H_{18}O + C_2H_3O_2NH_4 \rightarrow C_{10}H_{20}O_2 + NH_3 + H_2O$
2	$C_8H_{18}O + C_3H_5O_2NH_4 \rightarrow C_{11}H_{22}O_2 + NH_3 + H_2O$
	CHEA Pathway
1	$C_8H_{18}O + C_2H_4O_2 \rightarrow C_{10}H_{20}O_2 + H_2O$
2	$C_8H_{18}O + C_3H_6O_2 \rightarrow C_{11}H_{22}O_2 + H_2O$

Hydrogenolysis

The next step on both the CEA and the CHEA pathway is hydrogenolysis which converts the octyl esters into alcohols. For this project only ethanol and propanol were considered. Hydrogenolysis is a simple reaction that breaks an ester into two alcohols with the addition of hydrogen [8]. The reaction was modeled in a stoichiometric reactor at 220°C. The reactions for the CEA and CHEA route were the same and are shown in Table 7. The alcohol stream is then sent to a distillation column that recovers the lower boiling point alcohols from the octanol. The octanol is then recycled with additional octanol added since the distillation is not a perfect separation. Figure 3 displays the processes discussed for the CEA route which occur after the dewatering step. The units are labeld color coded and correspond to their equivalent steps in the CKA route. Esterification – including reaction and distillation – is matched to the thermal deoxygenation and the hydrogenolysis – including reaction and distillation – is matched with hydrogenation. The new step, ammonia conversion, is a different color and will be costed as a separate unit.

Table 7: Hydrogenolysis reactions for both the CEA and CHEA pathways.

Reaction Number	Stoichiometry
1	$C_{10}H_{20}O_2 + 2 \cdot H_2 \rightarrow C_2H_6O + C_8H_{18}O$
2	$C_{11}H_{22}O_2 + 2 \cdot H_2 \rightarrow C_3H_8O + C_8H_{18}O$

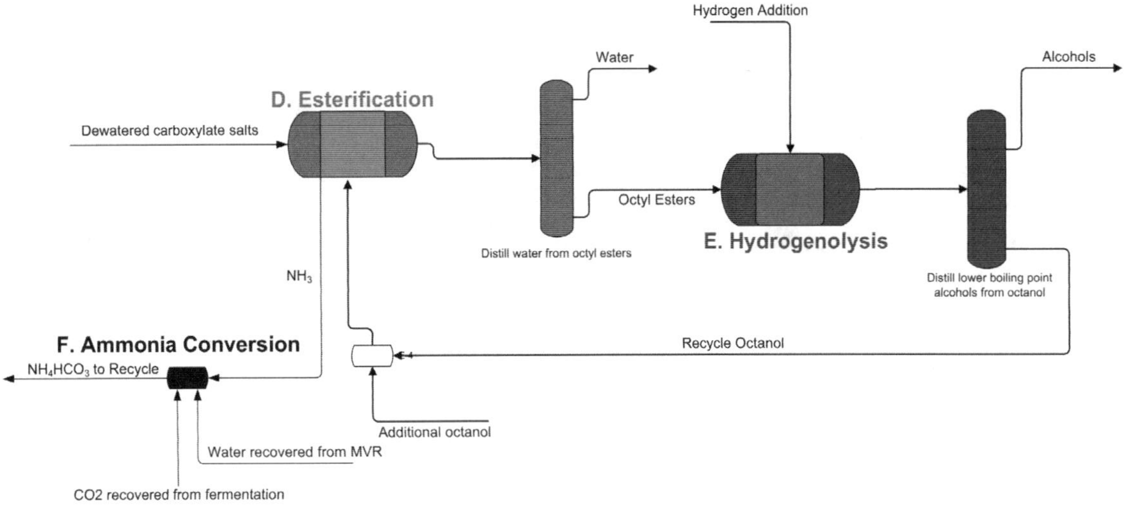

Figure 3: Configuration of units specific to the CEA route starting after the dewatering unit. Inlet stream to the w esterification reactor is the same as the outlet stream from the mechanical vapor recompression unit in Figure 1.

Ammonium Conversion

The ammonia made during distillation is recovered and recycled throughout the process. The ammonia is combined with the CO_2 that was vented from the fermentation with the water that was recovered during the dewatering step to form ammonium bicarbonate. This occurs in a stoichiometric reactor at 40°C. The ammonium bicarbonate recovery is shown in Figure 3.

Liquid-Liquid Extraction

The major difference between the CEA and the CHEA route is the addition of a liquid-liquid extraction step between dewatering and esterification. This step begins with the addition of carbon dioxide and water to release the calcium from the carboxylate salt [9]. The reactions were modeled with a stoichiometric reactor at 40°C. The acids produced are then recovered from the water stream using ethyl acetate as the extracting solvent. The liquid-liquid extraction was modeled with an extraction column and the split of the acids between the oulet streams is shown in Table 8.

Table 8: Resulting component split for extraction of acetic and propionic acid from water using ethyl acetate.

Component	Water-Rich Stream	Acid Rich Stream
Water	0.5527	0.4473
Acetic Acid	0.0080	0.9920
Propionic Acid	0.0001	0.9999
Ethyl Acetate	0.0790	0.9210

The acid rich stream is then distilled to remove the ethyl acetate and water which is then recycled back into the liquid-liquid extraction tower. Additional ethyl acetate is required since the distillation step is not a perfect separation. The acids are then sent to the esterificaton unit where they are combined with octanol to form octyl esters. Figure 4 illustrates the CHEA route after the dewatering step. The new portion is labeled as 'acid extraction'.

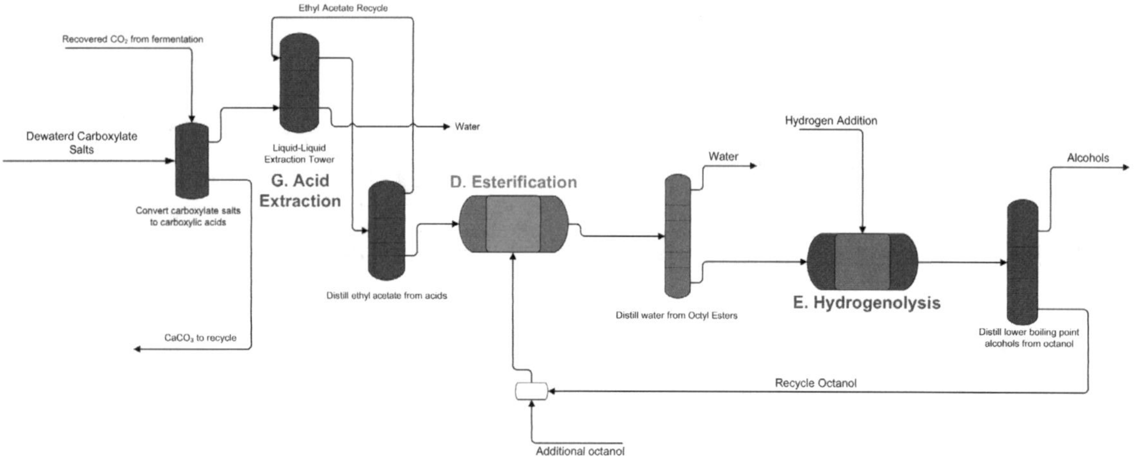

Figure 4: Configuration of units specific to the CHEA route starting after the dewatering unit. Inlet stream to the acid extraction unit is the same as the outlet stream from the mechanical vapor recompression unit in Figure 1.

Results & Discussion

Sizing

The material and energy balances simulated in Aspen were used in CAPCost to determine the size of the processing units and the capital costs for each section. The energy balances were also used to determine the operating costs for each pathway. Table 9 is a breakdown of the processing areas discussed in the previous section. The size of the liquid-liquid extraction column was manually determined using the mass balances and the density of the streams to calculate the diameter based on flooding calculations [10].

Table 9: Processing units from Aspen used for calculating the capital cost with CAPCost. Special notes for the the processing units are included where deemed necessary. The table is color coded to match the key areas in Figures 1 through 4.

Process Unit	Reactor	Units Included in ASPEN	Special Notes
A	Evaporation	1 flash tank	Considered a single effect evaporator
B	Fermentation	4 stoichiometric reactors – staged for easier troubleshooting	Broken down to 4 tanks for costing purposes Heater added for capital and operating costs Same for all scenarios
C	Dewatering	2 heat exchangers, 1 compressor and 1 flash tank	Stream needed to be heated prior to flash, heat from cooling used in flash tank
D	Thermal Deoxygenation (CKA) Esterification (CEA, CHEA)	1 stoichiometric reactor , TDO requires heat exchanger to cool stream, Esterification required distillation column	Thermal deoxygenation needs high heat All reactors have T_R = 35 min
E	Hydrogenation (CKA) Hydrogenolysis (CEA, CHEA)	1 stoichiometric reactor Hydrogenolysis requires cooling unit and distillation column	Hydrogenation at 40 °C and 10 bar Hydrogenolysis at 220°C and 62 bar All reactors have T_R = 35 min
F	Ammonia Conversion	Single stoichimetric reactor	Only required for CEA route T_R = 10 min
G	Acid Extraction	1 stoichiometric reactor, 1 extraction column and 1 distillation column	Only required for CHEA route Ethyl acetate is fully recycled

Costing

The capital cost was determined from the bare module cost given by CAPCost and is shown, separated by unit areas, in Figure 5. The largest cost area common to all three pathways is the dewatering step. The removal of water is an expensive step for any process, but it is essential because if the water were to remain in the stream the downstream processing units would be much larger and more expensive and water may play an inhibitory role in the downstream reactions.

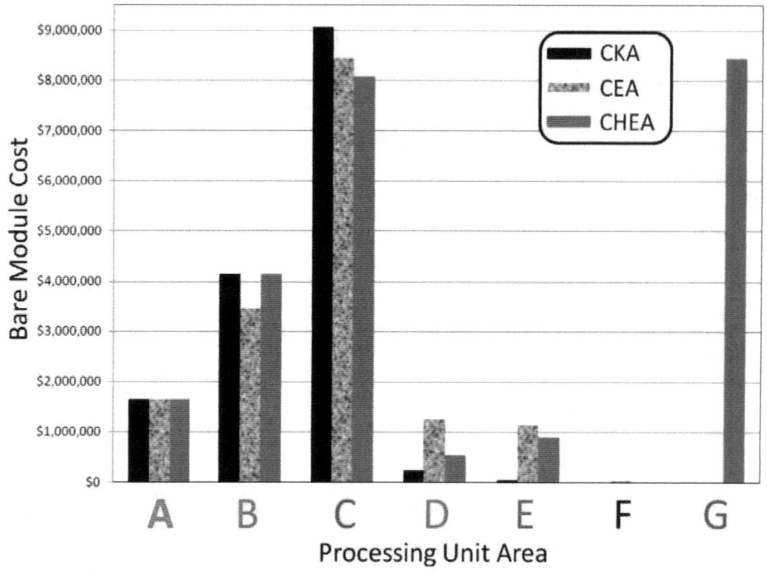

Figure 5: Bare module cost of each processing unit area as assessed by CAPCost using the material balances from the Aspen simulations.

The operating costs were determined using the energy balances from Aspen in the CAPCost program. The utilities used were low pressure steam ($13.28/GJ), cooling water ($0.534/GJ) and electricity ($16.8/GJ). The processing unit was matched with the required utility based on the energy balance attained from Aspen. The overall annual operating cost for each processing unit area is shown in Figure 6. Note that the fermentation for the CEA route required less energy to maintain the fermentation at 40°C because the reactions with ammonium bicarbonate are more exothermic. The addition of distillation to both the CEA and CHEA pathways cause an increase in operating costs for the hydrogenation/hydrogenolysis area but does not have much of an effective difference for the esterification area since the equivalent for the CKA route – thermal deoxygenation – requires high temperatures to decompose the salts.

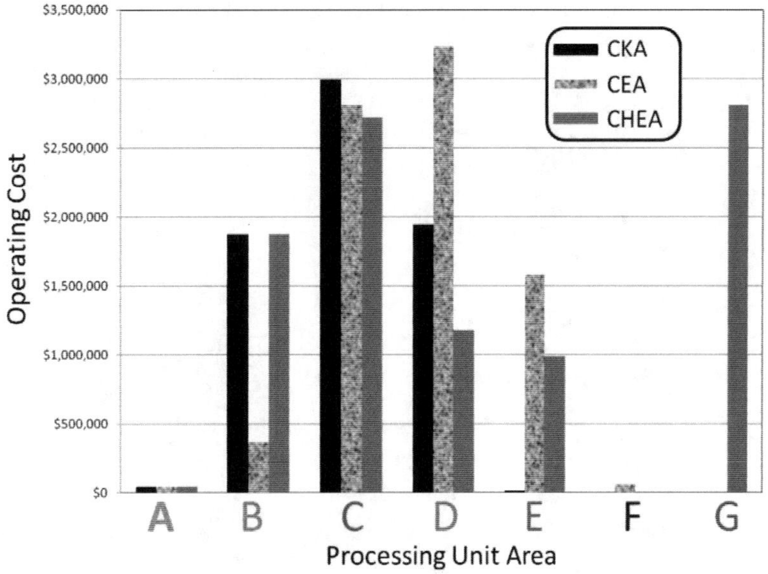

Figure 6: Annual operating cost of each processing unit area as assessed by CAPCost using the energy balances from the Aspen simulations.

Alcohol production

One way to compare the three pathways is based on the yearly annual production of alcohols. The CKA route, for this example, produces isopropanol and pentanol and the CEA and CHEA routes produce ethanol and propanol. Table 10 displays the annual alcohol production (in gallons) for each pathway. The CKA route has the least amount of annual alcohol production because during the TDO reaction two acetic acid groups are combined to form one acetone group. This differs from the other two pathways which each, in a sense, have one acid group forming the alcohol. The CHEA route produces less alcohol than the CEA route because some of the acetic acid is lost during the liquid-liquid extraction and recovery.

Table 10: Annual alcohol production in gallons for each upgrading pathway.

Alcohol Produced	CKA	CEA	CHEA
Ethanol	-	7,386,958	7,183,582
Propanol	-	2,709,230	2,708,452
Isopropanol	4,895,922	-	-
Pentanol	1,963,779	-	-
Total (gallons)	6,859,701	10,096,188	9,892,034

Another way of comparing the alcohol production of each of the three pathways is to look out how much energy is possible from combustion of each of the alcohols produced minus the energy of the hydrogen required to produce the alcohols. The CEA and CHEA pathways require two moles of hydrogen for the reactions while the CKA route only requires one mole for alcohol formation. Figure 7 displays the annual potential energy of combustion (in terajoules) for the alcohols, the energy of the hydrogen required to produce the alcohol and the difference in the energy. This will be considered the net energy of the process and does not include the operating costs. The net energy calculation shows that the CKA route has the most energy potential even though it produced the lowest volume of alcohol.

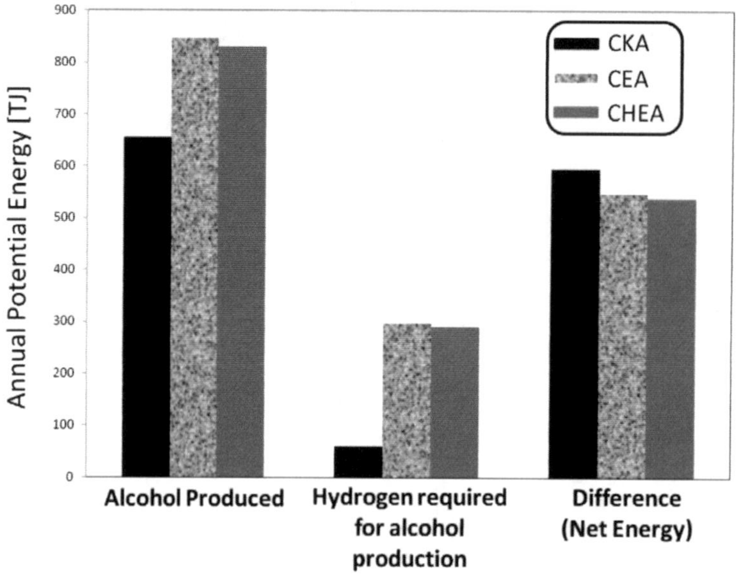

Figure 7: Net energy (in terajoules) of each of the pathways. The net energy is the potential energy in the annual alcohol production minus the energy in the hydrogen required to produce the alcohols.

Production Costs

Comparing the capital cost to the amount of alcohol produced and to the net potential energy gives an insight as to how the capital of each process weighs against the amount of product made. The total capital cost and the resulting $/gallon and $/MJ is shown in Table 11. This comparison illustrates that the CEA route is the least expensive based on the volume of alcohol produced and that the CKA route has the most potential for energy based on the capital required to produce the alcohols.

Table 11: Production cost for each pathway based on the annual volume of alcohol produced and on the net potential energy of the alcohol production.

Upgrading Pathway	Capital Cost	$/gallon	$/MJ
CKA	$15,226,700	2.22	0.026
CEA	$16,323,900	1.62	0.030
CHEA	$18,182,700	1.84	0.034

A similar analysis can be performed using the operating cost of each pathway. Table 12 shows the comparison of annual operating cost to the volume and energy potential of the products. The results follow the same trend as the capital cost per gallon and MJ, which is useful when comparing the three pathways.

Table 12: Comparison of the annual operating cost for each pathway based on the annual volume of alcohol produced and the net potential energy of the alcohol production.

Upgrading Pathway	Annual Operating Cost	$/gallon	$/MJ
CKA	$6,869,300	1.00	0.012
CEA	$8,088,900	0.80	0.015
CHEA	$9,621,900	0.97	0.018

Conclusions

The results of this project demonstrate that the CKA route is the least expensive in terms of potential energy for both capital and annual operating cost but that the CEA route is least expensive based on the annual volume of alcohol produced. Thus, to the extent that fuels are sold on a volume rather than energy content basis, the CEA route may be preferable. The cost calculations are also very sensitive to the cost of hydrogen, particularly for the esterification routes. Results from this model represent the process optima for these different pathways, since all reactions are modeled as giving 100% yield. Doubtless more representative yields will raise costs for all three pathways and could also change the economic calculations in favor of different pathways.

Acknowledgements:

This work was funded in part by award DE-FG36-08GO18165 from the Golden Field office of the DOE and by Department of Defense DARPA through Logos Tech. BAA 08-07 #HR0011-09-C-0075 SUB226-UME.

References

1. Holtzapple, M.T., et al., *Biomass conversion to mixed alcohol fuels using the MixAlco process.* Applied Biochemistry and Biotechnology, 1999. **77-9**: p. 609-631.
2. Mao, H., et al., *Technical economic evaluation of a hardwood biorefinery using the "Near-Neutral" hemicellulose pre-extraction process.* Journal of Biobased Materials and Bioenergy, 2008. **2**(2): p. 177-185.
3. Al-Dajani, W.W. and U.W. Tschirner, *Pre-extraction of hemicelluloses and subsequent kraft pulping Part I: alkaline extraction.* Tappi Journal, 2008. **7**(6): p. 3-8.
4. Granda, C.B., et al., *Carboxylate platform: the MixAlco process part 2: process economics.* Applied Biochemistry and Biotechnology, 2009. **156**(1-3): p. 537-554.
5. McCabe, W.L., J.C. Smith, and P. Harriott, *Unit Operations of Chemical Engineering*. 6th ed. 2001, Boston: McGraw Hill. 1114.
6. Ardagh, E.G.R., et al., *Distillation of acetate of lime.* Industrial and Engineering Chemistry, 1924. **16**: p. 1133-1139.
7. Pham, V., M. Holtzapple, and M. El-Halwagi, *Techno-economic analysis of biomass to fuel conversion via the MixAlco process.* Journal of Industrial Microbiology & Biotechnology, 2010: p. 1-12.
8. Adkins, H. and K. Folkers, *The catalytic hydrogenation of esters to alcohols.* Journal of the American Chemical Society, 1931. **53**(3): p. 1095-1097.

9. Chan, W. and M. Holtzapple, *Conversion of municipal solid wastes to carboxylic acids by thermophilic fermentation.* Applied Biochemistry and Biotechnology, 2003. **111**(2): p. 93-112.
10. Seader, J.D. and E.J. Henley, *Separation Process Principles*. 1998, New York: John Wiley & Sons, In.

(226f) Optimizing a Separate Hemicellulosic and Cellulosic Conversion Process Design for Producing Ethanol

Alexandre Chapeaux, Nancy Dowe and Daniel J. Schell, National Bioenergy Center, National Renewable Energy Laboratory, Golden, CO

Dilute-acid pretreatment is used to convert hemicelluloses present in lignocellulosic feedstocks into fermentable sugars. It also enhances the susceptibility of cellulose for conversion to glucose by enzymatic hydrolysis. Improving the conversion of sugars to ethanol may be possible by separately fermenting the hemicellulosic-sugar-rich stream and the glucose stream derived from enzymatic hydrolysis of cellulose. Fermenting the hemicellulosic sugars separately improves sugar conversion yields because the inhibitory effect of high ethanol concentrations is not present. Also, higher glucose yields from enzymatic cellulose hydrolysis are achieved by reducing inhibition of the enzyme by soluble sugars produced in pretreatment.

However, additional costs are incurred due to the presence of costly solid/liquid separation equipment, the additional capital equipment needed to operate two separate fermentation trains, and the potential additional water needed to dilute the cellulosic solids stream. This process configuration was investigated on pretreated corn stover produced in a 1 ton/d pilot-scale continuous reactor. The pretreated corn stover slurry was separated into a liquid and solid fraction using a custom hydraulic press. The hemicellulosic sugar stream was diluted with water to various concentrations, and then the sugars were fermented to ethanol. The cellulosic stream was re-suspended in either fresh water or spent broth from fermentation of the hemicellulosic sugar stream. The cellulose was converted to ethanol using cellulase enzymes and the yeast Saccharomyces pastorianus in either a simultaneous saccharification and fermentation (SSF) or separate enzymatic hydrolysis and fermentation (SHF) process.

Conversion yields were calculated from initial and final concentrations of sugars and ethanol and used to model process economics as a function of the total solids (TS) loading (based on untreated biomass). For the hemicellulosic sugar stream, we found that neither conditioning nor neutralizing the liquor adversely affected the yields; and we observed a sharp drop after 20% equivalent TS loading. For the cellulosic sugar stream, preliminary data suggests that there is no difference in the yield between the SSF and the SHF process. Further, we saw no significant effect on the yields even at 25% equivalent TS loading. Further results will be presented detailing the impact that the interaction of solids loading, liquor dilution, and recycling the spent broth has on conversion yields and ethanol production.

See more of this Session: Integrated Processes for Biochemical Conversion of Renewable Feedstocks to Fuels and Chemicals I

(252a) Biology In 4 Dimensions: Dynamic Hydrogel Niches for Tissue Regeneration

Kristi Anseth, Department of Chemical and Biological Engineering, University of Colorado, Howard Hughes Medical Institute, Boulder, CO

Methods for culturing mammalian cells in a biologically relevant context are increasingly needed to study cell and tissue physiology, expand and differentiate progenitor cells, and to grow replacement tissues for regenerative medicine. Two-dimensional culture has been the paradigm for in vitro cell culture; however, evidence and intuition suggest that cells behave differently when they are isolated from the complex architecture of their native tissues and constrained to petri dishes or material surfaces with unnaturally high stiffness, polarity, and surface to volume ratio. As a result, biologists are often faced with the need for a more physiologically relevant 3D culture environment, and many researchers are realizing the advantages of hydrogels as a means of creating custom 3D microenvironments with highly controlled chemical, biological and physical cues. Further, the native ECM is far from static, so ECM mimics must also be dynamic to direct complex cellular behavior. In general, there is an un-met need for materials that allow user-defined control over the spatio-temporal presentation of important signals, such as integrin-binding ligands, growth factor release, and biomechanical signals. Developing such hydrogel mimics of the ECM for 3D cell culture is an archetypal engineering problem, requiring control of numerous properties on multiple time and length scales important for cellular functions. New materials systems have the potential to significantly improve our understanding of how cells receive information from their microenvironment and the role that these dynamic processes may play in controlling the stem cell niche to cancer metastasis. This talk will illustrate our recent efforts to advance hydrogel chemistries for 3D cell culture and dynamically control biochemical and biophysical properties through orthogonal, photochemical reaction mechanisms.

See more of this Session: Professional Progress Award Lecture
See more of this Group/Topical: Liaison Functions

(257a) Investigation of Microbial Behavior of Yeasts in Solid State and Submerged Liquid Fermentations for Ethanol Production From Sweet Sorghum

Lei Zhang[1], Shi-zhong Li[2] and Min Zhang[1], (1)National Bioenergy Center, National Renewable Energy Laboratory, Golden, CO, (2)Institute of Nuclear and New Energy Technology, Tsinghua University, Beijing, China

Comparing to conventional submerged liquid fermentation (SLF), the solid state fermentation (SSF) has not found widespread application at large scale due to the barrier of heat and mass transfer, and limited understanding of behavior of microorganisms in SSF. The SSF, however, has the advantage of reducing the process cost of substrate pretreatment, lower water usage and negative impact by the waste of fermentation on environment. A novel fermentor using rotary drum concept for SSF was developed and recently, the demonstration of bioethanol production from sweet sorghum using this novel fermentor of 127 m3 in China was successfully achieved. Saccharomyces cerevisiae TSH1, applied in the fermentation, showed high efficiency of ethanol production during the course of SSF, which completed sugars utilization within 24 h using sweet sorghum containing a total sugar content of 10~14% (w/w). This study focuses on investigating the robust character of the TSH1 as compared to other yeast strains in SSF vs SLF and further offers an insight into a new approach to understand the microbial behavior in the process of SSF.

See more of this Session: Biological Conversions and Processes for Renewable Feedstocks II
See more of this Group/Topical: Sustainable Engineering Forum

(257b) Effect of Inhibitors in Corn Stover Hydrolysate On Fermentation of Glucose and Xylose by Zymomonas Mobilis

Min Zhang, National Bioenergy Center, National Renewable Energy Laboratory, Golden, CO

One of the biggest challenges is to achieve high sugar yields resulting from pretreatment and enzymatic saccharification of biomass while ensuring that the sugar streams generated are not overly toxic to fermentation organisms in order to achieve higher ethanol yields in fermentation at process concentrations. Biomass pretreatment is necessary to reduce the recalcitrance of lignocellulosic feedstocks to enzymatic saccharification and to allow the liberation of soluble sugars from plant cell wall carbohydrates, but pretreatment can also release a number of undesirable compounds into the hydrolyzate that can inhibit the growth and ethanol productivity of fermentation organisms. Inhibitors vary in range and amount depending on the feedstock, pretreatment methods and process conditions. We conducted systematic studies by 1) developing a quantitative, high throughput biological growth assay using an automated turbidometer to obtain the inhibitory kinetics for individual compounds (aldehydes, organic and inorganic acids along with various cations and alcohols) on growth of Z. mobilis 8b in glucose and xylose as well as in real hydrolyzates, 2) investigating the toxicity basis of the corn stover hydrolysates, and 3) examining the impact of the conditioning process on cell growth and fermentation through analysis of chemical composition of the conditioned hydrolyzates with detailed chemical identification and quantitation of potential inhibitory compounds. Research findings from these studies will be presented.

See more of this Session: Biological Conversions and Processes for Renewable Feedstocks II
See more of this Group/Topical: Sustainable Engineering Forum

(257c) Cellulase Binding to Insoluble Plant Cell Walls Is Substrate Dependent

Dahai Gao[1], Shishir PS Chundawat[1], Nirmal Uppugundla[1], Bruce E. Dale[2] and **Venkatesh Balan**[1], (1) Department of Chemical Engineering and Material Science, Michigan State University, Lansing, MI, (2) Department of Chemical Engineering and Material Science, Michigan State University, East Lansing, MI

Enzyme binding to insoluble cell walls is the first step towards effective deconstruction to monomeric sugars. However, the relationship between cellulase binding and synergistic activity as well as non-productive binding versus productive binding is unclear. We have applied a novel liquid chromatography based method to individually quantify the extent of cellulase (CBH I, CBH II and EG I) binding during the course of enzymatic hydrolysis for various cellulosic substrates. The binding isotherms for single, binary and ternary cellulase mixtures were determined for untreated, Ammonia Fiber Expansion (AFEX) treated and dilute-acid treated corn stover. Other isolated cellulosic substrates such as cellulose Iβ, liquid ammonia treated cellulose (cellulose III), sodium hydroxide treated cellulose (cellulose II) and phosphoric acid swollen cellulose (amorphous) were also studied for their cellulase binding affinities. EG I enhanced CBH binding for AFEX corn stover but not for untreated and dilute-acid pretreated corn stover. The total cellulase binding capacity for pretreated corn stover was higher than untreated corn stover. However, for the more digestible cellulose III, lesser binding was observed compared to native cellulose I. These results have important implications on our understanding of enzyme accessibility and productive binding for pretreated plant cell walls.

See more of this Session: Biological Conversions and Processes for Renewable Feedstocks II
See more of this Group/Topical: Sustainable Engineering Forum

(257d) Hydrolysis of Biomass with Plant Expressed Cell Wall Degrading Enzymes

Dongcheng Zhang, Amy VanFossen, Ryan Pagano and Jeremy C. Johnson, Agrivida, Medford, MA

Significant amounts of cell wall degrading enzymes are required to degrade lignocellulosic biomass into its component sugars. One approach to reducing that requirement is to produce the enzymes in planta. For this work, various cell wall degrading enzymes were expressed in the green tissue of maize. Following growth and dry down of the plant, the material was tested to determine the impact of the enzymes on the production of glucose and xylose from the corn stover. The hydrolysis experiments were performed after a moderate pretreatment and involved the addition of various amounts of exogenous enzymes. The results demonstrate that for plants expressing certain cell wall degrading enzymes, the monomer sugar yields after hydrolysis of the transgenic plant are higher than those of a control plant. This suggests that the production of enzymes in plants could lead to both reduced pretreatment and enzyme costs in cellulosic bioprocessing.

See more of this Session: Biological Conversions and Processes for Renewable Feedstocks II
See more of this Group/Topical: Sustainable Engineering Forum

(257e) Process Simplification for Cellulose Conversion

Michael Ladisch[1], Don Dimasi[2], Keven Wenger[3], David Hogsett[4], Youngmi Kim[5], Eduardo Ximenes[5] and Nathan S. Mosier[5], (1)Lorre, Purdue University, W. Lafayette, IN, (2)Process R&D, Mascoma Corporation, Lebanon, NH, (3)SR. Director PRD, Mascoma Corporation, Lebanon, NH, (4)VP External R&D, Mascoma Corporation, Lebanon, NH, (5)Agricultural and Biological Engineering & Laboratory of Renewable Resources Engineering (LORRE), Purdue University, West Lafayette, IN

The production of advanced biofuels from non-food cellulosic feedstocks requires the biomass to be pretreated and hydrolyzed to sugars that are fermented to ethanol or other molecules. The single major cost, other than feedstock, is the cost of enzyme. Consolidated bioprocessing (CBP) reduces the amount of enzyme needed, and therefore reduces enzyme costs through use of a genetically modified, pentose fermenting yeast that generates its own cellulolytic enzymes. CBP consolidates biochemical activities and fermentation capability into a single microorganism, and combines hydrolysis and formation of ethanol into a single unit operation, resulting in a simplified process with reduced costs. However, success with this approach also requires that biocatalyst inhibition due to phenolic compounds generated during the processing ligno-cellulosic biomass be overcome, and challenges caused by viscous slurries of pretreated cellulosic biomass be addressed. This talk presents chemical engineering applied to the direct fermentation of pretreated wood to ethanol and the engineering of processes that realize the benefits of CBP.

See more of this Session: Biological Conversions and Processes for Renewable Feedstocks II
See more of this Group/Topical: Sustainable Engineering Forum

(257f) Production of Yeast Biomass From Wood Molasses Using Non-Recombinant Xylose-Utilizing Saccharomyces Cerevisiae

Arthur Kollaras, Philip Bell and Paul Attfield, Microbiogen Pty Ltd, Lane Cove, NSW, Australia

The yeast Saccharomyces cerevisiae has been used for over 200 years in industrial applications for production of bread, beer, wine, fuel ethanol, protein-rich animal feeds, flavors, amino acids, vitamins, enzymes and various other products. The world consumes over 3 million tonnes of S.cerevisiae per annum, just for the manufacture of leavened bread alone. This yeast therefore represents a very important part of the human food chain. The major cost component for yeast manufacture is the substrate molasses or syrups that are rich in easily metabolized six carbon sugars such as glucose and fructose. The cost and availability of such substrates for yeast biomass production are under increasing pressure as more of these are used for fuel ethanol production. Therefore the yeast manufacturing industry seeks alternative substrates for manufacture of the yeast biomass.

Hemicellulose is present as 20 to 40% dry weight of all wood and grass derived materials. Hence, agricultural residues, forestry materials, paper/pulp byproducts and even organic fractions from municipal wastes could all be hydrolyzed to produce xylose-rich concentrates known as "wood molasses" and these could represent substrates for manufacture of yeast biomass. The problem with wood molasses as an alternative substrate though is that native S. cerevisiae is generally regarded as being unable to utilize the five carbon sugar xylose, which is the predominant sugar within hemicellulose. However, we have used natural genetic procedures, without the use of any genetic engineering, to obtain non-recombinant strains of S. cerevisiae that grow aerobically on xylose as well as other carbon present in wood molasses.

Wood molasses was produced from sugarcane bagasse using a dilute acid, low temperature pretreatment to disrupt the crystalline structure. This allowed for the hydrolyzed hemicellulose components to be separated as a liquid fraction from the still solid lignin-cellulose fibre. The xylose-rich hemicellulose fraction was concentrated to produce a wood molasses that was also high in acetate, furans and other compounds. This wood molasses was used to grow S. cerevisiae strain MBG3248, using an aerobic fed-batch process similar to that currently employed in the bakers yeast manufacturing industry. Yeast doubling times, yield efficiencies on substrate fed and productivity rates approached those achieved industrially when molasses is used as the substrate, although the source of the wood molasses did impact on these parameters.

The ability to grow non-recombinant S. cerevisiae on wood molasses opens up several opportunities in lignocellulosic bioconversion. Most obviously it will be possible to make cellulosic ethanol and yeast biomass from the same feedstock, which means that we are addressing issues of food versus fuel and making it possible to vastly improve the productivity of land currently used for biofuel crops. It will even be possible to manufacture protein-rich feed yeast from land used for energy crops such as switch grass. The value of the non-recombinant S. cerevisiae can range between 20 cents and $7 per kg, depending on its application, and this would be beneficial to a cost effective cellulosic ethanol biorefinery.

See more of this Session: Biological Conversions and Processes for Renewable Feedstocks II
See more of this Group/Topical: Sustainable Engineering Forum

Estimation of the cost impacts of retrofit biorefinery implementation using operations-driven costing

Eemeli Hytönen [1], Richard B. Phillips [2] and Paul R. Stuart [1]

[1] École Polytechnique de Montréal, Montréal, QC, Canada

[2] North Carolina State University, Raleigh, NC, United States

Abstract

Increased competition from low cost pulp and paper producer nations and decreased market demand of pulp and paper products have made it imperative for North American forestry companies to enhance the cost competitiveness of their production facilities. An effort has been made to achieve this by planning to change and diversify the product portfolio with biorefinery products such as biofuels and bio-chemicals either through mill retrofit or repurposing. Risk-adjusted financial justification of many of these strategies is challenging, partly due to uncertainties in the market place for both existing and new products, raw materials in the longer term, and technological viability of potential new biorefining processes in the short term.

We introduce a novel early-stage process design method targeted at estimating the changes in cost-efficiency of existing processes in different retrofit biorefinery implementation scenarios, for screening out non-promising alternatives. The method uses process-based data, process simulation and cost estimation based on the principles of activity-based costing to derive transparent and understandable direct and indirect cost information of an integrated system of continuous processes. Stochastic risk analysis (Monte Carlo) is integrated into the cost modelling framework to be able to analyse the effect of price and performance uncertainties on the cost-efficiency.

Several kraft pulp mill retrofit strategies, including mill modernization and biorefinery implementation projects with different production capacities are evaluated using the proposed method, and the results are presented. Because the method first allocates the costs to activities based on resource consumption and then to end products, the true costs of production can be better estimated. This leads to an understanding of the impacts of process integration on the individual product production costs in the integrated system. Differences – and the reasons for these differences – in cost-reduction potential and overall margin improvement, and therefore also in project profitability and mill's economic performance can be seen and analysed between retrofit strategies. The difference in the level of uncertainty, analysed in a systematic manner, shows that screening out retrofit strategies requires the use of multiple decision making criteria.

Introduction

A forest biorefinery (FBR) through retrofit modification of existing pulp and paper mills has been in focus of many recent biorefinery design studies. For example, Mao et al. and Amidon et al. have reported latest developments in near-neutral and hot water extraction of hemicellulose prior to pulping (VPP) –processes [1, 2], and Larson et al. have published a comprehensive study considering thermochemical black liquor conversion into biofuels and pulping chemicals [3]. This FBR approach is aiming at concurrently maintaining the core business profitable and generating new revenues by integrating bioproduct production into the P&P mill. The benefit compared to greenfield FBR plants is considered to come from lower capital cost requirement because existing pulp mill systems can be utilized instead of investing into all utility and service activities. Also the credit from possibly decreased core business production costs is expected to make the biorefinery perform better financially. Another approach for forest biorefinery has also been presented: Phillips et al. have analysed the economic potential of repurposing a shut-down kraft pulp mill into a biochemical bioethanol plant [4]. This presents the same capital cost benefits, even higher capital cost reductions can be realized since all old pulp mill equipment is available.

Retrofit biorefinery implementation will compete with traditional capital projects such as pulp & paper mill production rate ramp up or replacement of old boilers. The potential of advanced costing methods such as activity-based costing (ABC) –like operations-driven costing has been demonstrated in retrofit design decision making context by Janssen et al. and Laflamme-Mayer et al. [5, 6]. They used this method for analysis of traditional retrofit projects for newsprint mills in order to have better cost information for design decision making and supply chain (SC) optimisation. In both cases one product was produced at time (paper), hence the full potential of ABC in multi-product -costing was not needed. This is naturally the case of a retrofitted forest biorefinery which produces several products concurrently.

The current state of biorefinery technology development is still in some cases far from commercial implementation. This gives rise to additional uncertainties besides the external uncertainties of market place – even though ethanol and other biofuels have high demand their prices have shown detrimental volatility in the past; also the nature of emerging biorefinery industry might substantially increase the cost of biomass. These uncertainties have commonly been considered in biorefinery process design mainly using sensitivity and scenario analysis, some authors also report the use of stochastic method (Monte Carlo analysis) in design analysis [7, 8].

Objectives

Overall objective of this study is to demonstrate the use of operations-driven costing and stochastic risk analysis in retrofit biorefinery implementation analysis using a real Kraft pulp mill as a case study. More specifically the goals are 1) to illustrate how better understanding of integration impacts on core business manufacturing costs can be obtained in order to clearly see which product (pulp and/or bioproduct) is responsible of generating the profits, and 2) to show how the obtained credit from lower core business manufacturing costs impacts the overall profitability of the retrofit projects.

Methodology

A four step methodology was applied to the case study:

1. Development of the base case kraft pulp mill:
    - Definition of analysis boundaries
    - Development of a steady-state simulation model of existing processes

- Validation of the process model
- Development of operations-driven cost model of the base case operations
- Validation of the cost model using mill financial data
- Definition of productivity/efficiency and cost/price development over the analysis period

2. Development of retrofit design scenarios:
   - Definition of retrofit projects (feedstock, product, production scale)
   - Modification of the simulation model (addition of new process equipment/departments, integration of processes)
   - Addition of cost model structural units for new process departments in the mill cost model
   - Definition of efficiency, cost/price development over the analysis period related to new processes
3. Economic analysis:
   - Process simulation → Process parameter values (resource and activity drivers of operations-driven cost model)
   - Cost model simulation → Economic performance of the retrofit alternatives
   - NPV & IRR of the investment project
   - Core business impact measured with pulp production cost reduction
4. Risk analysis:
   - Risky cost model parameter identification
   - Definition of probability distributions of risky parameters
   - Monte-Carlo analysis → probability distribution of economic indicators
   - Downside project profitability (expected value - 2σ → 97.5% certainty that profitability is better than this value)

Case study

Base case P&P mill and analysis boundaries

The case study base case is a North American hardwood Kraft pulp mill producing slush and wetlap pulp (~50% dryness) for fine paper production. The paper mill (integrated into the pulp mill) is not considered in this case study, since the actual intermediate product pulp is assumed to be saleable with adjusted market pulp price for the economic analysis purpose.

Since the mill in question has also options for modernization as a pulp and paper producer, a second "Modernization Case" is set forth as another basis for evaluating different biorefinery options.

More detailed mill description and raw material costs delivered to the case mill are reported by Hytönen and Stuart [8, 9]. Based on their conceptual design level process analysis, three design alternatives were selected for the pre-feasibility level design analysis using more detailed analysis procedures. These biorefinery alternatives are described in the next part.

The analysis considers pulp mill process and all retrofit design alternatives without waste water treatment plant which is not located at the pulp mill site. In addition to waste water treatment, cooling water and oxygen are considered as fixed price utility to the analysed processes, consumption is based on simulation model. Electricity consumption is estimated based on mill's historical power consumption at process department level and estimated changes due to process changes in retrofit, biorefinery process electricity demand is calculated from reference estimates. Capital cost estimates are outside battery limit (OSBL) cost estimates, thus infrastructure required for operation is also accounted for.

Retrofit design alternatives

Table 1 lists the retrofit design alternatives considered in this study. One of these design scenarios is the "Modernization Case" alone (first alternative). All biorefinery alternatives have two design scenarios: integration into base case mill or modernized mill case. This results in total of 11 retrofit design scenarios.

Table 1. Retrofit process alternatives and their specifications (based on [8, 9])

Feedstock	Process description	Products	Design capacity	Feedstock capacity
Pulpwood	Modern Kraft pulping process and chemical recovery cycle utilizing maximum amount of existing pulping process equipment	***Kraft pulp***	1650 BDT pulp/day (35% increase from base case capacity)	1.5 million BDT/year
Hemicellulose extract	Value-prior to pulping (VPP): • Near-neutral green-liquor extraction • Acid hydrolysis • Liquid-liquid separation • Fermentation & distillation	***Ethanol*** Acetic acid Furfural	base case - 6.1 MMGPY	10% of pulp wood
			modernized - 8.0 MMGPY	10% of pulp wood
Corn stover Co-processed with	Biochemical lignocellulosic ethanol:	***Ethanol*** Organic	25 MMGPY	0.25 million BDT/year

kraft pulp, using mill infrastructure to maximum extent	• Dilute acid pre-treatment • Enzymatic hydrolysis • Fermentation & distillation	solid residue	100 MMGPY	1 million BDT/year
Forest residues Bark co-processed with kraft pulp, using mill infrastructure to maximum extent	Thermochemical Fischer-Tropsch: • Drying & grinding • Steam reforming • Syngas cleaning and compression • FT-synthesis	***FT-liquids*** Energy	37 500bdt/year	0.25 million BDT/year
			150 000 bdt/year	1 million BDT/year

Further key process design assumptions are given in Table 2, the listed references are used as the basis for process designs and the capital cost estimates. Mass and energy balances of the processes are developed using steady-state process simulator.

Table 2. Process assumptions

Retrofit option	Key process assumptions	References
VPP	10% extraction rate (of pulp wood) Extraction enabled with and additional extraction vessel Anthraquinone added to preserve same overall pulp yield and pulp properties (dose 0.05% of chips) Pulp production rate (in base case and modernized mill case) maintained	[1, 10, 11]
Corn stover-to-bioethanol (25 MMGPY)	In base case pulp mill configuration solid residue replace coal and added steam demand met by natural gas boiler steam • Power boilers solids load constrained • Natural gas boiler has excess capacity • Boiler investment avoided In modernized mill configuration solids burned in power boilers that have excess capacity	[12]
Corn stover-to-bioethanol (100 MMGPY)	In base case pulp mill configuration new boiler and turbine capacity installed In modernized mill configuration solids burned in power boilers that have excess capacity	[12]
FTL-process (37 500bdt/year)	Single-pass FT synthesis, tail gas burned in lime kiln to replace natural gas and excess burned in natural gas boiler Existing turbines used Steam surplus adjusted by lowering coal use in power boilers	[3, 13, 14]

FTL-process (150 000 bdt/year)	Single-pass FT synthesis, tail gas burned in lime kiln to replace natural gas and excess burned in natural gas boiler Additional turbine installed in base case (modernized mill scenario has excess turbine capacity installed) to utilise generated high pressure steam and excess steam generated from tail gas Coal use decreased to lower steam generation	[3, 13, 14]

In Table 3 the total project investment costs of considered retrofit alternatives are presented. Values are total installed capital costs of required new process equipment and the potential capital cost reduction based on the mill's existing infrastructure and equipment constraints described in Table 2 are considered. The reference capital costs are escalated to 2010-$ using Chemical Engineering Plant Cost Index (CEPCI) and scaled-up using the method and factors described by Hytonen and Stuart [8]. Values are assumed to be n*th* plant cost estimates.

Table 3. Total project investment cost of retrofit alternatives

Retrofit alternative	Capital cost (M$)	Remarks
Mill modernization	285	New evaporator plant, recovery boiler, lime kiln and additional steam turbine Power boilers, causticizing and fiberline retrofitted
Base case – VPP	47	Additional extraction vessels (2) in front of continuous Kraft digesters Sugar stream conditioning (liquid-liquid extraction, liming)
Base case – corn stover-to-ethanol (25 MMGPY)	58	No additional boiler
Base case – corn stover-to-ethanol (100 MMGPY)	233	Includes additional boiler and turbo-generator
Base case – FTL (37 000 bdt/year)	117	
Base case – FTL (150 000 bdt/year)	332	
Modernized mill – VPP	57	
Modernized mill – corn stover-to-ethanol (25 MMGPY)	58	No additional boiler
Modernized mill – corn stover-to-ethanol (100 MMGPY)	151	No additional boiler and turbine required
Modernized mill – FTL (37 000 bdt/year)	117	
Modernized mill – FTL (150 000 bdt/year)	302	

Accuracy of the capital cost estimate is assumed to be -10%...+15% and -15%...+25% for the mill modernization and biorefinery projects correspondingly. This reflects the higher uncertainty in new biorefinery process technologies compared to traditional pulp mill processes/process parts. Even though the level of detail in the reference studies for capital cost estimates was not the same, the assumed ranges were selected to represent all design scenarios. Investment schedule for mill modernization is assumed to be three years (20/30/50) and for biorefinery projects two years (30/70), both starting at 2010, thus simultaneous retrofit is considered in modernized mill cases.

Assumptions used in operations-driven costing

Table 4 describes the cost allocation rules including variable, fixed and investment cost categories. Variable costs are traced by the cost model based on resource and activity drivers defined in the table. Fixed costs and investment costs are allocated and assigned using the given basis through dedicated cost drivers correspondingly.

Table 4. Cost allocation bases used in operations-driven cost model

Category	Cost type	Allocation basis
Fixed	Labour	Current mill employees shared, labour costs allocated based on headcount • Feedstock preparation (yard & mobile equipment) shared up to the level that is sufficient for pulp process requirement • Maintenance crew and supervisors shared • Other shift operators, laboratory technicians and supervisors shared New employees hired if design requires and all allocated to biorefinery
	Maintenance	Annual budget set for pulp mill and biorefinery departments separately Maintenance supplies for biorefinery – 2% of initial capital investment
	SG&A, benefits	Headcount
	Other fixed (operating supplies, outside services)	Annual budget set for pulp mill and biorefinery departments separately
Variable	Water	Clean water consumption
	Energy	Fuel cost shared based on steam demand Fuel production (solid residues, tail gas) credited
Investment	Investment	Energy system (additional boiler and utility system) capital investment cost shared based on steam consumption Mill modernization capital costs in commonly used departments allocated fully to pulp mill

The allocation rules listed in Table 4 are not the only possible rules, for example energy costs could be allocated based on source of the energy: pulp mill produces as by-products black liquor and bark which are used in steam and electricity generation. These low resource cost activities could also be directly traced to the pulp. Moreover, the excess steam and electricity can be "sold over-the-fence" to external processes such as paper mill and biorefinery with a fixed price. If the integrated facility requires more steam and/or electricity than can be generated from the black liquor and bark, the costs of fossil fuels and purchased electricity should be allocated based on the energy demand. Compared to above allocation rule, this would increase the cost of steam for biorefinery and paper mill and at the same time lower energy costs of pulping process.

In Table 5 all parameters that are expected to have different future trend than consumer price index (CPI) or producer price index (PPI) are listed. Also the uncertainty in the trends is given for most important parameters. The biorefinery product prices are assumed to follow oil price trend (all products are fuels).

Table 5. Future trends of input parameters

Parameter	Annual Inflation (%)	Trend standard deviation (Normal distribution)	Remarks [reference/data source]
Prices & costs			Price/cost trends in real-$
• Pulp	-0.5	0.5	[15]
• Oil	2.0	1.0	[16]
• Pulp wood	1.5	0.75	
• Forest residues & agro residues	2.5	1.25	
	0.25	-	
• Labour cost	0.5	-	
Annual capital expenditures	CEPCI	-	25% to development (cost reduction) and 75% to replacement (regulatory, quality and reliability)
Productivity & efficiency			Annual development capital expenditures with 2 years payback time and division 50% fibre, 40% chemicals and 10% energy cost reduction result in different productivity gain in every scenario
• Fibre productivity			
• Energy efficiency			
• Chemical productivity			
• Labour productivity	0.5	-	Annual replacement capital expenditures achieve additionally labour productivity and production reliability increases
• Pulp production reliability	0.25	-	

In addition to long term trend uncertainty, many prices have short term uncertainty (volatility). In this retrofit design assessment following price volatility is considered:

- Raw materials – normal price distribution, 5% standard deviation from expected price at the mill gate is assumed

- Gasoline (ethanol) – Weibull distribution (β = 1.336, γ = 0.846 and μ = 1.14 in $/gal) [17]
- Diesel (FT-liquids) – Weibull distribution (β = 1.125, γ = 0.819 and μ = 1.19 in $/gal) [17]
- Pulp – normal price distribution, 10% standard deviation from expected price [15]
- Natural gas – truncated exponential price distribution (λ = 0.192, min = 2.1 and max = 13.7 in $/MMBTU) [18]
- Electricity – Weibull distribution (β = 1.55, γ = 0.752 and μ = 5.46 in ¢/kWh) [19]

Results and discussion

Manufacturing costs

In Figure 1 and Figure 3 the production costs of pulp and biofuels in all design scenarios, using the above described cost allocation rules and analysis method, are presented. Costs are divided to variable and fixed production costs of mill areas or process departments, where pulp mill area costs are aggregated costs of several departments for clarity reasons (e.g. fiberline consists of total 8 departments).

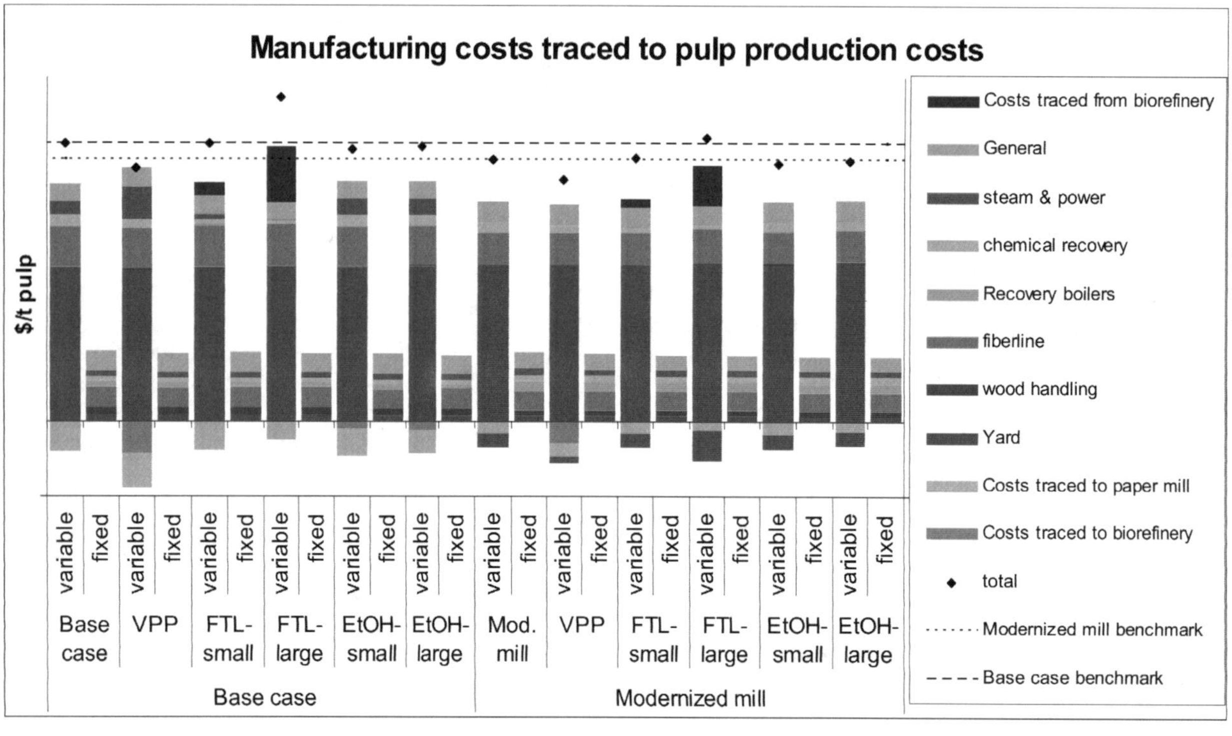

Figure 1. Variable and fixed pulp production costs in pulp mill areas and the total costs traced to pulp in considered design scenarios

Total costs of pulp mill areas that are traced to pulp manufacturing costs vary significantly between design alternatives. Highest pulp production costs are seen in both pulp mill configurations with high production rate FTL process followed by small capacity FTL process alternatives. Lowest costs are observed in both VPP retrofit scenarios; corn stover-to-ethanol alternatives present comparable pulp production costs with their corresponding base case.

The total fixed cost of pulp mill activities (including labour, maintenance, business unit and corporate overheads and depreciation) do not change substantially between retrofit scenarios. This is the result of combined effects of increased depreciation costs and decreased other costs: New asset implementation increases depreciation costs whereas when absolute amounts of business overhead and labour costs are not assumed to change, their cost per ton is decreased. In addition, the fixed cost allocation to new biorefinery processes seems to lower only slightly the overall fixed costs that are traced to pulp manufacturing costs. The decrease from base case is the largest in high biofuel production volume cases, about 10$/bdt of pulp.

In variable pulp production costs substantial variation is observed. The main changes occur in steam & power, recovery boilers and chemical recovery areas. In Figure 2 more detailed cost breakdown of steam & power area is shown including variable and fixed costs and also the traced costs between this area and pulp mill and biorefinery. It reveals the impacts of different available bio-based fuels on fossil fuel requirement for meeting the process steam demand – the costs of fossil fuels and credits from net electricity production change substantially between design scenarios (pulp mill and biorefinery electricity demand is already reduced from total electricity generation and not shown in Figure 2). The net steam & power area costs traced to end product pulp vary from base case -14$/bdt pulp to highest value of 74 $/bdt pulp in small FTL design scenario implemented into base case mill and to lowest value of -34 $/bdt pulp in modernized mill scenario. Highest costs can be explained by lower available black liquor amount due to hemicellulose extraction and increased process steam demand, which in turn increases the demand for natural gas based steam (coal also used to its maximum extent). The lowest area costs in all modernized mill scenarios on the other hand are result of higher energy efficiency in pulping process (no natural gas is needed and excess electricity is generated) and excess of bio-based fuel (either tail-gas from thermochemical process or organic solids from bioethanol process). Also coal demand varies depending on the scenario.

Similarly other process areas with substantial cost variations between scenarios can be analysed. The impact of selected allocation rules can be examined by drilling down to traced costs.

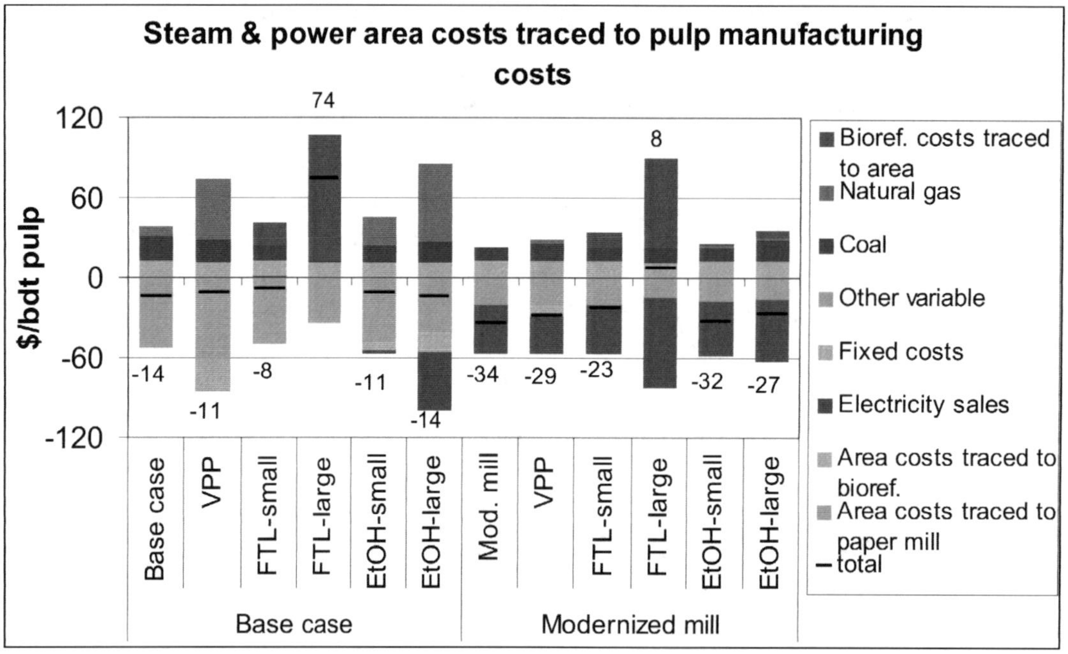

Figure 2. Variable costs of steam & power area at the pulp mill

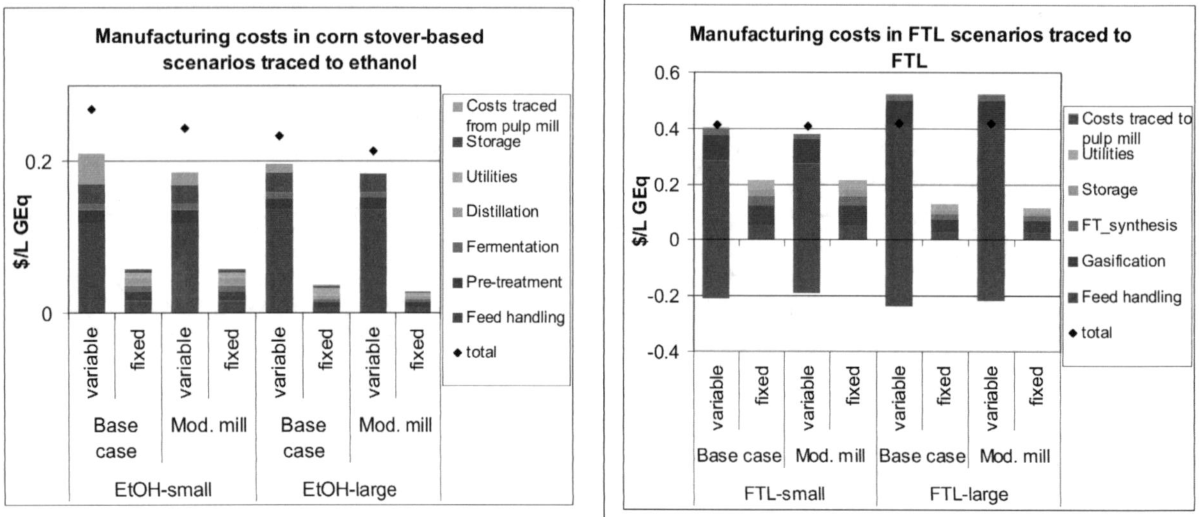

a) b)

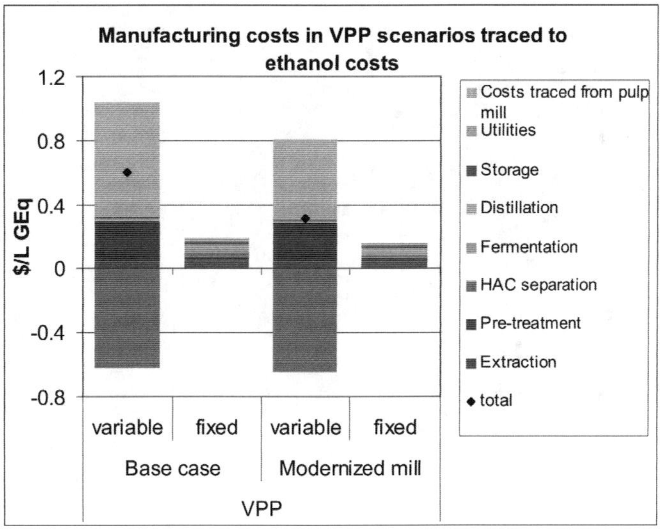

c)

Figure 3. Biofuel production costs in gasoline equivalent litres. a) corn stover ethanol design scenarios, b) FTL design alternatives and c) VPP design alternatives

Biofuel production costs vary between 0.22 and 0.6 $/GEg litre, the corn stover based ethanol processes having lowest production costs and VPP alternative implemented in base case mill the highest production costs. Important to notice is the magnitude of the costs traced from/to pulp mill: the VPP processes have substantially lower production capacity but relatively high energy demand (selected allocation basis: all steam needed in extraction is traced to biorefinery process department) and cost transfer is therefore higher from steam & power area. Contrary to that, FTL process designs generate substantial amount of tail-gas that is a key credit for the biorefinery process. In VPP design scenarios additional credit is obtained from sold acetic acid and furfural.

Profitability and risk analysis

The profitability of a retrofit project is defined here to include all credits from changed pulp profits compared to the corresponding base case. In modernized mill scenarios the changes in profit from current mill operations due to mill modernization are not included, hence, the profitability is evaluated for biorefinery projects alone (mill modernization also assessed separately as a retrofit alternative). Two measures for profitability were applied, net present value (NPV) using 11% rate of return, and internal rate of return (IRR).

Results of Monte Carlo (MC) analysis are used for obtaining both project profitability and risk level, expected and downside profitability respectively. The downside profitability represents the value above which the profitability is expected to be with ~95% certainty, mathematically defined as expected profitability minus two times standard deviation. In MC analysis 200 iterations was conducted (Monte Carlo analysis using operations-driven cost model is computationally heavy), more iterations were tested and no significant change in either target measure, expected or downside profitability, was observed. The results are shown in Figure 4.

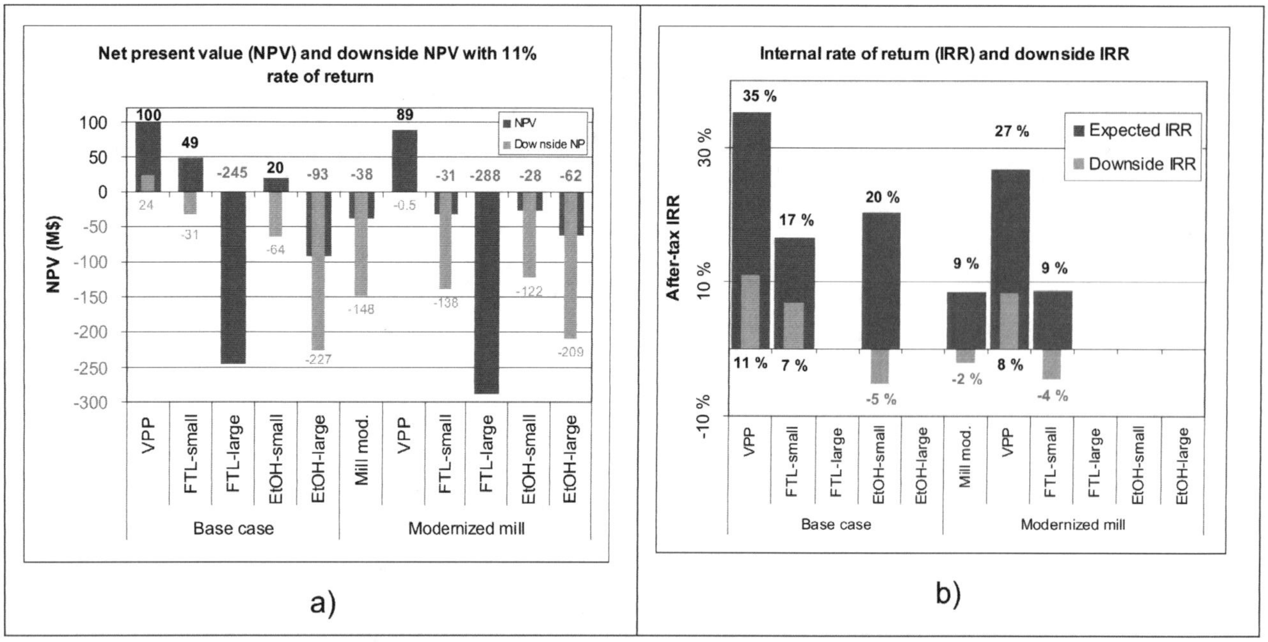

Figure 4. Profitability of analysed retrofit projects. a) Net present value and downside NPV, and b) internal rate of return and downside IRR

From the results it can be seen that many of the retrofit project show low and even highly negative profitability if it is measured with fixed rate of return of 11% and NPV. However, many of the projects have positive expected IRR. The expected IRR and downside IRR are presented only for projects that have positive expected IRR, thus for all projects having positive NPV with 0% rate of return. Using this break-even performance as screening heuristic (instead of e.g. 11% weighted average cost of capital (WACC) which is commonly considered to be minimum rate of return for strategic investment projects) makes sure that we do not screen out a possibly promising retrofit alternative but also is able to remove projects that most probably will not be profitable. Moreover, this heuristic lowers the probability to obtain such annual discounted cash flow series that change sign several times over the project lifetime. Therefore it enhances the reliability of the IRR estimate. However, in small ethanol retrofit project and mill modernization project scenarios ~25% of the iterations generated such cash flow series that do not have IRR solution and therefore these samples were not considered in the probability distributions shown in Figure 4 b).

The VPP projects appear to have highest expected IRR in both mill configurations, however it must be noted that in modernized mill case the value does not include the mill modernization project's impacts. This profitability value differs notably from reported value by Mao et al. [1]: differences between assumptions (costs traced outside the analysis boundaries, annual productivity/efficiency improvements and inflation for different resources) lead to substantial pulp production cost reduction and higher profit in this study compared to no annual cost reduction in the study by Mao et al. In addition, the changes in all common activity costs have influence beyond the analysis boundaries, namely on the papermaking costs. These impacts are not explicitly considered in the profitability measures of this study.

Downside NPV is positive only for the VPP alternative, downside IRR values show that the mill modernization project alone as retrofit project has lowest risk (standard deviation smallest) but it has slightly negative downside IRR value. VPP alternatives have because of their high expected profitability the highest absolute value of the downside IRR. All analysed biorefinery projects have relatively similar risk, and the small FTL project has the only other positive downside IRR in addition to VPP projects.

Compared with a design alternative pre-screening analysis using the same retrofit alternatives and case mill conducted by Hytönen and Stuart (conceptual level design analysis considering integration into base case mill with no impacts on pulp production costs estimated and fixed prices for common utilities and services [8]), all expected IRR values obtained with the presented method are somewhat improved. On the other hand their risk level seems to be relatively much higher, this can result from the additional uncertainties considered in this study compared to the pre-screening – project investment cost uncertainty and pulp price variation – and the fact that also the changes in pulp production costs are credits/additional costs to the retrofit projects.

Conclusions

A pre-feasibility level techno-economic assessment of different biofuel production scenarios at a North American hardwood pulp and paper mill was conducted using operations-driven costing combined to Monte Carlo risk analysis. Activity based costing –like cost accounting method was based on validated steady-state simulation model of the case study pulp mill and models of considered biorefinery process alternatives integrated into the pulp mill model.

Using the presented set of allocation and tracing rules, substantial manufacturing cost differences between retrofit design alternatives are observed for both pulp and the biofuels. These differences can be further analysed by drilling-down to individual resource consumption and fixed costs assigned to each process department because of the structure of the costing method that was used. If other allocation bases would be utilised, these differences would naturally be different, however the important fact to notice is that the core business production costs are changed and differences in this change between different biorefinery processes exist. The impacts of allocation rules on the final results should also be assessed before selecting the appropriate allocation basis.

The combined effect of manufacturing costs of the two concurrent products and their respective sales prices on project profitability showed also significant differences between analysed retrofit projects. Clearly the most promising project among the analysed cases is the near-neutral VPP alternative. Another key conclusion is that small capacity processes seem to have better profitability compared to same processes with higher capacity. This can be partly explained by the relatively higher capital cost due to required additional boiler and turbine capacity.

The presented measures for biorefinery retrofit implementation project performance analysis, changes in core business manufacturing costs and project profitability and downside profitability, all show similar behaviour: Highest expected IRR project (VPP) is clearly the best retrofit project if NPV is used as a measure, however the relative risk in that project is high compared to the traditional retrofit project (mill modernization). In addition, based on the operations-driven cost analysis, this project relies on high revenues from by-products (acetic acid and furfural). Hence, it can be concluded that it is imperative to analyse retrofit biorefinery projects using a set of criteria instead of only one singe indicator of performance to enable better informed decisions in capital appropriation process. Especially the analysis of the changes in the performance of core operations seems to have an important effect on the profitability of the retrofit projects. However, it is important to notice that in this specific case of integrated pulp and paper mill, the integration impacts reach also beyond the pulp mill and that these impacts, especially the changes in energy costs, should not be neglected in the decision making. This issue should be studied in future.

In addition, in the case of stochastic risk analysis the suitability of the metric must be carefully addressed: calculation of IRR from sign changing cash flow series that are possibly obtained as a results of Monte Carlo sampling for relatively low profitability design projects, poses a risk of having several solutions. Other measures can be used in these cases, for example NPV or modified internal rate of return (MIRR).

Acknowledgements

This work was supported by the Natural Sciences and Engineering Research Council of Canada (NSERC) Environmental Design Engineering Chair at École Polytechnique de Montréal, and by a research project in the BioRefine Technology Programme of the Finnish Funding Agency for Technology and Innovation (TEKES). The industrial partner of this project supplying the case specific data is gratefully acknowledged

References

1. Mao, H., et al., *Technical Economic Evaluation of a Hardwood Biorefinery Using the "Near-Neutral" Hemicellulose Pre-Extraction Process* Journal of Biobased Materials and Bioenergy, 2008. **2**(2): p. 1-9.
2. Amidon, T.E., et al., *Biorefinery: Conversion of Woody Biomass to Chemicals, Energy and Materials.* Journal of Biobased Materials and Bioenergy, 2008. **2**(2): p. 21.
3. Larson, E.D., et al., *A Cost-Benefit Assessment of Gasification-Based Biorefining in the Kraft Pulp and Paper Industry*. 2006, U.S. Department of Energy. p. 365.
4. Phillips, R.B., H. Jameel, and E.C. Clark Jr. *Repurposing a Kraft Pulp and Paper Mill to the Production of Ethanol.* in *92nd Annual Meeting of the Pulp and Paper Technical Association of Canada*. 2010. Montreal, Canada.
5. Janssen, M., P. Naliwajka, and P. Stuart, *Development of an Operations-Driven Cost Model for Continuous Processes Part I: Retrofit process design application.* 2008: p. 17.
6. Laflamme-Mayer, M., et al., *Development of an Operations-Driven Cost Model for Continuous Processes Part I: Framework for Design and Operations Decision Making.* International Journal or Production Economics, 2008. **In press**: p. 18.
7. Richardson, J.W., et al., *Risk Assessment in Economic Feasibility Analysis: The Case of Ethanol Production in Texas.* 2006, Agricultural and Food Policy Center The Texas A&M University System. p. 16.

8. Hytönen, E. and P.R. Stuart, *Biofuel Production in an Integrated Forest Biorefinery—Technology Identification Under Uncertainty.* Journal of Biobased Materials and Bioenergy, 2010. **4**(1): p. 1-10.

9. Hytönen, E. and P.R. Stuart, *Integrating Bioethanol Production into a Kraft Pulp Mill - Technology Assessment.* Pulp & Paper Canada, 2009. **110**(5): p. 58-65.

10. Wooley, R., et al., *Lignocellulosic Biomass to Ethanol Process Design and Economics Utilizing Co-Current Dilute Acid Prehydrolysis and Enzymatic Hydrolysis current and Futuristic Scenarios.* 1999, NREL National Renewable Energy Laboratory. p. 130.

11. van Heiningen, A., *Converting a kraft pulp mill into an integrated forest biorefinery.* Pulp and Paper Canada, 2006. **107**(6): p. 38-44.

12. Aden, A., et al., *Lignocellulosic Biomass to Ethanol Process Design and Economics Utilizing Co-Current Dilute Acid Prehydrolysis and Enzymatic Hydrolysis for Corn Stover.* 2002. p. 154.

13. Phillips, S., et al., *Thermochemical Ethanol via Indirect Gasification and Mixed Alcohol Synthesis of Lignocellulosic Biomass.* 2007, NREL National Renewable Energy Laboratory. p. 132.

14. Phillips, S.D., *Technoeconomic Analysis of a Lignocellulosic Biomass Indirect Gasification Process To Make Ethanol via Mixed Alcohols Synthesis.* Industrial & Engineering Chemistry Research, 2008. **46**(26): p. 11.

15. FOEX Indexes Ltd. *The PIX Pulp EUROPE Benchmark Indexes.* 2010 [cited 2010 June 10]; Available from: http://www.foex.fi/.

16. DOE/EIA, *Annual Energy Outlook 2009.* 2009, DOE/EIA. p. 230.

17. U.S. Energy Information Administration. *Petroleum Navigator.* 2009 [cited 2009 September 2]; Available from: http://www.eia.doe.gov/.

18. U.S. Energy Information Administration. *Natural gas navigator.* 2009 [cited 2009 January 9]; Available from: http://www.eia.doe.gov/.

19. U.S. Energy Information Administration. *Wholesale electricity market data.* 2009 [cited 2009 September 2]; Available from: http://www.eia.doe.gov/.

(285b) Fully Integrated Lignocellulosic Biorefinery with On-Site Production of Enzymes and Yeast

Jarno Kuijvenhoven[1], Hans Kroon[1], Reinder Hamstra[1] and Manoj Kumar[2], (1)DSM White Biotechnology B.V., Delft, Netherlands, (2)White Biotechnology, DSM Innovation Center Inc, Parsippany, NJ

Lignocellulosic biomass is the most abundant, least expensive renewable natural biological resource for the production of bioenergy, biofuels and biobased products via biorefining. As such, lignocellulosic biomass and biorefineries will enable the sustainable development of human civilization in 21st century.

It is a must to fully integrate the design, construction and operation of a biorefinery to make it sustainable and cost effective, when producing fermentable sugars and biofuels such as cellulosic ethanol from lignocellulosic biomass. In this way many value added features can be realized, like improved cellulosic ethanol productivities and yields, lower operational and capital costs and an improved eco-footprint. DSM estimates based on LCA analysis that on-site production of biochemical conversion technology products such as cellulase enzymes and a C5/C6 sugars co-fermenting ethanol yeast generates much less CO_2/ton of enzyme and yeast bioproducts in comparison to a similar off-site production facility. A fully integrated biorefinery operation will provide a) the overall lowest cost option for sourcing bio-conversion components with increased security of supply, b) an enzyme mix continuously tuned to the specific substrate and process conditions at hand and c) increased total process stability and versatility.

In this presentation, we will provide DSM's design efforts and project approach in developing a fully integrated on-site production platform of biochemical conversion technology products in a biorefinery. It will also be evident from this presentation that on-site production of enzymes and yeast (E&Y) for a biorefinery producing cellulosic ethanol will help make bioconversion technology products more effective and efficient by a) reduced down-stream processing costs including water, b) minimal transportation and packaging costs, c) minimal E&Y process losses via direct transfer to the cellulosic ethanol process, d) reduced capital expenditure costs vs. dedicated (remote) facilities, e) lower cost carbon source for the E&Y production fermentations, f) direct availability of induction and adaptation components for E&Y, g) direct access to utilities and energy sources within the biorefinery and h) continuous E&Y output tuning and improvement opportunities directly within the biorefinery.

Royal DSM (www.dsm.com) creates innovative products and services in Life Sciences and Materials Sciences that contribute to the quality of life. DSM White Biotechnology is part of the Innovation Center of DSM and strives to develop new business in the area of industrial biotechnology. Key areas of interest are Next Generation BioFuels and BioBased Chemicals, where DSM is developing new production processes based on renewable feedstocks and BioProducts, where DSM is developing enzymes and fermentation organisms for biofuels that do not compete in the food chain.

See more of this Session: Integrated Processes for Biochemical Conversion of Renewable Feedstocks to Fuels and Chemicals II
See more of this Group/Topical: Sustainable Engineering Forum

(285c) Improving Simultaneous Saccharification and Co-Fermentation (SSCF) Performance by Optimizing Biomass and Enzyme Feeding Strategies

Jian Shi, Center for Environmental Research and Technology, Bourns College of Engineering, University of California, Riverside, Riverisde, CA, **YI Jin**, Chemical and Environmental Engineering, Center for Environmental Research and Technology (CE-CERT), Riverside, CA, Mirvat Ebrik, College of Engineering-Center for Environmental Research and Technology, University of California, Riverside, Riverisde, CA, Taiying Zhang, Bourns College of Engineering---Center for Environmental Research & Technology, University of California, Riverside, Riverside, CA and Charles Wyman, College of Engineering-Center for Environmental Research and Technology, University of California, Riverside, Riverside, CA

Low enzyme usage and high ethanol yields and concentrations are the keys to cost effective cellulosic ethanol production. The simultaneous saccharification and co-fermentation (SSCF) configuration is often favored due to its lower end product inhibition and better performance than separate hydrolysis and fermentation. However, further improvement has slowed primarily because enzymatic hydrolysis is rate limiting, especially at low enzyme loadings. An important challenge is to better match enzyme properties and enhance performance of ethanol fermentation. In this study, we focused on testing biomass and enzyme feeding strategies to improve SSCF performance by using stream exploded corn stover and recombinant C5-C6 co-fermenting *Zymomonas mobilis*. Fed batch operation was evaluated to process high solids and yield high ethanol concentrations and compared to hybrid operations based on hydrolysis with just enzyme at higher optimal temperatures in the initial stage followed by cooling to allow introduction of a fermentative organism for the SSCF approach in later stages. We also looked at enzyme feeding strategies such as supplementation with xylanase and beta-xylosidase and staged addition of enzymes in conjunction with SSCF to enhance performance. The effect of pretreatment severity on enzyme usage and fermentation inhibition was also evaluated and used to suggest integration strategies for these essential operations.

See more of this Session: Integrated Processes for Biochemical Conversion of Renewable Feedstocks to Fuels and Chemicals II
See more of this Group/Topical: Sustainable Engineering Forum

(285d) Integrated Design of the MixAlco Process for the Conversion of Biomass to Hydrocarbon Fuels

Viet Pham, Mark Holtzapple and Mahmoud M. El-Halwagi, Chemical Engineering, Texas A&M University, College Station, TX

MixAlco is a robust process that converts biomass to fuels and chemicals. To make hydrocarbon fuels, the MixAlco process has the following steps: (1) pretreatment with lime, (2) fermentation with a mixed culture of acid-forming microorganism to obtain carboxylate salts, (3) dewatering with a high-efficiency vapor-compression evaporator, (4) thermal conversion of salts to ketones, (5) hydrogenation of the ketones to mixed alcohols, and (6) oligomerization of alcohols to hydrocarbons using zeolite catalysts. This work focuses on process synthesis, simulation, and integration for optimal designs of the MixAlco process. For the fermentation which is the key feature of the process, reactor design alternatives (steel tank, pile, pond) were compared. The fermentation operating conditions (conversion, residence time, product concentration) that strongly affect the dewatering cost were optimized. Integration was performed at plant-wide level to reduce the fresh usages of raw materials, minimize the demand of steam and power, and determine the strategies of supplying hydrogen for the hydrogenation. Cost estimation for various scenarios was also performed to demonstrate the merits of the integration.

See more of this Session: Integrated Processes for Biochemical Conversion of Renewable Feedstocks to Fuels and Chemicals II
See more of this Group/Topical: Sustainable Engineering Forum

(285e) Consolidated Bioprocessing (CBP) of AFEX Treated Biomass by *Clostridium Phytofermentans*

Mingjie Jin, Dahai Gao, Bruce Dale and **Venkatesh Balan**, Department of Chemical Engineering and Material Science, Michigan State University, East Lansing, MI

Our previous publications showed that high sugar conversion and high ethanol yield can be achieved by using Ammonia Fiber Expansion (AFEX) pretreatment, commercial enzymes, and Saccharomyces cerevisiae 424A(LNH-ST) in separate hydrolysis and fermentation (SHF) and simultaneous saccharification and co-fermentation (SSCF). Some of the promising finding using those studies include: 1) AFEX pretreatment greatly improves biomass digestibility for commercial enzymes; 2) AFEX treated preservers the nutrients in biomass for yeast fermentation; 3) the inhibition effect of degradation products produced during AFEX is mild for yeast fermentation. When compared to SHF, SSCF, Consolidated Bioprocessing (CBP) which performs enzymes production, enzymatic hydrolysis and fermentation in one bioreactor is strongly believed to be the most economic way to produce biofuel in a biorefinery.

Clostridium phytofermentans (ATCC 700394) is a promising CBP microbe, which can degrade cellulose and hemicellulose at the same time and produce ethanol, acetate, lactate, formate and hydrogen. We will report a detail study on the enzyme secreted by this microbe using proteomic analysis. The CBP performance of *C. phytofermentans* (ATCC 700394) on various AFEX treated biomass including corn stover, switchgrass, rice straw, cane leaf and bagasse will be discussed.

See more of this Session: Integrated Processes for Biochemical Conversion of Renewable Feedstocks to Fuels and Chemicals II
See more of this Group/Topical: Sustainable Engineering Forum

(285f) Reconfigurable Bioenergy Production

Henry Y. Wang, University of Michigan, Ann Arbor, MI

A flexible technology is proposed for enabling localized biofuel production to fit the uncertain energy demands of the near future. Key characteristics of the present biofuel infrastructure include centralized processing, long-distance transportation (>25mi) of biomass, and a reliance on agricultural crops. An alternative to this model is a system of mobile biofuel processing units, or MPUs. These MPUs would be designed to fit on a standard container and be easily transportable by truck. They could be linked with other MPUs to accomplish a specific task and they would be designed for safe, predictable operation. Early attempt at engineering shall focus on a micro combined heat & power (microCHP) processing unit and a photobioreactor system for high cell density microalgae cultivation. The key unit operations and design principals are outlined along with an initial economic analysis of the operating costs. The results of this research will be used to examine commercial viability and compare lifecycle energy usage and emissions with alternative methods of biofuel production.

See more of this Session: Integrated Processes for Biochemical Conversion of Renewable Feedstocks to Fuels and Chemicals II
See more of this Group/Topical: Sustainable Engineering Forum

(305a) Production of High Quality Biodiesel Fuel From Waste Cooking Oil Using Anion-Exchange Resin Catalyst

Naomi Shibasaki-Kitakawa, Takahiro Tsuji, Masaki Kubo and Toshikuni Yonemoto, Department of Chemical Engineering, Tohoku University, Sendai, Japan

1. Introduction

The anion-exchange resin has a high catalytic activity for the transesterification of triglycerides to produce fatty acid esters, used as biodiesel fuel[1]. The principal advantage of the production process of biodiesel fuel using the anion-exchange resin catalyst can be summarized as follows: 1) The alkali soap, contaminant for the fuel, is never formed by saponification[2]. 2) The impurities contained in the raw oils, such as free fatty acid (FFA) and dark pigment, are removed by adsorption on the resin[2,3]. 3) The complete conversion can be achieved even at the stoichiometric molar ratio of 3:1[2].

By combining the catalytic activity and adsorption ability of the anion-exchange resin, the high quality biodiesel fuel would be produced without the pretreatment of the raw oils and the purification of the products.

2. Experimental methods

By supplying the mixed-solution of waste cooking oil and methanol to the expanded-bed reactor packed with the anion-exchange resin, the continuous production experiment of the fatty acid methyl esters (FAME) was performed. The effluent from the reactor was analyzed to check the quality of the biodiesel fuel.

3. Results and discussion

Figure 1 shows the time courses of the triglyceride residual ratio and the FAME concentration in the effluent from the column. The starting time to supply the reaction solution to the column was set to zero. The open and filled symbols were the results using the original resin and the reused one after regeneration, respectively. Both symbols almost overlapped each other and the resin could be repeatedly used without any loss in the catalytic activity. The FAME concentration increased after 3 h and became constant after 10 h, when the steady state was attained. The triglyceride residual ratio was almost zero up to 25 h. so that the transesterification was found to completely proceed.

The result to check the quality of the biodiesel fuel is listed in Table 1 compared with standards of Europe (EN14214) and USA (ASTM D6751). The component properties are picked up. The contents of impurities (FFA and water), unreacted reactants (methanol and glycerides), and the byproduct (glycerin) in the effluent from the reactor fully satisfied with both standards. Only the FAME content was a little lower than that in EN14214, but it was not limited in ASTM D6751. From these results, water and glycerin as well as FFA and pigment were found to be removed by the adsorption on the resin. The distillation process to remove the residual methanol from the product can be omitted in this system.

4. Reference

1) N.Shibasaki-Kitakawa *et al., Bioresour. Technol.,* **98, 416 (2007)**

2) T.Tsuji *et al., Energy Fuels,* **23**, 6163 (2009)

3) N.Shibasaki-Kitakawa *et al., Energy Fuels,* (2010) in printing

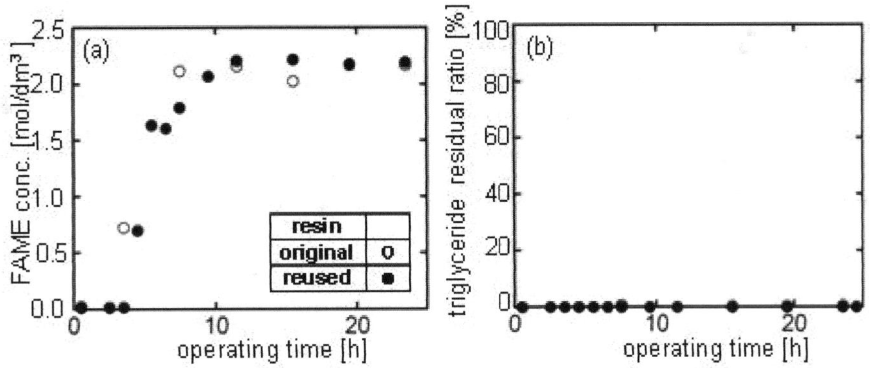

Fig. 1 Time courses of FAME concentration (a) and triglyceride residual ratio (b) in effluent from reactor

Table 1 Component properties of effluent from reactor compared with standards of biodiesel fuel

property	units	effluent	standards	
			EN14214	ASTM D6751
acid value (FFA)	[mg-KOH/g]	0.08	<0.5	<0.8
water	[mg/kg]	190	<500	-
methanol	[wt%]	0.03	<0.20	-
monoglyceride	[wt%]	<0.25	<0.80	-
diglyceride	[wt%]	<0.05	<0.20	-
triglyceride	[wt%]	<0.05	<0.20	-
free glycerin	[wt%]	<0.005	<0.02	<0.02
total glycerin	[wt%]	<0.10	<0.25	<0.24
FAME	[wt%]	91.5	>96.5	-

See more of this Session: Sustainable Diesel Fuel From Renewable Resources

See more of this Group/Topical: Sustainable Engineering Forum

(305b) Electrochemical Conversion of Fatty Acids to Produce Green Diesel

Charles J. Coronella, Chemical Engineering, University of Nevada, Reno, Reno, NV, Cody Wagner, University of Nevada, Reno, Reno, NV and Rainer Busch, Pacific Propellants, Reno, NV
Electrochemical conversion of fatty acids to produce green diesel (aka renewable diesel)

We are developing an electrochemical process to convert fatty acids to green diesel. Under appropriate conditions of temperature, current density, solvent mixtures etc., free fatty acid salts can be decarboxylated in a specialized electrolysis cell. The resulting alkyl fragments form hydrocarbons with a carbon chain shortened by one atom, producing $C_{15}$and C_{17} hydrocarbons from typical, lipid-based fatty acid feedstocks.

The chemistry is relatively straightforward. An unsaturated fatty acid, e.g. oleic acid, subjected to this process is transformed into unsaturated C17-hydrocarbons:

$$CH_3(CH_2)_7CH{=}CH(CH_2)_7COO^- \rightarrow CH_3(CH_2)_7CH{=}CH(CH_2)_5CH{=}CH_2 + CO_2 + H^+ + 2e^-$$

Traditional refining and upgrading processes, such as "hydrotreating", are commonly being run at 300-500 Celsius as well as thousands of psi of pressure. The process under development permits the stereospecificity (i.e. cis-double bonds) in oleic and other unsaturated acids to be retained. This contributes to the excellent low-temperature performance of the hydrocarbons so produced. Thus, this fuel has properties that easily meets or even exceeds those of fossil-derived diesel fuel.

As shown in the equation above, the process also produces molecular hydrogen as well as carbon dioxide as byproducts. Hydrogen is a valuable byproduct, useful in other areas of a biorefinery, or directly as a fuel or salable commodity. The carbon dioxide can be easily separated and used as is, or fed back to an algae colony or other CO_2-consumer.

The talk presents results from experiments with a prototypical reactor, that has processed pure oleic acid, palmitic acid and commercially available mixtures to produce high quality hydrocarbon fuel products. The effects of pH, temperature, and current density are evaluated.

See more of this Session: Sustainable Diesel Fuel From Renewable Resources
See more of this Group/Topical: Sustainable Engineering Forum

Parametric Study of Biodiesel Quality and Yield Using a Bench-Top Processor

Rebecca Wilson and Ihab H. Farag, Chemical Engineering Department, University of New Hampshire (UNH), Durham NH 03824-3591

Abstract

A 1.5 liter bench-top biodiesel processor was constructed to study the effect of feedstock and operating conditions on the conversion and the biodiesel quality. Both the one-stage base process and the base-base two-stage production process were tested. The two-stage process exhibits greater conversion, lower cost and improved biodiesel quality.

1- Background

1.1 Introduction

Energy is consumed constantly, and is often taken for granted. There are finite resources accessible to us, and future generations will have to deal with the consequences when petro fuels are no longer available. Most sources of energy (such as petro fuels, wood, oil, coal and natural gases) release undesirable emissions, such as sulfur dioxide, carbon dioxide, hydrocarbons, and others that affect air quality. This has provided an incentive to research renewable, ecofriendly fuel sources such as biodiesel.

1.2 Biodiesel

Biodiesel is a plant-derived fuel that can be used to replace petro diesel (Wikipedia). Biodiesel is essentially CO_2 neutral, biodegradable and much less toxic than petro diesel. Figure 1 (Farag 2008) shows the environmental benefits of biodiesel relative to petro diesel. Significant emission reductions are observed in CO, polycyclic aromatic hydrocarbons (PAH) and nitro PAH (nPAH) (both PAH and nPAH are potential cancer causing substances), unburned hydrocarbons (largely responsible for the formation of smog), particulate matter (a health hazard when inhaled) and sulfates (a major component of acid rain). Clearly biodiesel is an environmentally friendly alternative to petro diesel.

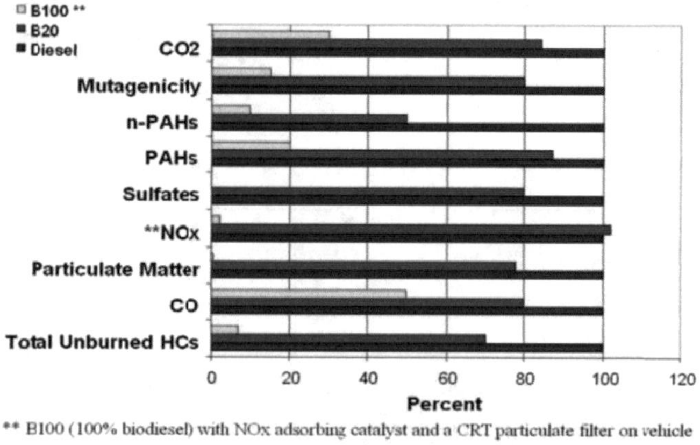

Figure 1: Emission Reduction of Biodiesel (pure or B100, and a mixture of 20% Biodiesel and 80% diesel) relative to petro diesel (Basis petro diesel emission of each component is 100).

1.3 Biodiesel Production

Biodiesel is made by the transesterification reaction of renewable resources like a vegetable oil or animal fat with an alcohol (methanol or ethanol,) in the presence of a catalyst (sodium hydroxide, NaOH, or potassium hydroxide, KOH), which also neutralizes the free fatty acid in the oil. The oils used in the present work include virgin vegetable oil (VVO), canola oil and waste vegetable oil (WVO). Canola oil is also known as Rapeseed oil. WVO is essentially free, ecofriendly and an excellent resource when making biodiesel. VVO is

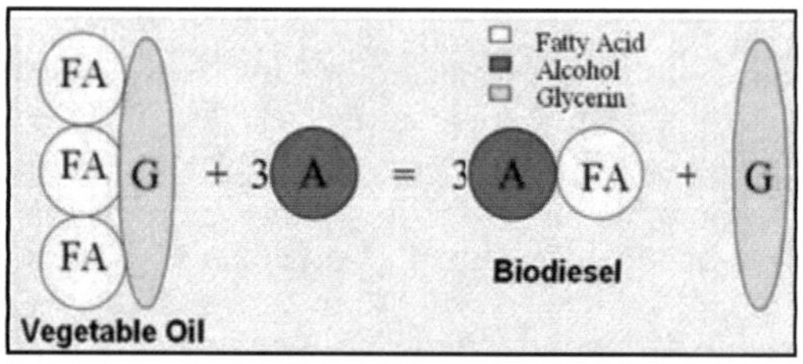

an arduous process to produce, starting with growing the crop, harvesting it, extracting the oil, and then transporting the oil to a facility. This is energy intensive and costly. WVO can both be transported to a biodiesel facility and then cleaned, or can be sent to a cleaning company then transported to the biodiesel facility. The alcohols investigated were methanol and ethanol. Both have pros and cons (Mulumba and Farag 2010). Methanol was used because it is less expensive and is a more efficient reactant. Important parameters that affect the transesterification are ratio of reactants, reaction temperature and reaction time (Briggs, Pearson and Farag 2004, Math 2007). One major problem is soap formation due to the reaction of the base catalyst (KOH in the present study) and the free fatty acids (FFA) which are present mainly in WVO and some VVO, i.e.,

Free Fatty Acids in Oil + KOH = Soap (undesirable)

Residual soap in Biodiesel may plug the fuel filters. Also, if soap is burned in a diesel engine it forms ash residues which result in higher levels of soot out the tailpipe.

1.4 Bench-Top processor

A major constraint in the wide scale use of biodiesel is the production cost. Optimizing biodiesel production is very important in reducing cost and increasing production efficiency. This involves studying and testing different parameters and feedstocks used to make biodiesel. One option that has been the subject of several studies is the use of low cost feedstocks such as WVOs. However, the challenge in using WVO is the high content of free fatty acids (FFA). In order to experiment with different feedstocks and process parameters, our UNH Biodiesel group designed and constructed the bench-top processor shown in Figure 2. It is a scaled down partially transparent processor (1.5 L) that allows for testing runs that would be more difficult and expensive on a full-sized industrial processor.

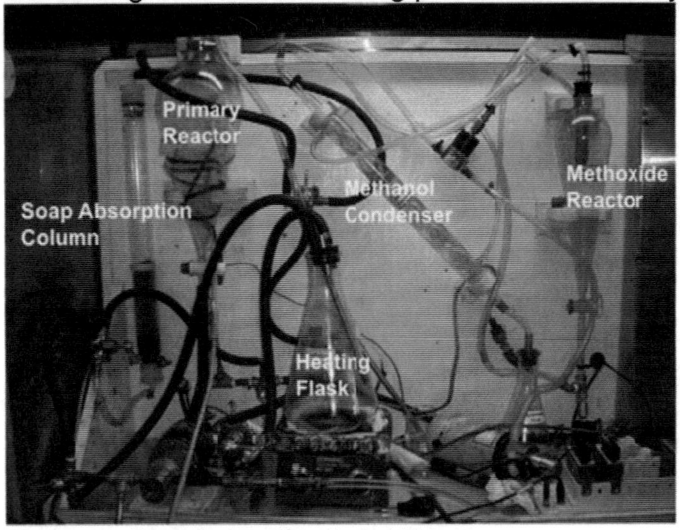

Figure 2: Bench-top Biodiesel Processor

The bench-top biodiesel processor includes 5 major parts (Wilson and Farag, 2010): 1- Methoxide Reactor: KOH and methanol are mixed to form methoxide, 2- Primary Reactor: the oil and methoxide are fed and the transesterification reaction takes place, 3- Soap adsorption column where Amberlite™ BD10DRY™ is used to adsorb soap from the biodiesel product of the reactor, 4- Heating flask: the excess methanol is removed from the biodiesel, and 5- Methanol condenser where the evaporated methanol is recovered for possible reuse.

1.5 Base-Base Two-Stage processing

The simplest method to produce biodiesel is the one-stage base processing. This involves adding all the methoxide to the oil in the reactor, run the reaction at about 135-140° F for 2 hours and let the biodiesel and glycerin products separate for at least four hours. After draining the glycerin, the biodiesel goes through soap adsorption and methanol recovery.

In the base-base two-stage process only some of the methoxide (for example, 80%) is mixed with oil in the primary reactor. The reaction proceeds for 1.5 hours, and the glycerin is drained. The rest of the methoxide (20%) is then added and the reaction proceeds for another 1 hour. After separating the glycerin the biodiesel is cleaned of soap and methanol. The base-base two-stage process studied most in depth was one in which first 80% methoxide was used, and then 20% (this is denoted an 80/20 process).

The base-base two-stage process is intended to increase the conversion of the oil to biodiesel. Removing the glycerin at the end of the first stage will reduce the likelihood of a reverse transesterification reaction. The reaction will proceed forward after more methoxide has been added. It is also possible to reduce the amount of methanol used, to achieve the same conversion as the one-stage process. The base-base two-stage process also reduces the amount of soap formed because the oil is in contact with less KOH in each stage.

2. Goal/Objectives

2.1 Goal

The overall goal of this project is to develop a bench-top biodiesel processor and use it to investigate different feedstocks and operating conditions that result in the highest yield of good quality biodiesel.

2.2 Objectives

The specific objectives of this research project are; 1- construction and flow testing of the bench-top biodiesel processor unit in the UNH Biodiesel Lab. 2- Establish that the processor unit is working properly before employing it with vegetable oil and other chemicals. 3- Run a base case for the production of biodiesel in the bench-top processor to establish operating parameters, 4- Do a parametric investigation to study the effect different variables/operating conditions have on oil conversion and biodiesel quality, 5- Compare the one-stage process with the base-base two-stage process.

3. Methodology/Approach

3.1 Approach

Our first objective to accomplish is to complete the construction and testing of our bench-top biodiesel processor. Operating conditions investigated are reaction temperature, ratio of reactants and the processing time. The quality of biodiesel is assessed by measuring properties such as viscosity, density, the heating value and the soap impurities. The metrics to evaluate optimal biodiesel yield are the oil conversion to biodiesel and the production cost.

3.2 Transesterification

The oil feedstocks used are VVO, canola oil and WVO. The alcohol used was methanol (MeOH), and the catalyst used was KOH in pellet form. Excess methanol was used to increase conversion because the transesterification reaction is reversible. Additional KOH was needed to neutralize the FFAs in the oil. The FFA content in the oil was determined by titration with KOH solution.

3.3 Soap Removal

The Amberlite™ BD10DRY™ purification ion exchange was used for soap adsorption. There are other techniques, such as the water washing or using a synthetic magnesium silicate called Magnesol XL. Water washing is not cost effective because it generates wastewater and is energy intensive. Magnesol XL can be used once, and needs a paper filter to be used. Amberlite™ BD10DRY™ can be used until saturated with soap, then cleaned with methanol, and reused, typically 5-6 times. Its lifespan is about 6 months.

4. Metrics and Measurement Methods

In order to determine the quality of the biodiesel being produced, the percent conversion, the viscosity, the heat of combustion, and the leftover soaps were measured.

4.1 Percent Oil Conversion: 25/225 Test

The percent conversion is measured by dissolving 25 mL of biodiesel in 225 mL of methanol. Biodiesel is soluble in methanol, while vegetable oil is not. If there is 100% conversion, then the biodiesel will be completely dissolved. If conversion is less than 100% than there is some vegetable oil in the biodiesel that will settle at the bottom. For every 1mL of vegetable oil settled there was 4% oil that was not converted.

4.2 Viscosity

Viscosity is a viable measure of quality biodiesel because it is an indicator of the resistance of biodiesel to flow. When the fuel is injected into a combustion chamber, it will atomize and combine with air to combust. If the ignition temperature is low and the fuel is too viscous, then it may not completely combust, and it will form engine deposits. This is why vegetable oil is not typically used in a diesel engine. Vegetable oil has a much higher kinematic viscosity between 36 mm^2/s and 48 mm^2/s (or centistokes, cSt), increasing the probability of engine deposits. Biodiesel typically has a kinematic viscosity between 4 mm^2/s and 6 mm^2/s at 40°C. Diesel viscosity range at 40°C is 1.9 mm^2/s to 4.1 mm^2/s (Knothe, Van Gerpen and Krahl, 2005). In the present study, the kinematic viscosity was measured using a Parr Ball capillary viscometer, which required the density of the biodiesel or oil being tested.

4.3 Heat of Combustion/Heating Value

The heating value is a measure of the potential power obtained from combustion. It was measured using a Parr bomb calorimeter with benzoic acid pellets as the standard.

4.4 Residual Soap: 50/50 Test

Residual soap can be detected by mixing 50 mL of water with 50 mL of biodiesel. If there is clear separation between the water and the biodiesel, then there is no soap left in the biodiesel. If the separation is not clear then the soap made during the reaction was not completely removed by the Amberlite™ BD10DRY™ ion exchange resin. This may indicate that there was probably excess potassium hydroxide (KOH) in the reactants.

5. Results

5.1 Percent Conversion

Figure 3 shows the percent conversion of the biodiesel samples, using the 25/225 test. The results show that the percent conversion increased with the 80/20 process, compared to the one-stage process. The WVO runs show that the percent conversion increased with the 80/20 process, as well as with increased KOH catalyst. WVO contains more FFAs than VVOs so more potassium hydroxide is needed to neutralize the free fatty acids. The conversion results of the vegetable oil show that using less methanol decreases the conversion of the vegetable oil to biodiesel. It also shows that using the 80/20 two-stage process increased the conversion using the same reactants as the one-stage process, and using less methanol the 80/20 process resulted in the same conversion as the one-stage process. This supports one of the benefits of the two-stage process that less alcohol is needed in the two-stage process to achieve the same conversion as the one-stage process.

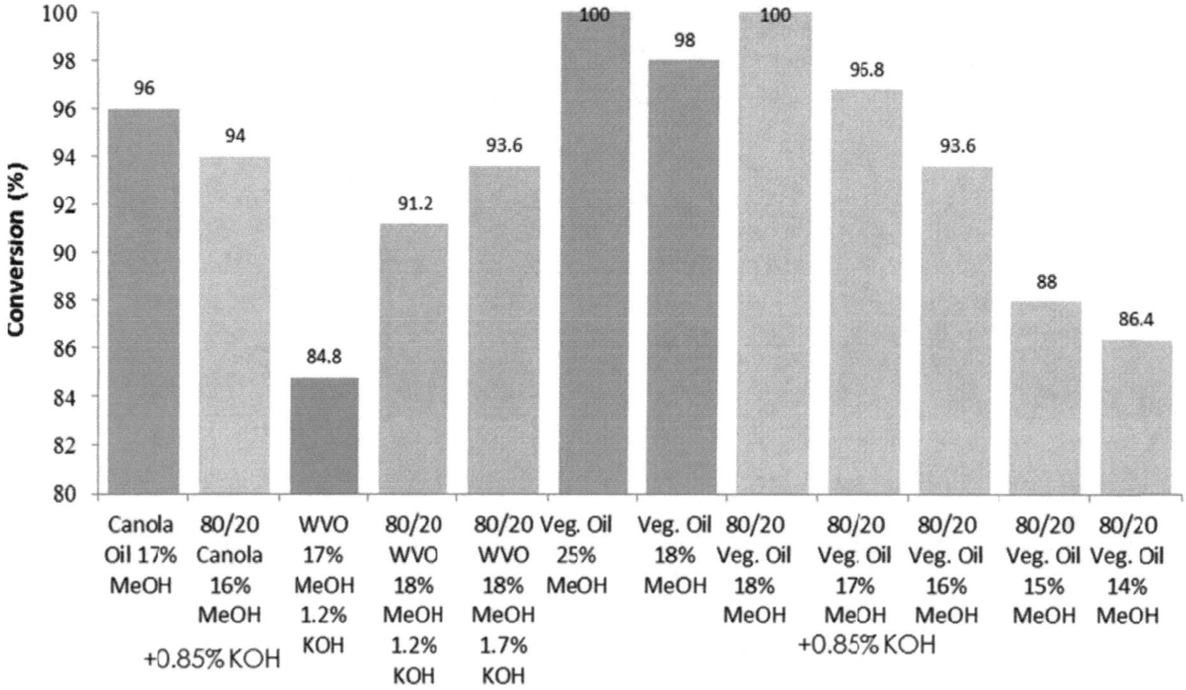

Figure 3. Percent conversion of oil, labeled with the reactants used to make the biodiesel and percent by weight relative to oil, e.g., WVO 17% MeOH 1.2% KOH has 100 g WVO + 17 g MeOH + 1.2 g of KOH pellets. All canola oil and vegetable oil batches were mixed with 0.85% KOH:oil ratio by weight of oil, i.e., 100 g oil + 0.85 g KOH.

5.2 Viscosity

Figure 4 shows that the kinematic viscosity of all biodiesel samples produced in our bench-top processor was in the usual range of 4 mm^2/s to 6 mm^2/s. The WVO and VVO feedstocks have a much greater viscosity (36 – 48 mm^2/s). The biodiesel produced in the 80/20 two-stage process typically had a lower viscosity than the one-stage process. The lower viscosity in the 80/20 two- stage process biodiesel is due to the higher conversion and reduced vegetable oil in the biodiesel, which will raise the viscosity.

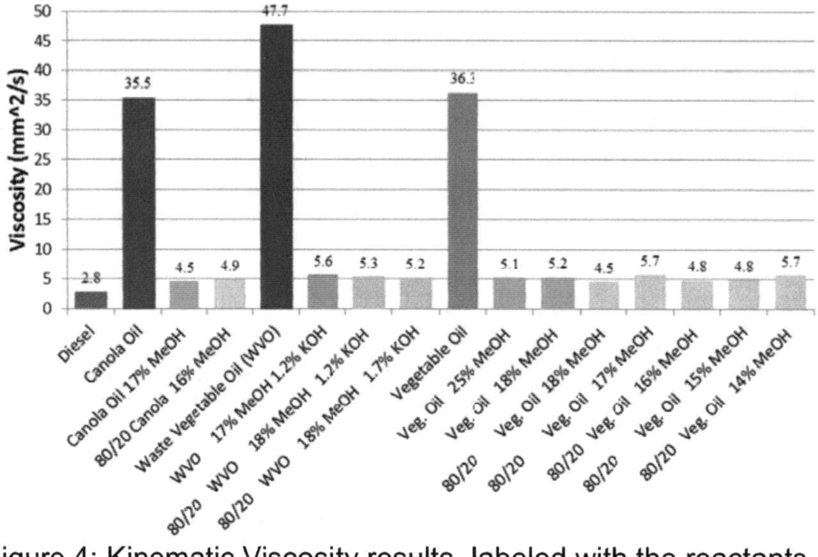

Figure 4: Kinematic Viscosity results, labeled with the reactants used to make the biodiesel and percent by weight. All canola oil and vegetable oil batches were mixed with 0.85% KOH by weight.

5.3 Heat of Combustion

Figure 5 shows that the heat of combustion (HOC) of diesel fuel is greater than that of biodiesel or feedstock oil. The HOC of pure vegetable oil is lower than the biodiesel made from the respective oil. Virgin vegetable oil (VVO) and waste vegetable oil (WVO) show upward trend in the HOC from the one-stage to the 80/20 two-stage process.

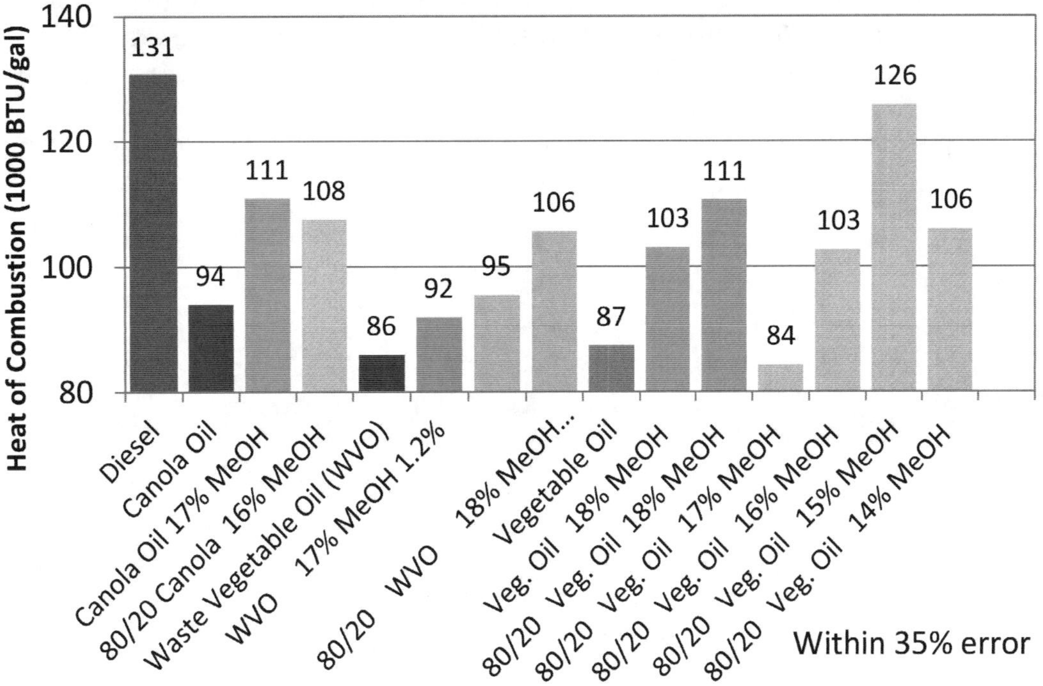

Figure 5. Heat of Combustion of Biodiesel (within 35% error), labeled with the reactants used to make the biodiesel and percent by weight. All canola oil and vegetable oil batches were mixed with 0.85% KOH by weight relative to oil.

5.4 Residual Soap

Figure 6 displays pictures from the soap tests. The 80/20 two-stage processes show less residual soap than the one-stage processes with the same reactants (up to 17% MeOH). The WVO 80/20 process using 0.5% more KOH formed more soap due to the increase in KOH. Also, the vegetable oil one-stage process using 25% MeOH had comparable separation as the 80/20 two-stage process using 18% MeOH/oil ratio.

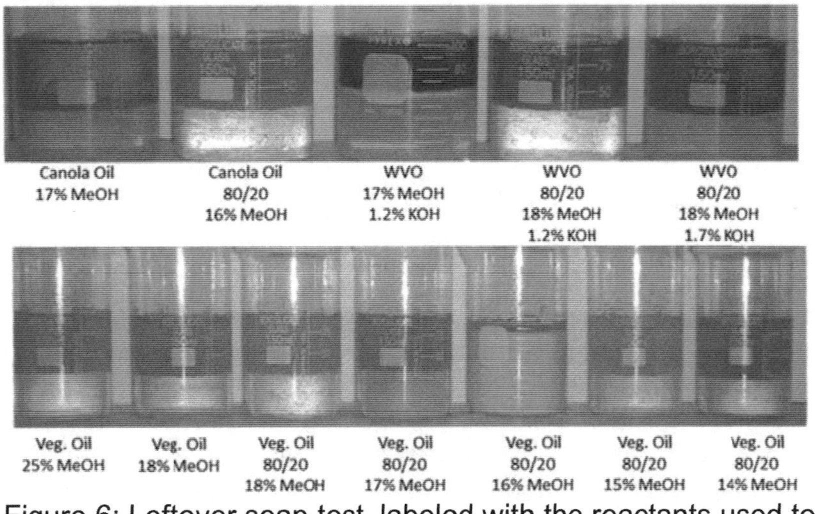

Figure 6: Leftover soap test, labeled with the reactants used to make the biodiesel and percent by weight. Canola oil and vegetable oil batches were mixed with 0.85% KOH by weight.

5.4 Effect of Methanol/oil Ratio

The effects of the methanol/oil ratio on the oil conversion, the viscosity, the heat of combustion and the residual soaps have been discussed in the previous sections. The specific results of the methanol/oil ratio are presented in this section. Figure 7 shows the effect in one-stage operation using vegetable oil. Figure 8 shows the 50/50 residual soaps test results for the same samples in Figure 7. Figures 7 and 8 show a drop in conversion with a decrease in the MeOH/Oil ratio from 25/100 to 18/100, and a slight increase in residual soap. The same effects are found in the base-base 80/20 two-stage process using vegetable oil, as shown in Figures 9 and 10. These results show that using a methanol/oil ratio of more than 18/100 in a two-stage process of vegetable oil is uneconomical because it will not improve the conversion.

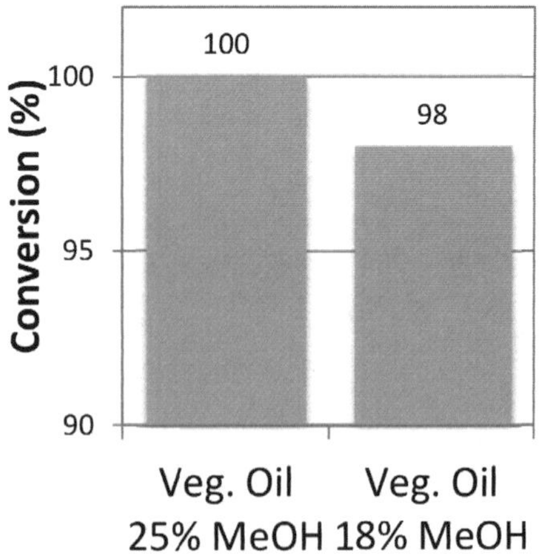

Figure 7. Effect of Methanol/Oil ratio on conversion of vegetable oil, one-stage process

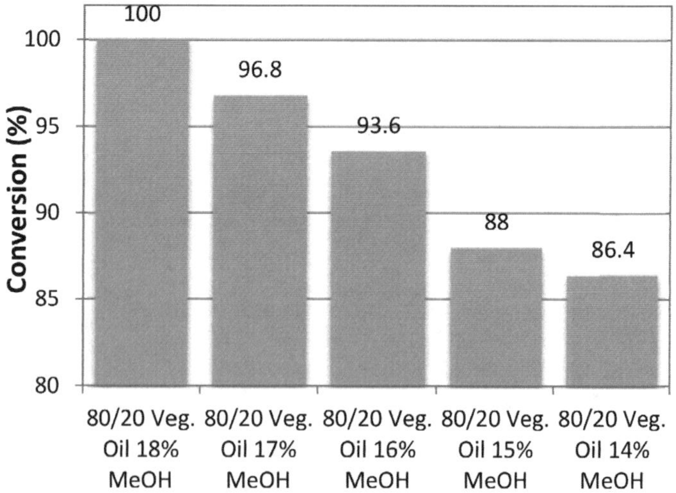

Figure 9. Effect of Methanol/Oil ratio on conversion of vegetable oil, base-base two-stage process

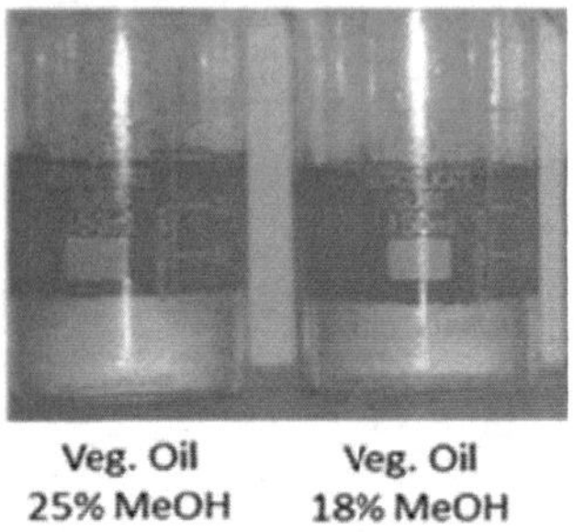

Figure 8. Effect of Methanol/Oil ratio on residual vegetable oil, one-stage process

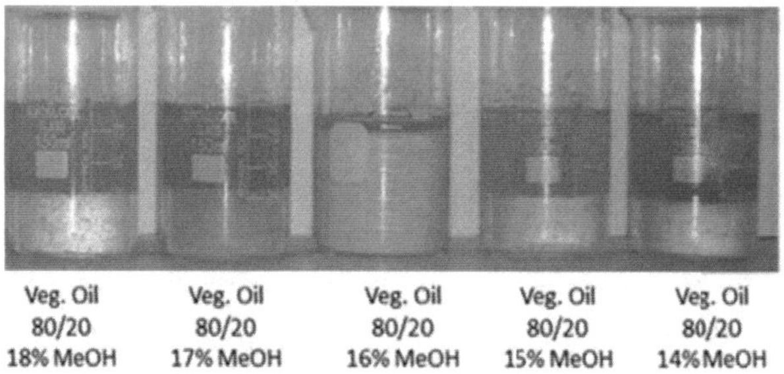

Figure 10. Effect of Methanol/Oil ratio on the residual vegetable oil, base-base two-stage process

5.5 Effect of Catalyst (KOH)/Oil Ratio

Figure 11 shows the effect of the amount of catalyst (KOH in the present study) on the conversion of waste vegetable oil (WVO). The data was obtained from the case of an 80/20 base-base two-stage run. Increasing the catalyst (KOH)/oil ratio from 1.2:100 to 1.7:100 caused an increase in the oil conversion. It also caused an increase in the heat of combustion, as shown in Figure 12. However, increasing the KOH also increased the residual soap, shown by Figure 13.Further experiments are needed to fully investigate the effect of KOH. Based on our results we recommend the use of WVO in a two-stage process using 18% MeOH/oil ratio and 1.2% KOH/oil ratio because of the lower residual soap relative to the case of 1.7% KOH/oil ratio even though this compromises the oil conversion and the heat of combustion.

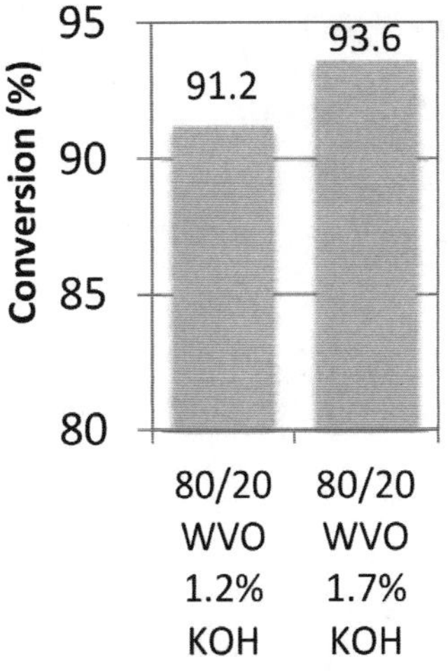

Figure 11. Effect of KOH/oil ratio on conversion of WVO base-base two-stage process

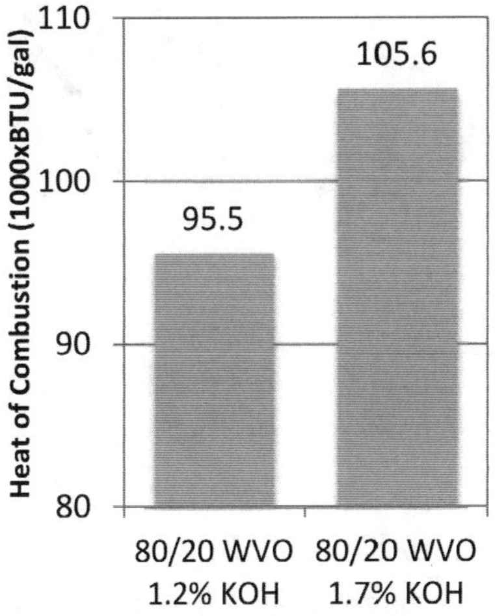

Figure 12. Effect of KOH/oil ratio on heat of combustion of biodiesel made from WVO base-base two-stage process

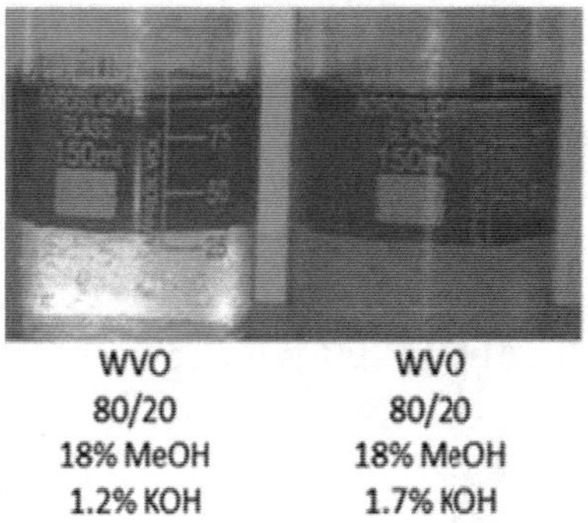

Figure 13. Effect of KOH/oil ratio on leftover soap in biodiesel made from WVO base-base two-stage process

5.5 One-Stage versus Two-Stage Operation

Figure 14 shows the layers of biodiesel and water after mixing. The one-stage process (left beaker) evidently has more residual soap than the two-stage process (right beaker). This demonstrates that the base-base two-stage process produces less soap than the one-stage process. Figure 15 indicates that for the same oil/ methanol ratio the two-stage process results in a higher oil conversion than the one-stage process.

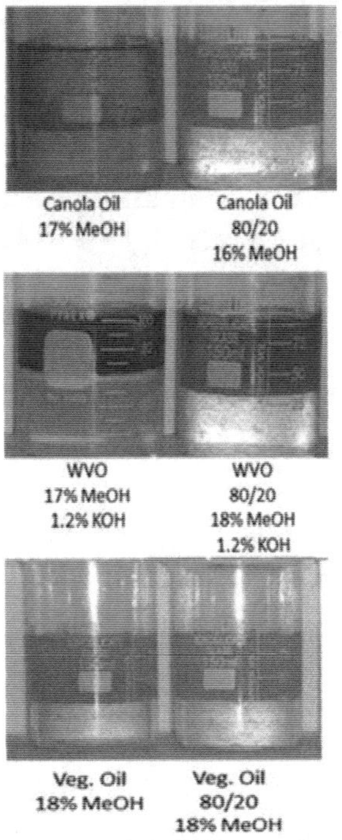

Figure 14. Effect of one-stage versus two-stage process on residual soap in biodiesel.

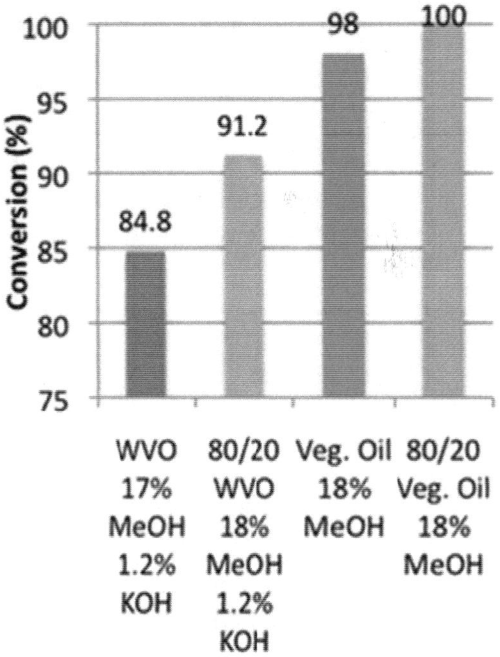

Figure 15. Effect of one-stage versus two-stage process on percent conversion of biodiesel made from WVO and Veg. Oil.

25% MeOH/oil ratio gave the same conversion (100%) as using 18% MeOH/oil ratio in an 80/20 batch. This is shown in Figure 16. It is evident from these results that the base-base two-stage process achieves the same oil feedstock conversion as the one-stage process, but with smaller methanol/oil ratio. This makes the base-base two-stage process more economically viable, as well as environmentally friendly.

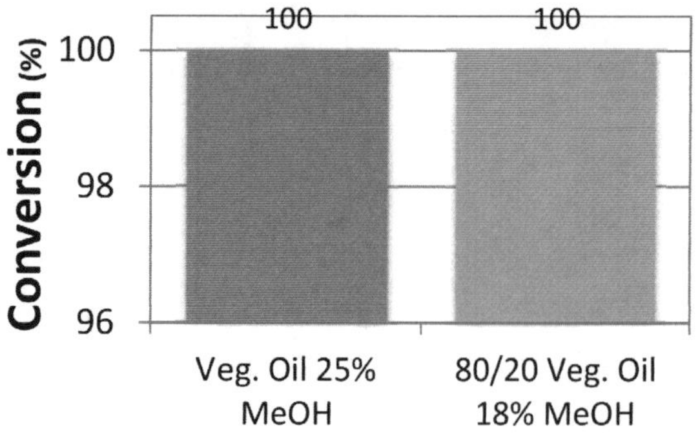

Figure 16. Effect of one-stage versus two-stage process on percent conversion of biodiesel made from Virgin Vegetable Oil (VVO).

6-Conclusions

It is evident from our study that an 80/20 base-base two-stage process is more economically viable because it uses less methanol while producing the same quality biodiesel as a one-stage process. Our results also show that using waste vegetable oil (WVO) is comparable to virgin vegetable oil. It is evident that an 80/20 base-base two-stage process using 18% MeOH/oil ratio and 1.2% KOH/oil ratio is the optimal choice for producing biodiesel. This is the least expensive option, and it had a 91% oil conversion and the biodiesel produced had a viscosity of 5.2 mm^2/s, a reasonable heat of combustion (92,000 BTU/gal) and no residua soap. While better quality biodiesel was produced using a virgin vegetable oil (such as the 80/20 vegetable oil batch using 18% MeOH/oil ratio), the waste vegetable oil is much less expensive and overall better for the environment, so we feel it is the optimal choice.

Acknowledgment

The authors would like to acknowledge the support of the UNH Hamel Center for Undergraduate Research, the UNH Dairy Research Center, Nancy Whitehouse and Sue Chase for use of their equipment.

References

- http://en.wikipedia.org/wiki/Biodiesel last Modified: February 16, 2010.
- Briggs, M., Pearson, J., Farag, I.H., (2004) “Biodiesel Processing,” UNH Biodiesel Group. http://www.unh.edu/p2/biodiesel
- Farag, I.H. (2008), “Biodiesel: Challenges and new Fronts”, Keynote Paper, International Biodiesel Workshop, National Research Center, March 23-25, 2008, Cairo, Egypt
- Knothe, Gerhard; Van Gerpen, Jon; Krahl, Jurgen (2005), ““The Biodiesel Handbook,” AOCS Press, Champaign, Ill.
- Math, M.C. and G. Irfan (2007), “Optimization of Restaurant Waste Oil Methyl Ester Yield,” Journal of Scientific and Industrial Research, 66, 772-776.
- Mulumba, N and I.H. Farag (2010), “Production of Biodiesel from Microalgae,” M.S. Thesis, Chemical Engineering Dept., University of New Hampshire (UNH), Sep. 2010.
- Wilson, R. Grohs, J and I. H. Farag (2010) “Optimal Biodiesel Production Through Bench-top Processor Testing,” Northeast Biomass Conference and Expo, August 4-6, 2010, Westin Copley Plaza, Boston.

(305d) Disruption of Microalgal Cells for Biodiesel Production

Ronald Halim, Michael Danquah and Paul Webley, Chemical Engineering, Monash University, Melbourne, Australia

The production of biodiesel from microalgae involves lengthy processing steps. Cell disruption is an integral part of this downstream protocol as it facilitates the release of the intracellular lipids that will be used for biodiesel production. This study examines different mechanical methods for the disruption of microalgae cells, including High Pressure Homogenization or HPH, ultrasonication and bead beating. The performance of cell disruption method was evaluated in terms of two key indicators: reduction in the intact cell counts per unit volume (C/C0) and reduction in the average diameters of microalgal colonies (D/D0). The most effective disruption was obtained using HPH (optimum C/C0 = 0.06, D/D0 = 0.35) followed by bead beating (optimum C/C0 = 0.55, D/D0 = 0.25). Even though ultrasonication failed to disrupt the microalgal cells (optimum C/C0 = 0.91), it still managed to disintegrate cellular agglomerations (optimum D/D0 = 0.23). It is hoped that the finding from this study can be used for future microalgal biofuel research.

See more of this Session: Sustainable Diesel Fuel From Renewable Resources
See more of this Group/Topical: Sustainable Engineering Forum

(305e) Production of Biodiesel From Alternative Feedstocks – a Bubble Reactor for Converting Oils Containing Free-Fatty-Acids to Biodiesel

Richard A. Cairncross[1], Cory Melick[2] and Minhazuddin Mohammed[2], (1)Chemical and Biological Engineering, Drexel University, Philadelphia, PA, (2)Chemical & Biological Engineering, Drexel University, Philadelphia, PA

Biodiesel is a diesel substitute that is typically produced by transesterification of refined vegetable oils (soybean oil in USA) with methanol; this chemistry is simple and effective with base catalysts and refined oils. However, crude vegetable oils and many alternative oil feedstocks contain significant (> 1%) concentrations of free-fatty-acids (FFA); for example waste fryer oil typically is 10-15% FFA and trap grease is >85% FFA. Free-fatty-acids cause low conversion and difficult separation with the traditional chemistry because they form soaps with base catalysts. There are several alternative techniques for producing biodiesel from high-FFA oils including two-step reactions, high pressure reactions, and supercritical reactions. In this research a two-phase bubble column reactor is used to esterify FFA at moderate temperatures (100-150°C) and atmospheric pressure. Methanol added to the reactor boils at these temperatures and produces bubbles that rise through the reactor. Despite having lower methanol concentrations in the reacting liquid phase, acceptable reaction rates and high conversions are achieved with this configuration. Our research has also shown that bubble reactors are effective at overcoming equilibrium limitations of the esterification reaction. Because the bubbles remove by-product water, the reactor is much more robust for feedstocks containing water, and conversions approaching 100% are achieved with up to 10% water in methanol. Similar results have been obtained for reactions with ethanol/water mixtures showing that it is not necessary to break the ethanol-water azeotrope when purifying ethanol for this reactor. A model of the reactor has been developed that accurately predicts trends of the experimental data and can be used to design improved reactors. Feasibility studies of full-scale production of biodiesel from trap-grease have been performed using the bubble reactor and two other reaction schemes – supercritical reactors and medium pressure CSTR's in series. The bubble reactor process has the highest return on investment and lowest greenhouse gas emissions.

See more of this Session: Sustainable Diesel Fuel From Renewable Resources
See more of this Group/Topical: Sustainable Engineering Forum

High-Pressure Viscosity of Bio-diesel from Soybean, Canola, and Coconut Oil

Andrew M. Duncan[1], *Azita Ahosseini*[1], *Reece McHenry*[1], *Christopher D. Depcik*[2,4], *Susan M. Stagg-Williams*[1-3,*] *and Aaron M. Scurto*[1-3,*]

[1] *Department of Chemical & Petroleum Engineering*
[2] *DOT-KS Transportation Research Institute (TRI)*
[3] *Center for Environmentally Beneficial Catalysis*
[4] *Department of Mechanical Engineering*

1. Introduction

Biodiesel, as an alternative fuel, has undergone significant research, development, and large-scale distribution over the past years. Biodiesel is a mixture of fatty acid esters and is commonly used as an alternative to, or blended with, No. 2 petroleum diesel. The utilization of biodiesel as an alternative fuel has been driven by the fact that, compared to petroleum based diesel, biodiesel is: domestically produced, renewable, has low sulfur and aromatic content, has low particulate emissions, has high energy content, and can be used in conventional Compression-ignition (diesel) engines with relatively little or no modification.[1]

Typical biodiesel is produced via the transesterification of fats and oils with an alcohol such as methanol. The composition of the fat or oil feedstock determines the fatty acid methyl ester profile and ultimately the chemical and physical properties of the resulting biodiesel. For example, biodiesel produced from soybean oil typical contains methyl ester profiles rich in oleate (C18:1) and linoleate (C18:2), while biodiesel from coconut oil is rich in laurate (C12:0) and myristerate (C14:0). These differences in compositions can affect properties such as cloud point, viscosity, density, oxidative stability, etc.[2, 3]

Current fuel injection systems for compression ignition (diesel) engines reach pressures approximately 1600 and 1400 bar for passenger car and commercial vehicle systems respectively.[4] Future systems are targeting even higher pressures in excess of 2000 bar while also expanding the possible number of injections per cycle from five to nine.[5] Under these pressures, the liquid viscosity of the fuel can increase up to approximately ten times atmospheric levels.[6] Upon injection of the fuel in the cylinder, the large de-pressurization of the fuel results in a significant gradient of the viscous properties of the fluid. A proper understanding of these values is therefore required for modeling the discharge coefficient and flow rate through the injector. These parameters are key in ensuring the proper rate shaping of the fuel injection spray in order to optimize performance while minimizing emissions.[7]

Generally speaking, biodiesel and biodiesel blends from a range of feedstocks are considered to have lower emissions of carbon monoxide (CO), total hydrocarbons (THC) and particulate matter (PM) than standard diesel fuel, but equal or slightly higher levels of nitrogen oxides (NO_x).[8] In some cases, recalibration of engines based on blending of biodiesel with other fuels has resulted in decreasing NO_x concentrations.[9, 10] Part of the change in NO_x relates to the inadvertent advancing of injection timing caused by the rapid transfer of the pressure wave from the fuel injection pump to the fuel injector causing it to open earlier. This pressure wave is a function of the properties of the fuel. Choi *et al.* attributed the change in fuel injection timing to the higher viscosity of biodiesel [11]; however, others conclude that viscosity does not affect timing and the bulk modulus, which effects the speed of sound as the main factor.[12-15] In any case, viscosity does affect the pressure dynamics within the fuel lines and during the spray break-up process inside the cylinder. This is a concern for engine manufacturers while designing the fuel injection systems for these engines under the range of potential pressure and temperature conditions (up to 3500 bar and 273.15 K to 373.15 K).[16, 17]

2. Experimental High-pressure and ambient viscosity measurement

A Cambridge Applied Systems (currently Cambridge Viscosity, Inc.) high pressure viscometer was utilized for these measurements (ViscoPro 2000 System 4- SPL-440 with Viscolab software). The apparatus has been described in high detail in Ahosseini and Scurto[18], but an overview will be given here. The apparatus consists of a temperature-controlled oven (±0.1 K) which houses the high-pressure viscometer sensor, a precision pressure transducer (PT), and a resistance temperature detector (RTD) (±0.05 K). The viscometer is connected to a manual high-pressure syringe pump purchased from High Pressure Equipment Company (HIP) (Model No. 50-575-30; 30,000 psi, capacity of 10 cm^3 with PolyPak) to pressurize the diesel samples. The viscometer utilizes the principles of annular flow around an axially oscillating piston[19] and has recently been given an ASTM certification.[20] The sensor is capable of measurements from 0.2 to 10,000 cP (mPa·s), at a maximum pressure of 137.9 MPa and in a temperature range of 233.15 K to 463.15 K. The diameter of the inner chamber is 0.314 inches. Various size pistons are used according the nominal viscosity range suggested by the manufacturer: 0.25-5 cP (0.3085 inch diameter), 1-20 cP (0.3055 inch diameter) and 2.5-50 cP (0.3025 inch diameter).

The apparatus was used to find the viscosity of soybean, Vistive (Monsanto's low-linlolenic soybean), canola, used canola, coconut-based biodiesels, and #2 ultra low sulfur diesel fuel. The viscosities were measured over a temperature range of 283.15 K to 373.15 K and a pressure range of 0.1 MPa to 130 MPa.

2. Correlation and Results

Various theoretical models and empirical expressions can be found in the literature that can be used to represent the viscosity of liquids (η) as functions of pressure and temperature. Litovitz[21,22] has suggested an empirical equation, which has been used at a single pressure over wide temperature ranges

$$\eta = A\exp(B / RT^3)$$

where R is the gas constant and A and B are fitted parameters.

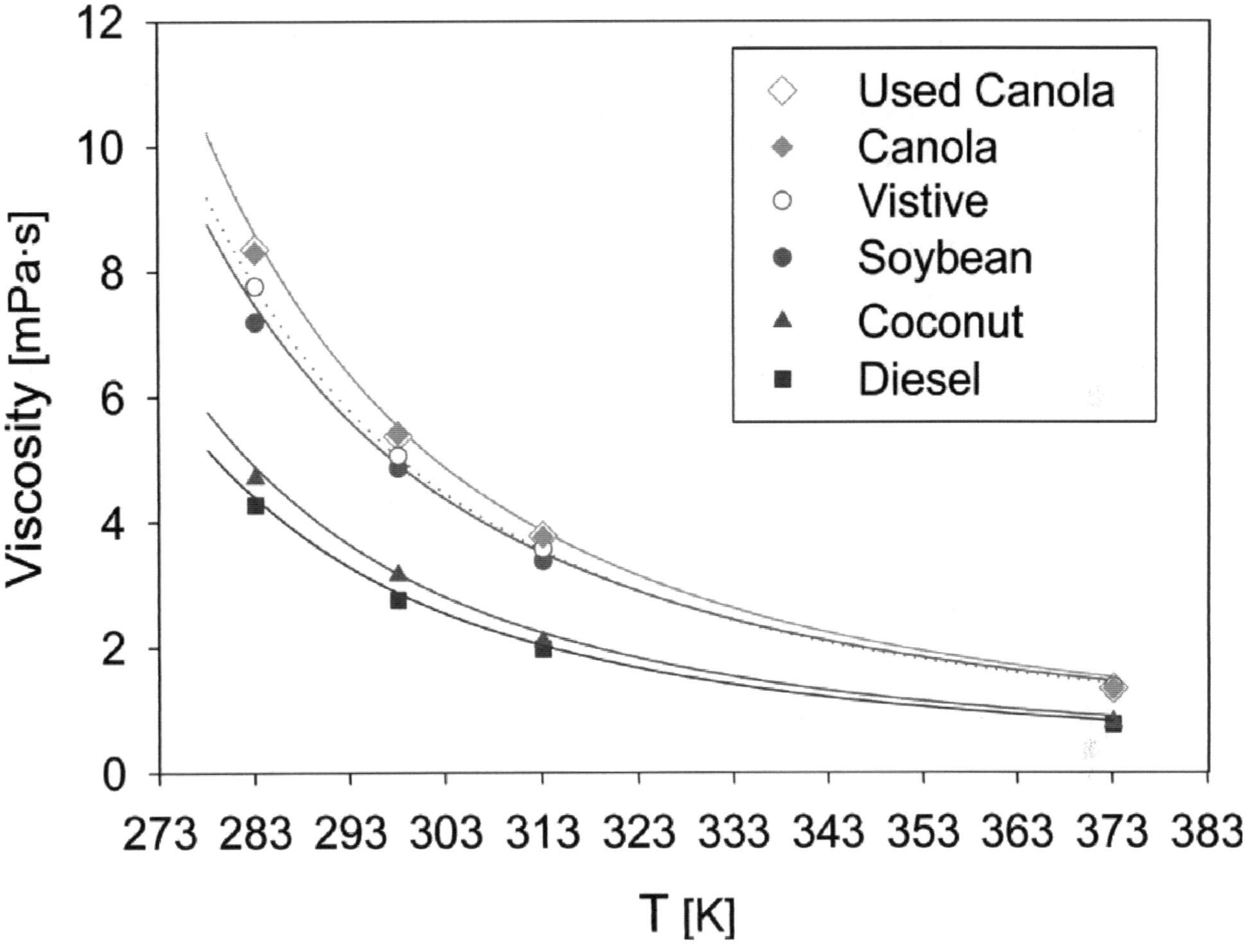

Figure 1. Ambient-pressure viscosity with the temperature for the number 2 diesel and biodiesel samples. The lines represent the correlation from the Litovitz equation from parameters regressed from all data.

Figure 1 above shows the results of the regression of the ambient pressure viscosities of each fuel source tested in this investigation to the Litovitz parameters. The data, correlate very well to the Litovitz equation—represented by the solid or dashed lines. The soybean and canola fuels had very similar behavior, with a very sharp increase in viscosity with decreasing temperature. The coconut biodiesel had very similar viscous behavior to the petroleum-based fuel.

After the ambient pressure values were fit to a Litovitz equation, the high-pressure data were fit to a Tait equation. The Tait equation is well-known to represent the pressure volume-temperature relationship of many liquids.[23,24]

$$\ln\left(\frac{\eta_P}{\eta_o}\right) = E\ln\left(\frac{D+P}{D+0.1}\right)$$

where η_p and η_0 are the viscosities at a pressure P and at 0.1 MPa, respectively. This equation contains only two fitted parameters and has yielded good correlation with experimental data as shown by Kashiwagi and Makita 4 for aromatic hydrocarbons and cyclohexane up to 110 MPa.

This investigation will utilize a hybrid Tait-Litovitz equation at elevated pressures for the viscosity data for a series of biodiesels:

$$\eta_P = A\exp(B'/T^3)((D+P)/(D+0.1))^E$$
$$B' = B/R$$

The Tait-Litovitz hybrid equation was applied and regressed for each of the six fuels tested. The results of this hybrid equation can be seen by the solid lines in Figures 2 and 3. The accuracy of the correlation in these figures was typical of all fuel samples.

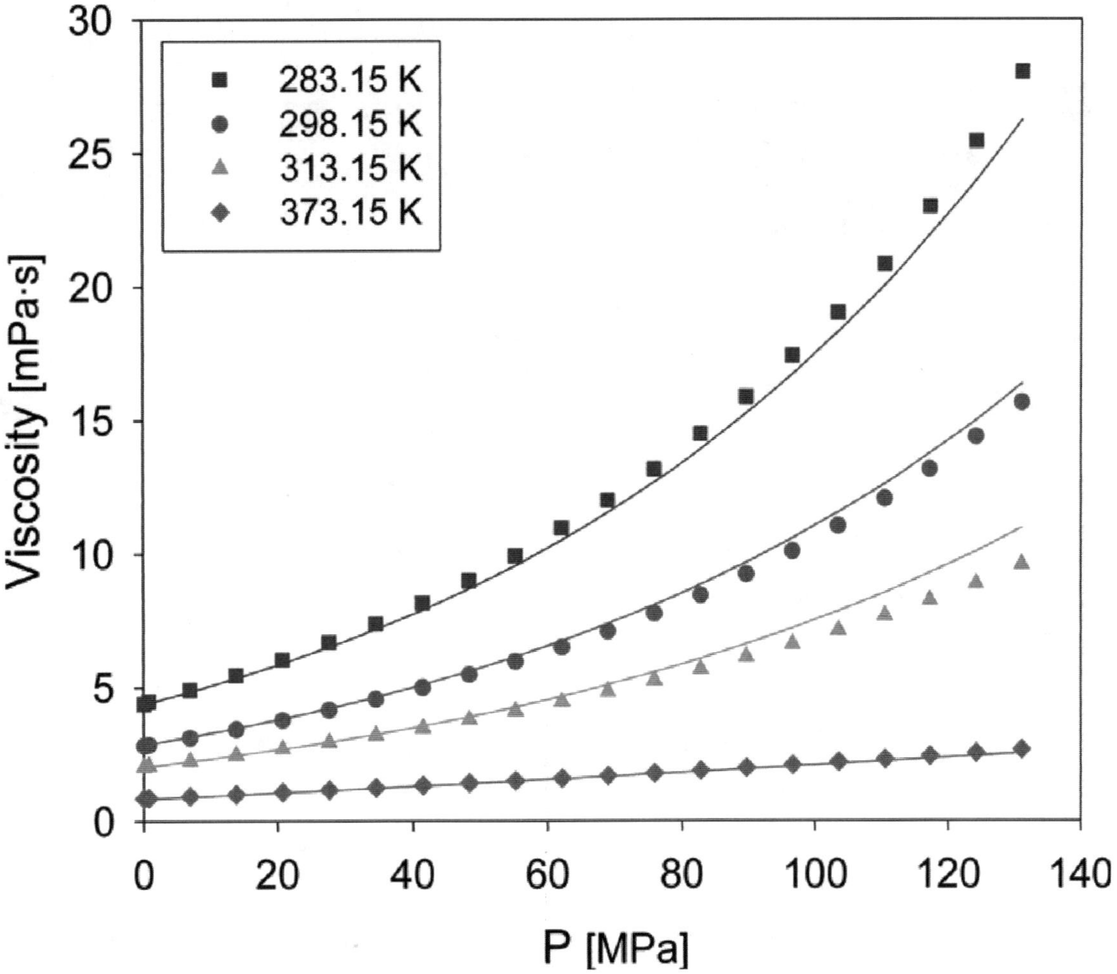

Figure 2. Viscosity with pressure for petroleum-based diesel (number 2) at four isotherms: 283.15, 298.15, 313.15, and 373.15 K. Lines represent the Tait-Litovitz equation from temperature dependent parameters regressed from all four isotherms.

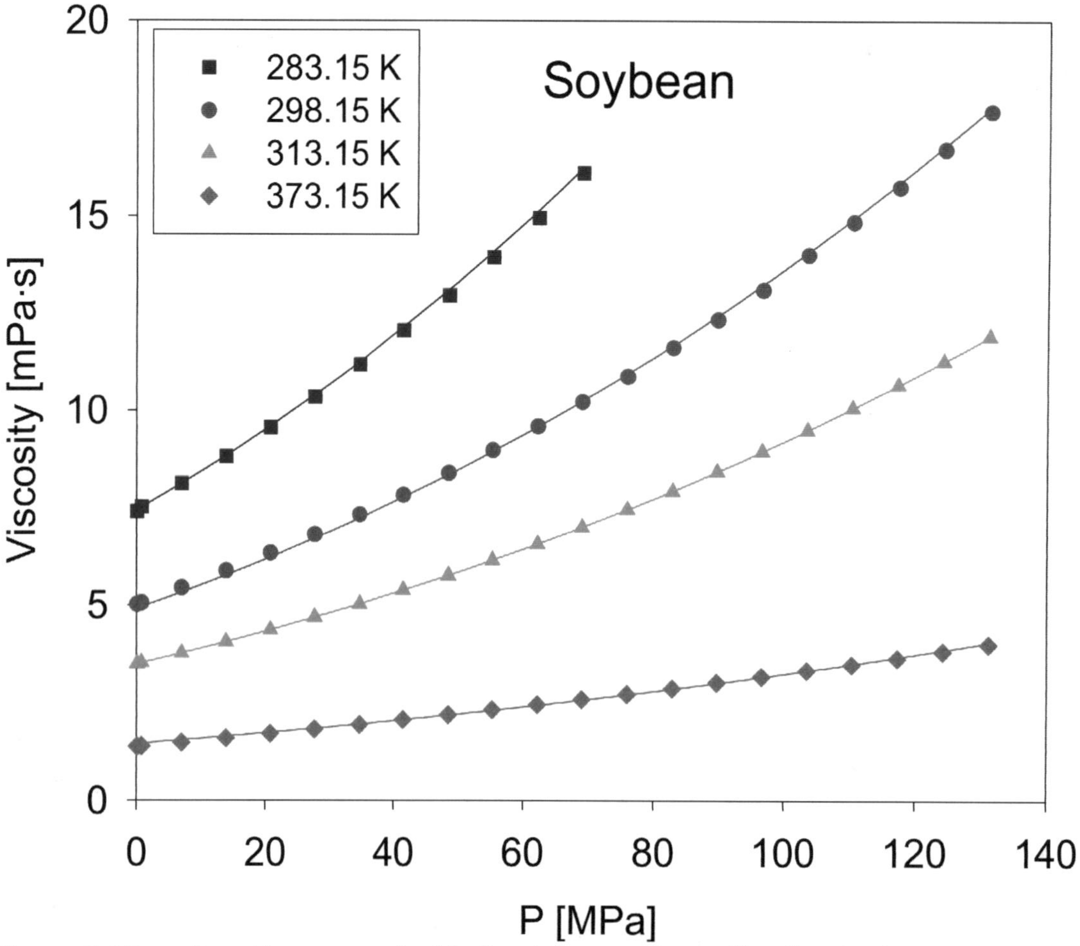

Figure 3. Viscosity with pressure for biodiesel from soybean oil at four isotherms: 283.15, 298.15, 313.15, and 373.15 K.

While the viscosities of the soybean biodiesel shown in Figure 3 is higher compared to the diesel isotherms shown in Figure 2, the percentage change for each respective isotherm was found to be greater for the diesel fuel. This percentage change was typical for the biodiesels compared to the petroleum-based fuel.

Conclusions

The viscosity of a series of biodiesels from various biorenewable sources was measured to determine the effect of high pressures (up to 130 MPa) and temperatures from 283.15 to 373.15 K. These data are the first to explore comprehensively the effect of temperature, pressure, and vegetal source on the viscosity of biodiesel. The viscosity change that a biodiesel would experience in a fuel injector would be up to several hundred percent greater than the ambient pressure viscosity. While differences in ambient pressure viscosity exist among the different biodiesels, the effect of pressure is relatively similar. The viscosity data have been correlated very well with a hybrid Tait-Litovitz equation with average absolute relative deviation percentages (AARD%) from 0.9% for coconut derived biodiesel to 2.1% for canola derived biodiesel. The correlation could be used for a wide range of simulation and design.

References:

1. Graboski, M.; McCormick, R., Combustion of fat and vegetable oil derived fuels in diesel engines. *Prog. Energy Comb. Sci.* **1998,** 24, (2), 125-164.
2. Rodrigues, J.; Cardoso, F.; Lachter, E.; Estevão, L.; Lima, E.; Nascimento, R., Correlating chemical structure and physical properties of vegetable oil esters. *J. Am. Oil. Chem. Soc.* **2006,** 83, (4), 353-357.
3. Ramos, M.; Fernandez, C.; Casas, A.; Rodriguez, L.; Perez, A., Influence of fatty acid composition of raw materials on biodiesel properties. *Bioresource Tech.* **2009,** 100, (1), 261-268.
4. Bosch, *Bosch Automotive Handbook.* 7th ed.; Robert Bosch GmbH: 2007.
5. Tomishima, H.; Matsumoto, T.; Okl, M.; Nagata, K., The Advanced Diesel Common Rail System for Achieving a Good Balance Between Ecology and Economy. *SAE Paper 2008-28-0017* **2008**.
6. Giannadakis, E.; Gavaises, M.; Theodorakakos, A., The Influence of Variable Fuel Properties in High-Pressure Diesel Injectors. *SAE Paper 2009-01-0832* **2009**.
7. Schmidt, D. P.; Corradini, M. L., The internal flow of diesel fuel injector nozzles: a review. *Int. J. Engine Res.* **2001,** 2, (1), 1-22.
8. Agency, E. P. *A Comprehensive Analysis of Biodiesel Impacts on Exhaust Emissions; Draft Technical Report*; EPA420-P-02-001; Office of Transportation and Air Quality: Washington, DC, 2002.
9. Kegl, B.; Pehan, S., Influence of biodiesel on injection, fuel spray, and engine characteristics. *Therm. Sci.* **2008,** 12, (2), 171-182.
10. Lu, X. C.; Ma, J. J.; Ji, L. B.; Huang, Z., Simultaneous reduction of NOx emission and smoke opacity of biodiesel-fueled engines by port injection of ethanol. *Fuel* **2008,** 87, (7), 1289-1296.
11. Choi, C. Y.; Bower, G. R.; Reitz, R. D., Effect of Biodiesel Blended Fuels and Multiple Injections on D.I. Diesel Engines. *SAE Paper 970218* **1997**.

12. Arcoumanis, C.; Gavaises, M.; Yamanishi, M.; Oiwa, J., Application of a FIE computer model to an in-line pump-based injection system for diesel engines. *SAE Paper 970348* **1997**.
13. Rakopoulos, C. D.; Hountalas, D. T., A simulation analysis of a DI diesel engine fuel injection system fitted with a constant pressure valve. *Energy Conv. Managm.* **1996,** 37, (2), 135-150.
14. Szybist, J. P.; Boehman, A. L., Behavior of a Diesel Injection System with Biodiesel Fuel. *SAE Paper 2003-01-1039* **2003**.
15. Szybist, J. P.; McFarlane, J.; Bunting, B. G., Comparison of Simulated and Experimental Combustion of Biodiesel Blends in a Single Cylinder Diesel HCCI Engine. *SAE Paper 2007-01-4010* **2007**.
16. Gault, R. In *Biodiesel Outlook - An Engine Manufacturer's Perspective*, Diesel Engine-Efficiency and Emissions Research (DEER), Dearborn, MI, August 7, 2008; Dearborn, MI, 2008.
17. Beyer, T. The Measurement of Diesel Fuel Properties. Georgia Institute of Technology, Atlanta, GA, 1998.
18. Ahosseini, A.; Scurto, A. M., Viscosity of Imidazolium-Based Ionic Liquids at Elevated Pressures: Cation and Anion Effects. *Int. J. Thermophys.* **2008,** 29, 1222-1243.
19. Bird, R. B.; Stewart, W. E.; Lightfoot, E. N., *Transport Phenomena*. Wiley: New York, 1960.
20. *ASTM D7483 - 08 Determination of Dynamic and Kinematic Viscosity of Liquids by Oscillating Piston Viscometer.*; 2008.
21. Litovitz, T. A., Temperature Dependence of the Viscosity of Associated Liquids. *J. Chem. Phys.* **1952,** 20, 1088-1089.
22. Litovitz, T. A., Temperature Dependence of the Viscosity of Associated Liquids. *J. Chem. Phys.* **1952,** 20, 1980.
23. Dymond, J. H.; Malhotra, R., The Tait equation: 100 years on. *Int. J. Thermophys.* **1988,** 9, (6), 941-951.
24. Kashiwagi, H.; Makita, T., Viscosity of Twelve Hydrocarbon Liquids in the Temperature Range 298-348 K at Pressures up to 110 MPa. *Int. J. Thermophys.* **1982,** 3, (4), 289-305.

(305g) Activated Sludge Oil: Identification and Characterization of Components

Patrisha J. Pham, Rafael Hernandez, Emmanuel D. Revellame and William T. French, Dave C. Swalm School of Chemical Engineering, Mississippi State University, Mississippi State, MS

The demand for alternative sources to petroleum is growing. Recently, sewage sludge as a raw material for biodiesel production has been demonstrated in our group[1-2]. In this present study, activated sludge oil (ASO) extracted from activated sludge was characterized to provide information for activated sludge oil refinement. The refinement process aims for activated sludge to be considered as a feedstock for biodiesel production and as a possible source of value-added products. Fractionation of the ASO extract with solvents of varying polarities gave five ASO solvent fractions. The % composition of the ASO hexane (HX), chloroform ($CHCl_3$), acetone (ACE), methanol (MeOH) and methanol/ammonium hydroxide (HPC) fractions were 7.0, 26.0, 32.0, 31.0 and 4 %, respectively. The ASO HX fraction was found to contain a homologous series of n-alkanes and squalene. The ASO $CHCl_3$ fraction was found to contain free fatty acids (C14:0-C20:0), tocopherols, monoglyceride, free sterols, diglyceride, wax esters, steryl esters and triglycerides. The ASO ACE fraction contained two unknown glycolipids. Approximately 1.3% of the ASO ACE fraction is carbohydrates with glucose (57%) as the main glycosyl moiety present. Majority of the ASO ACE fraction are sterols or steryl derivatives. The ASO MeOH fraction was 35 % phospholipids based on ^{31}P NMR. Phosphatidylethanolamine at 18% was the major phospholipid component detected. Conversion into fatty acid methyl esters was at 38, 18 and 15 % for ASO $CHCl_3$, ACE and MeOH fractions, respectively. The fatty acid profiles show C14:0, C16:0, C16:1, C18:0 and C18:1 as the dominant fatty acids in all three major ASO solvent fractions. However, longer and unsaturated FAMEs were observed in the ASO MeOH fraction. ASO extract (13%) of dry AS is a source of biodiesel but simultaneously can be refined to obtain value-added products such as squalene and tocopherols. Although the utilization of activated sludge oil can reduce the cost for feedstocks, the actual process requires additional expense for refinement but may be compensated by the separation and purification of high value-added compounds of nutraceutical or biologically importance.

[1]Dufreche S, Hernandez R, French T, Sparks D, Zappi M, Alley E. 2007 *J Amer Oil Chem Soc* 84, 181-187.

[2]Mondala A, Liang K, Toghiani H, Hernandez R, French T. 2009 *Biores Technol* 100, 1203-1210.

See more of this Session: Sustainable Diesel Fuel From Renewable Resources
See more of this Group/Topical: Sustainable Engineering Forum

(307a) Introduction and Objectives of the Session

Jim Davis, Chemical and Biomolecular Engineering, UCLA, Los Angeles, CA and Thomas F. Edgar, Chemical Engineering, The University of Texas at Austin, Austin, TX

This session will examine plant, company and industry roles in energy, sustainability, environment, EH&S and economic performance across several areas of Smart Manufacturing and manufacturing innovation. Presentations and discussion are organized to identify key elements that involve cross application development, common approaches and shared infrastructure.

See more of this Session: The Application of Simulation-Based Engineering and Science to Industrial Operations and Engineering Management of Enterprise Process Systems: Invited Speakers
See more of this Group/Topical: Topical 2: Simulation Based Engineering and Science

(307b) Smart Process Manufacturing - Capital Programs and Industry Operations

James B. Porter Jr., DuPont, Wilmington, DE

Smart Process Manufacturing (SPM) is a dramatically intensified knowledge-enabled industrial enterprise in which ALL business and operating actions are executed to achieve substantially enhanced energy, sustainability, environmental, safety and economic performance. When SPM is applied comprehensively in terms of technology, methodology, operational assimilation and elevation to company and industry enterprises it becomes a disruptive / breakthrough use of knowledge and information technology.

See more of this Session: The Application of Simulation-Based Engineering and Science to Industrial Operations and Engineering Management of Enterprise Process Systems: Invited Speakers
See more of this Group/Topical: Topical 2: Simulation Based Engineering and Science

(307c) Unit Operations for Smart Nanomanufacturing

Roger T. Bonnecaze, Department of Chemical Engineering, The University of Texas at Austin, Austin, TX

There is a chasm between promising nanotechnological advances in the laboratory and commercial manufacturing. Many techniques have been developed to produce exquisitely controlled nanopatterned surfaces with amazing and potentially valuable properties. These methods, however, are almost always on the laboratory scale producing a few samples of the materials over small areas. To make them viable manufacturing processes, the methods must be scaled up to produce nanopatterned materials at high speeds over large areas, usually with low-defectivity. Currently, scale up is often accomplished by time consuming and expensive trial-and-error approaches. In this talk it is argued that the time has come to identify, model and simulate the unit operations of nanomanufacturing. Such models accelerate the design and optimization of nanomanufacturing processes from predominantly empirical approaches to more reliable and cost-effective ones based on computational simulations. Furthermore, these computational tools will enable greater innovation and early identification of untenable designs, thus mitigating risk. This research will transform manufacturing of nano-enabled devices in areas such as computing, communications, data storage, solid state lighting, energy storage, solar cells, biomedicine, and electronic displays. Examples of the benefits of such computational simulations will be given in the context of UV Imprint Lithography. The beginnings of a list of unit operations for nanopatterning will also be identified.

See more of this Session: The Application of Simulation-Based Engineering and Science to Industrial Operations and Engineering Management of Enterprise Process Systems: Invited Speakers
See more of this Group/Topical: Topical 2: Simulation Based Engineering and Science

(307d) Sustainability and Systems Engineering: Reducing the Carbon Footprint of Chemical Plants

Jeffrey J. Siirola, Eastman Research Division, Eastman Chemical Company, Kingsport, TN

Despite some carbon dioxide producing chemistries, Most CO2 emissions in the chemical industry result from the combustion of carbonaceous fuels. Reducing the carbon footprint of chemical plants then involves two strategies: reducing process energy requirements, and reducing the carbon emissions associated with whatever energy is required. This presentation will discuss process systems considerations to both reducing energy requirements and reducing carbon emissions highlighting capital cost, operating cost, and fitness-for-use tradeoffs, and operational, controllability, and other impacts that should be anticipated. The presentation will also discuss systems approaches to fuel substitution, use of non-fossil energy, energy system carbon capture and sequestration, and the modeling challenges required to help understand the full impact of alternative courses of action.

See more of this Session: The Application of Simulation-Based Engineering and Science to Industrial Operations and Engineering Management of Enterprise Process Systems: Invited Speakers
See more of this Group/Topical: Topical 2: Simulation Based Engineering and Science

(307e) Automation for Smart Manufacturing--Trends and Technologies

Tariq Samad, Honeywell Labs, Golden Valley, MN

In this presentation I provide a perspective on the future of smart manufacturing and the technologies and capabilities that are in development to support it. The paper will focus on technology enablers and advances and illustrate capabilities by describing global-scale mega projects. Taking a forward look at technology, I will discuss wireless and its benefits for industrial plants, the state of the art in process simulation, and supply chain optimization for manufacturing plants. The last part of the presentation will focus on energy: an analysis of energy efficiency opportunities, using refineries as an example, and the potential for improved demand-response with smart grids.

See more of this Session: The Application of Simulation-Based Engineering and Science to Industrial Operations and Engineering Management of Enterprise Process Systems: Invited Speakers
See more of this Group/Topical: Topical 2: Simulation Based Engineering and Science

(307f) Open Discussion On Smart Manufacturing

Thomas F. Edgar, Chemical Engineering, The University of Texas at Austin, Austin, TX and Jim Davis, Chemical and Biomolecular Engineering, UCLA, Los Angeles, CA

This will be an open discussion with the speakers and the attendees to identify key elements of high performance computing in Smart Process Manufacturing. An objective will be to define actions and approaches to support cross-application, common infrastructure or collective development requirements of Simulation Based Engineering and Science for Smart Process Manufacturing.

See more of this Session: The Application of Simulation-Based Engineering and Science to Industrial Operations and Engineering Management of Enterprise Process Systems: Invited Speakers
See more of this Group/Topical: Topical 2: Simulation Based Engineering and Science

(325a) Optimizing Pretreatment of Brown Midrib Maize Silage for Effective Fermentation at High Solids Loading

Arun Athmanathan, ABE/Lorre, Purdue University, West Lafayette, IN, Peter Friedemann, Dow AgroSciences, Homer, IL and Nathan Mosier, Agricultural and Biological Engineering & Laboratory of Renewable Resources Engineering (LORRE), Purdue University, West Lafayette, IN

Both corn grain and grain stover have been examined and utilized as biofuel feedstocks. Maize silage (wet stored, partially fermented maize stover plus immature grain) is an alternative that combines starch and cellulosic processing in a single feedstock. The commercial brown midrib (BMR marketed by Mycogen, wholly owned subsidiary of Dow AgroSciences) has lowered expression of caffeic acid O-methyl transferase, a key enzyme in the biosynthesis of S monolignols. We carried out a compositional analysis for two commercial varieties of maize silage (regular and brown midrib) for starch, cellulose, hemicellulose, and lignin content. Our results show that for the commercial varieties, the lignin content (Klason lignin plus acid soluble lignin) is indistinguishable. However, the BMR silage exhibits significantly higher cellulose enzymatic digestibility. Liquid hot-water pretreatment was optimized for each silage variant. Optimal pretreatment conditions were similar between BMR and regular silage, which was less severe than required for dry stover from similar maize varieties. Simultaneous saccharifications and fermentations were subsequently performed on pretreated whole silage and ground silage at 25% (w/v) total solids using Celluclast 1.5L and Novozyme 188 and the glucose/xylose co-fermenting yeast S. cerevisiae 424A(LNH-ST). The results show that the improved cellulose hydrolysis performance of BMR silage compared to regular silage is also seen in pretreated material, resulting in significantly higher yields of ethanol after SSF.

See more of this Session: Developments in the Pretreatment of Lignocellulosics for Bioconversion
See more of this Group/Topical: Sustainable Engineering Forum

(325b) Ionic Liquid-Solvent Mixtures: Enabling Cost-Effective Biomass Pretreatment

E. Kate Brown[1], Emma Sebeey[1], Xinglian Geng[2] and Wesley A. Henderson[1], (1)Department of Chemical & Biomolecular Engineering, North Carolina State University, Raleigh, NC, (2)Chemical & Biomolecular Engineering, North Carolina State University, Raleigh, NC

To date, only a handful of solvents are capable of solvating cellulose. Among the known solvent systems, ionic liquids (ILs) are exceptional cellulose solvents suggesting their use in a biomass pretreatment method in which the polysaccharides are dissolved and then regenerated prior to the subsequent hydrolysis step. This process greatly improves the enzymatic hydrolysis efficiency through dramatic changes in the structural properties of the regenerated biomass. Significant roadblocks exist, however, for the implementation of such a pretreatment process. Amongst these are the costly nature of the ILs, as well as practical processing limitations which exist due to the high IL viscosity. Although common solvents do not dissolve cellulose, we have discovered that relatively dilute solvent-IL mixtures readily dissolve significant amounts of cellulose. ILs have therefore been screened with a variety of molecular solvents for cellulose and corn stover dissolution. Through the addition of solvents, the solution properties can be fine-tuned, and a drastic reduction in the amount of IL is needed. Thus, this solvent-IL approach provides a viable route to a facile, cost-effective biomass pretreatment process.

See more of this Session: Developments in the Pretreatment of Lignocellulosics for Bioconversion
See more of this Group/Topical: Sustainable Engineering Forum

(325c) Conversion of Switchgrass Into Fermentable Sugars by Electrolyzed Water Pretreatment and a Two-Stage Pretreatment

Xiaojuan Wang, Chemical Machinery at DUT and Department of Food Science and Human Nutrition at UIUC, Dalian University of Technology (DUT) and University of Illinois at Urbana& Champaign (UIUC), Urbana, IL, Hao Feng, Department of Food Science and Human Nutrition, Energy Biosciences Institute, University of Illinois at Urbana-Champaign, Urbana, IL and Zhiyi Li, Department of Chemical Mashinery, Dalian University of Technology, Dalian, China

Biofuel is a carbon-based and renewable energy source that is originally from solar energy captured and stored in the form of carbohydrate in plants. Biofuel can be produced from monomeric sugars released from plants by fermentation into ethanol, butanol, and other transportation fuels. Switchgrass (Panicum virgatum) is a warm season perennial grass that is native to North America. Besides a high cellulosic content, switchgrass also has some significant advantages, such as being CO2-neutral, pest and disease resistance, natural wildlife habitat, drought and flood tolerance, among others, making it one of the best energy corps. Electrolyzed water is a relatively new pretreatment catalyst which was first proposed and tested by scientists at University of Illinois at Urbana-Champaign (UIUC) in 2006. A two-stage pretreatment method for deconstructing biomass with a purpose of partially removing lignin and hemicellulose to achieve a mild pretreatment was also recently developed at UIUC. In this study, the two methods were evaluated for conversion of switchgrass into fermentable sugars. The fermentablility of hydrolyzates obtained from the two-stage pretreatment was also examined. Acidic electrolyzed water (AEW) and alkaline electrolyzed water (ALEW) were used as pretreatment catalysts to pretreat switchgrass at three temperatures (170°C, 185°C, and 200°C) and 3 treatment times (5min, 15min, and 25min). A solid loading of 20% w/w was used (8 g switchgrass (dry basis) to make a 40 g slurry) in pretreatment. The enzymatic hydrolysis was performed in 100mL VITLAB flasks in a shaker water bath at 50°C and 200rpm for 120h with a solid loading of 8%. The enzyme loading was: 15FPU cellulase/ g glucan and 40U beta-glucosidase/ g glucan. In two-stage pretreatment, 1.0%, 2.0% and 3.0% H2O2 solutions were used to partially remove hemicellulose and lignin during the 1st stage in a shaker water bath at 50°C and 150rpm for 16h, 24h, and 32h with a solid loading of 5%. The supernatants of the 1st stage were used for hemicellulose precipitation. The remaining solids were washed with DI water into VITLAB flasks and subjected to a 2nd treatment at 121°C for 30min in an auto-clave with a solid loading of 12.5%. After the 2nd treatment, the obtained slurry was used for enzymatic hydrolysis at a solid loading of 10%. The other conditions of the enzymatic hydrolysis were the same as those in the electrolyzed water pretreatment experiments except that the hydrolysis time was 96h. The highest glucose yield in the electrolyzed water pretreated samples was 29.75g glucose from 100g dry switchgrass when using ALEW at 200°C and 25min. Over 50% hemicellulose was removed in the 1st stage treatment of the two-stage process, and most of hemicellulose in the liquid portion could be recovered after a further treatment. Over 30g glucose was released from 100g dry switchgrass in the two-stage process. The hydrolyzate obtained from the treatment of 3% H_2O_2 solution for 16h was fermented by Baker yeast in a shaker water bath at 30°C for 96h without any concentration and detoxification. The final ethanol concentration is 18.89g/L and the productive yield was 119% of the theoretical yield. A mass balance for the two-stage process showed that 18g ethanol was produced from 100g dry switchgrass.

See more of this Session: Developments in the Pretreatment of Lignocellulosics for Bioconversion
See more of this Group/Topical: Sustainable Engineering Forum

Isolation and characterization of lignin-modifying activity of bacteria from rainforest soils

Navaneetha Santhanam[a,b,e], Dayakar Badri[b], Jorge M. Vivanco[b,e], Dan K. Manter[c], Steve R. Decker[d,e], Kenneth F. Reardon[a,e]

[a]Department of Chemical and Biological Engineering, CSU,[b]Department of Horticulture and Landscape Architecture, CSU,[c]USDA-ARS Soil-Plant-Nutrient Research Unit, Fort Collins, CO,[d]Chemical and Biosciences Center, National Renewable Energy Laboratory, Golden, CO, [e]Colorado Center for Biorefining and Biofuels

The commercialization of cellulosic biofuels requires the cost-effective release of fermentable sugars from plant biomass. Since cellulose in plant cell walls is intricately interlinked with hemicellulose and lignin, access to the fermentable sugars can only be achieved by the removal of lignin. Current physicochemical pretreatment technologies involve high energy consumption and operating costs. Enzymatic "loosening" of the lignin matrix may be an environmentally benign and less expensive pretreatment alternative due to its potential for the production of value-added compounds from lignin. While wood-decaying fungi abound, their enzymes are often less robust in terms of thermal and pH stability and are more difficult to optimize by protein engineering than are bacterial enzymes. The Peruvian Rainforest of Tambopata supports a rich diversity of vegetation that grows in relatively nutrient poor soils. It is believed that the soil microbes in that region have developed fast turnover rates of organic matter which is credited as one of the main factors that supports plant mega diversity. In this study, bacterial strains isolated from rainforest soils were screened for the production of lignin-modifying enzymes, such as, laccases. Laccases are a class of oxidizing enzymes that can catalyze a multitude of reactions that include the degradation of polymers, oxidative coupling of phenolic compounds, aromatic ring cleavage and functionalization of polymers, to name a few. Their intrinsic versatility makes them industrially significant enzymes. ABTS (2,2-azino-bis (3-ethylbenzthiazoline-6-sulfonic acid) a non-phenolic substrate, which when oxidized by laccase produces a green radical cation, was used to select for bacteria that secrete laccases into the culture supernatant. To date, 140 isolates have been screened and the supernatants of 7 isolates grown in rich medium were found to show visible and measurable green product formation with ABTS. In order to test the ability of selected strains to modify lignin structure, a method is being developed using liquid chromatography/mass spectrometry (LC/MS), to detect monomers released from corn stover lignin or Kraft lignin after incubation with bacteria.

(325e) The Effect of Ferrous Ion On High Solids Dilute-Acid Low Severity Steam Explosion Pretreatment of Corn Stover

Xiaowen Chen[1], Erik Kuhn[1], Yun Ji[2] and Melvin Tucker[1], (1)National Biology Center, National Renewable Energy Lab, Golden, CO, (2)Chemical Engineering, University of North Dakota, Grand Forks, ND

Acid-catalyzed steam explosion pretreatment is capable of solubilizing significant amounts of hemicellulose from lignocellulosic biomass, producing more digestible solids for enzymatic hydrolysis. Dilute sulfuric acid is commonly used because it is less expensive than other acid catalysts. However, dilute-sulfuric acid pretreatment requires expensive acid resistant reactors, rapid degradation of desired monomeric sugars into fermentation inhibitors occurs during pretreatment, and addition of post pretreatment neutralization reagents are needed. Lowering the severity of pretreatment by lowering the sulfuric acid concentration can lower these costs, as well as the production of furfural and other degradation products. However lower yields in pretreatment and enzymatic hydrolysis result. A process must be developed that overcomes the limitations of low severity dilute-acid pretreatment. In this study, high solids (~45% w/w) dilute acid pretreatment with co-catalyst supplementation of ferrous ions was investigated using corn stover. Results show that dilute acid with ferrous co-catalyst pretreatment increased xylose yields in pretreatment by 1~7% (on xylose basis) compared to dilute-acid only pretreatment at the same conditions. However, the effect is highly dependent on the ferrous ion concentration and reaction temperature. Furthermore, the enzymatic hydrolysis of washed acid-ferrous co-catalyst pretreated solids indicate enhanced enzymatic yields over controls. The pretreatment results for xylan hydrolysis kinetics were fitted into a four-step linear sequence of first order irreversible reactions. Activation Energy (Ea) was derived from the rate constants and fitted into Arrhenius plots where a 30~50% reduction in Ea was found with the supplementation of small catalytic amounts ferrous ions into the low severity dilute-acid pretreatments.

See more of this Session: Developments in the Pretreatment of Lignocellulosics for Bioconversion
See more of this Group/Topical: Sustainable Engineering Forum

(325f) Characterization of Process-Relevant Operating Conditions for Alkaline Hydrogen Peroxide Pretreatment of Switchgrass

Daniel L. Williams, Chemical Engineering and Material Science, Michigan State University, East Lansing, MI, Tongjun Liu, Chemical Engineering and Materials Science, Michigan State University, E. Lansing, MI and David Hodge, Chemical Engineering and Materials Science, Michigan State University, East Lansing, MI

Alkali and oxidative treatments of lignocellulose are well-established in chemical pulping and bleaching. However, their application for pretreatment technologies optimized for liberating carbohydrates as a substrate for biofuels production is different in several regards and requires further research before they can be used in cellulosic biofuels processes. This work will present research results that characterize the efficacy of alkaline hydrogen peroxide pretreatment with regard to several process requirements that include quantifying the consumption of alkali and hydrogen peroxide during pretreatment, determining the potential inhibition of downstream hydrolysis and fermentation processes, investigating the potential separation of hemicellulose from hydrolyzate liquor ,and investigating the impact of Fenton's chemistry on the selectivity of lignin vs carbohydrate oxidation and its effect on enzyme digestibility.

See more of this Session: Developments in the Pretreatment of Lignocellulosics for Bioconversion
See more of this Group/Topical: Sustainable Engineering Forum

(341a) Directed Evolution of a Highly Efficient Arabinose/Xylose Utilization Pathway in *Saccharomyces Cerevisiae*

Jing Du, Byoungjin Kim and Huimin Zhao, Department of Chemical and Biomolecular Engineering, Institute for Genomic Biology, University of Illinois Urbana Champaign, Urbana, IL

Bioconversion of plant derived lignocellulosic materials into biofuels has drawn increasing attention for being an attractive candidate for replacement of fossil fuels. *Saccharomyces cerevisiae*, also known as baker's yeast, is considered one of the most promising organisms for ethanol production from lignocellulosic feedstock. Unfortunately, pentose sugars, which make up to 30% of biomass hydrolysate, cannot be utilized by *S. cerevisiae*. Heterologous pentose utilization pathways from bacterial and fungal resources have been introduced into *S. cerevisiae* to enable the assimilation of pentose sugars. However, pentose utilization of recombinant *S. cerevisiae* strains are inefficient due to the low expression level and activity of heterologous genes, redox imbalance resulted from different preference of cofactor for oxidation and reduction reactions, and suboptimal metabolic flux through different catalytic steps. A lot of research has been done to improve the pentose utilization by *S. cerevisiae* targeting different aspect of these issues, but it is very challenging to come up with a single strategy which can solve all three problems at the same time. Here we report the directed evolution of a highly efficient pentose utilization pathway in *S. cerevisiae*. Specifically, we chose the fungal five-step arabinose/xylose utilization pathway as our target pathway. For each catalytic step, 10 to 20 different enzyme homologues from various fungal species with different catalytic efficiency and cofactor preference have been cloned, and the whole pathway was assembled into a random library using our recently developed DNA assembler method. The resultant library exhibited a very high efficiency for correct assembly of the complete multi-gene pathway and good diversity of different homologues within the same catalytic step. Using the library generated, clones with better combination of enzyme homologues can be selected by faster cell growth. This method can also be applied to directly search for the pathway with a better fit in industrial yeast strains, which may have different metabolic flux patterns compared to the laboratory *S. cerevisiae* strains.

See more of this Session: Microbial Engineering of Microorganisms for Biofuels and Chemicals I
See more of this Group/Topical: Sustainable Engineering Forum

(341b) Towards Consolidated Bio-Processing: Engineering Saccharomyces Cerevisiae for the Expression and Evaluation of Novel Anaerobic Fungal Cellulases

Michelle A. O'Malley and Chris A. Kaiser, Department of Biology, Massachusetts Institute of Technology, Cambridge, MA

The most promising approach for harnessing the energy contained within cellulosic biomass is to break down the cellulose polymer into simple sugars that can be converted via fermentation into energy-rich chemicals such as ethanol. At present time, no single microorganism has been identified that can perform both cellulolytic and fermentation processes with suitable efficiency. Thus, industrial approaches to biomass conversion typically rely on co-culturing two different organisms or pre-treating cellulose-rich biomass with expensive chemicals to reduce lignin content prior to incubation with purified cellulolytic enzymes. More recently, advances in the identification of the genes responsible for cellulose breakdown in native cellulolytic organisms as well as advances in our ability to genetically modify a wide variety of microbial species has opened the way to genetic engineering of a single microbe that can efficiently carry out both cellulose breakdown and fermentation – a strategy referred to as Consolidated Bio-processing (CBP).

In order to facilitate CBP, we are developing the yeast *Saccharomyces cerevisae* as a robust expression platform for the production of novel cellulases derived from anerobic fungi that naturally inhabit the rumen of large herbivorous animals. Despite the powerful cellulolytic properties of rumen fungi, remarkably few enzymes have been characterized or cloned from these organisms due to difficulties associated with the isolation and culture of these microbes. Towards introducing these enzymes into yeast, we have developed unique methods to culture the anaerobic fungi *Piromyces equi*, which produces a collection of catabolically-regulated cellulases. Among these enzymes, most exist in multi-enzyme complexes termed cellulosomes, which tether cellulases together for the targeted, synergistic hydrolysis of cellulose. We will discuss the properties of these enzymes as they are expressed in *S. cerevisiae*, including their enhanced specific activities, thermostability, resistance to inhibitors, and their ability to cleave novel chemical linkages in cell wall material. Reconstitution, assembly, and catalytic activity of fungal cellulosomes produced in *S. cerevisiae* will also be discussed.

See more of this Session: Microbial Engineering of Microorganisms for Biofuels and Chemicals I
See more of this Group/Topical: Sustainable Engineering Forum

(341c) Improving Acetic Acid and Ethanol Resistance of S. Cerevisiae 424A(LNH-ST) During the Co-Fermentation of Glucose and Xylose

Nathan Mosier, Agricultural and Biological Engineering & Laboratory of Renewable Resources Engineering (LORRE), Purdue University, West Lafayette, IN, Miroslav Sedlak, ABE/Lorre, Purdue University, West Lafayette, IN and Nancy W. Y. Ho, ChemE/LORRE, Purdue University, West Lafayette, IN

Efficient conversion of hemicellulose-derived sugars to ethanol at high yields and titers are goals toward commercializing cellulosic ethanol production. S. cerevisiae 424A (LNH-ST) developed at Purdue University can efficiently ferment glucose and xylose. However, inhibitors present in cellulosic feedstocks (acetic acid) and the desired fermentation product (ethanol) reduce yeast growth rate and fermentation rates, especially during xylose fermentation.

Through adaptation we have developed new strains with improved xylose fermentation compared to the original strain. The new strain has 500% higher ethanol volumetric productivity on xylose in the presence of higher ethanol concentrations (above 6%) than the original strain. An acetic acid-resistant yeast strain co-fermenting glucose and xylose in the presence of acetic acid (10 g/L) when compared to the original strain has 3 times the rate of xylose utilization (1.05 g/L/h from 0.32 g/L/h) and results in a higher final ethanol titer (76.3 g/L from 61.2 g/L). We present the results from a system biology approach to analyzing differences between our original strain and newly developed strains. We focus not only on expression profiling (transcriptomics), but also report changes in metabolic intermediates and fluxes, and lipid membrane composition to elucidate the basis for improved yeast performance.

See more of this Session: Microbial Engineering of Microorganisms for Biofuels and Chemicals I
See more of this Group/Topical: Sustainable Engineering Forum

(341d) Systematic and Recursive Engineering of E. Coli Using TReMR for Desired Tolerance to Acidic pH

Younghwanjaoon Kim, Joseph Roy Warner and Ryan T. Gill, DEPT of Chemical ENG, UNIVERSITY OF COLORADO BOULDER, Boulder, CO

TReMR is a novel genetic tool that utilizes a systematic and trackable approach for improvement of strains with a desired trait under a specific environment. We can monitor the response of microorganisms under certain circumstances by a genetic signal that is marked on each gene of the genome. Using TMR/TReMR, expression of each gene can be controlled positively or negatively. If the expression of a gene is helpful to survival and growth of the host strain under certain conditions, then the frequency of the strain with a beneficial trait increases compared to other strains with undesired traits. Based on these strategies and recombineering technique, we engineered E. coli for a desired trait, specifically pH tolerance herein, by recursive and multiple recombineering with control of each gene in the chromosome. This approach made it possible to engineer the strain systematically and efficiently. Engineered strain showed relatively high tolerance to acidic pH compared to wild type strain.

See more of this Session: Microbial Engineering of Microorganisms for Biofuels and Chemicals I
See more of this Group/Topical: Sustainable Engineering Forum

(341e) Production of Biofuels Using Genetically Engineered Escherichia Coli

Seunghyun Ryu, Department of Chemical Engineering, Texas Tech University, Lubbock, TX and M. Nazmul Karim, Chemical Engineering Department, Texas Tech University, Lubbock, TX

A whole-cell biocatalyst system was developed in our research for the production of ethanol from cellulosic substrate in a single step. The whole cell biocatalyst was constructed with Escherichia coli LY01, which is the most ethanologenic strain. The cellulase genes, cel5A, cel9E and beta-glucosidase, were cloned from the mesophilic strain Clostridium cellulolyticum, and displayed on the surface of LY01 using the anchor protein PgsA. The pure cellulosic substrate phosphoric acid swollen cellulose (PASC) and the industrial substrate, dilute acid pretreated corn stover (PCS) were fermented. The yield of produced ethanol from PASC was 3.20 ± 0.15 g/L from 10 g/L PASC. 0.30 ± 0.02 g/L ethanol was produced from PCS equivalent of 1 g/L glucose. During the fermentation of the whole-cell biocatalyst, it was observed that the saccharification product glucose was uptaken immediately by the host cell and metabolized into ethanol. Currently, butanol has been considered as a potential liquid biofuel, and the butanol production has long been studied using clostridia strains. One of the difficulties of using clostridia strains is the formation of mixed-fermentation products, such as butyrate, acetate, acetone and ethanol. In this study, we engineered E. coli by introducing recombinant butanol pathway from multiple clostridia strains. The key enzyme coding genes: thiolase (thl), 3-hydroxybutyryl-CoA dehydrogenase (hbd), crotonase (crt), butyryl-CoA dehydrogenase (bcd), and the bifunctional butyraldehyde/butanol dehydrogenase (adhE), which are involved in the butanol production pathway in C. acetobutylicum were expressed in E. coli BL21(DE3). In the same E. coli, some of the carbon competitive pathway genes, e. g. phosphotransacetylase (pta), lactate dehydrogenase A (ldhA), bifunctional acetaldehyde/ethanol dehydrogenase (adhE), fumarate reductase (frdBC), and pyruvate oxidase B (poxB) were disrupted to maximize the carbon flux to butanol pathway with the expectation of increasing specific butanol productivity. Additionally, the transcriptional regulator FNR protein gene was also deleted. The engineered E. coli produced 830.1 mg/L butanol from 20 g/L glucose in 48 h fermentation.

See more of this Session: Microbial Engineering of Microorganisms for Biofuels and Chemicals I
See more of this Group/Topical: Sustainable Engineering Forum

(341f) High Titer Butanol-Producing Clostridium Beijerinckii Mutant Strain and Its Proteome Analysis

Jingbo Zhao, William G. Lowrie Department of Chemical and Biomolecular Engineering, The Ohio State University, Columbus, OH and Shang-Tian Yang, Department of Chemical and Biomolecular Engineering, The Ohio State University, Columbus, OH

Butanol received increasing interest due to its potential as an alternative fuel and expanding demand as an industrial chemical. Biological production of butanol is promising because it can use renewable biomass as feedstock and is environmentally friendly. However, butanol production by solventogenic Clostridia in conventional acetone-butanol-ethanol (ABE) fermentation is limited by its low yield, productivity and titer. To make the bioprocess more economically attractive, a mutant strain of Clostridium beijerinckii with high butanol tolerance and productivity was developed through in-process adaptation in a fibrous bed bioreactor (FBB) followed with extensive strain screening. Compared with the parental strain, the mutant showed several good traits, including improved butanol tolerance, autolysis resistance, robust adjustment of membrane fatty acid composition, more complete substrate utilization, and dramatically increased butanol titer. A high butanol titer of 28.2 g/L was achieved in batch fermentation with the mutant immobilized in a FBB. To investigate the molecular basis for the acquired traits of the mutant, the proteomic profiles of the mutant and its parental strain during early exponential phase and late stationary phase were analyzed using 2-D gel electrophoresis. Proteins with dramatic difference in their expression levels were identified and classified based on their potential roles. The comparative proteome analysis between the mutant and its parental strain suggested several novel gene targets for metabolic engineering of the solventogenic Clostridia to further improve butanol tolerance and production in ABE fermentation.

See more of this Session: Microbial Engineering of Microorganisms for Biofuels and Chemicals I
See more of this Group/Topical: Sustainable Engineering Forum

(341g) Exhaustive Screening of 3985 Isogenic Escherichia Coli Mutants for Enhanced Bacterial Hydrogen Production

Toshinari Maeda[1], Viviana Sanchez-Torres[2] and Thomas K. Wood[2], (1)Department of Biological Functions and Engineering, Kyushu Institute of Technology, Kitakyushu, Japan, (2)Artie McFerrin Department of Chemical Engineering, Texas A&M University, College Station, TX

Hydrogen is a promising fuel since it is renewable, efficient, and clean. Use of biological methods for hydrogen production should significantly reduce energy costs as these processes do not require extensive heating or electricity. We have used metabolic engineering with *Escherichia coli* to enhance hydrogen production by manipulating the key enzymes of the formate hydrogen lyase system (FHL) consisting of hydrogenase 3 and formate dehydrogenase-H (Maeda et al., Microb. Biotechnol. 1:30-39, 2008). To date, we have constructed metabolically-engineered strains that produce 141-fold higher hydrogen from formate by inactivating genes that encode enzymes for hydrogen uptake (*hyaB* and *hybC*), by inactivating the gene encoding the FHL repressor (*hycA*), by inactivating the gene for formate depletion (*fdoG*), and by producing an essential activator of FHL, FhlA. In addition, we have achieved 5-fold higher hydrogen production from glucose by inactivating the succinate synthesis pathway (*frdC*), the lactate synthesis pathway (*ldhA*), and pyruvate-consuming activity (*aceE*) (Maeda et al., Appl. Microbiol. Biotechnol. 77:879-890, 2007). Furthermore, our studies using random mutation approaches such as chemical mutagenesis and transposon mutagenesis have succeeded in generating several mutants in which hydrogen productivity has been improved which implies other unknown pathways are important for increasing hydrogen production. Here, we conducted an exhaustive search of all *E. coli* pathways for their impact on hydrogen production by screening the entire Keio mutant library using both glucose and formate as substrates; this search identifies mutations which alter hydrogen production while simultaneously identifying the key pathway. The screen was facilitated by use of chemochromic membranes (GVD Corp., Cambridge, MA) formed by a thin film of WO_3 covered with a catalytic layer of palladium; these plates were used to detect directly the amount of hydrogen gas by a colorimetric response from colonies grown anaerobically. We previously utilized this screen to engineer FhlA for improved hydrogen production (Sanchez-Torres et al., Appl. Environ. Microbiol. 75:5639–5646, 2009) and to perform the first protein engineering of hydrogenase 3 to improve hydrogen production (Maeda et al., Appl. Microbiol. Biotechnol. 79:77–86, 2008). As expected, using this powerful screen and the Keio Collection, deleterious mutations were identified in genes related to hydrogen production such as hydrogenase 3 (*hyc* genes), hydrogenase maturation proteins (*hyp* genes), formate dehydrogenase-H (*fdhF*), and the FHL activator (*fhlA*). In addition, hydrogen production was decreased by inactivating genes related to nickel transport (*nikABCDE*), selenocysteine synthesis and incorporation (*selAB*), molybdopterin biosynthesis (*moeAB* and *moaABCDE*), and ATP synthesis (*atp* genes). In contrast, several beneficial mutations were identified including those that inactivated several cold shock proteins, those in uncharacterized proteins, as well as a mutation in a metal ion transporter (*mntH*). Hence, this approach identifies genes which may be deleted as well as those that may be overexpressed to improve hydrogen production.

See more of this Session: Microbial Engineering of Microorganisms for Biofuels and Chemicals I
See more of this Group/Topical: Sustainable Engineering Forum

(365a) Framing Workshop Objectives for Determining AIChE Institute Roles In Simulation Based Engineering and Science

Bond Calloway, Alternative Energy, Savannah River National Laboratory, Aiken, SC

This opening presentation will describe plans in this workshop session to examine AIChE's roles as an institute in Simulation Based Engineering & Science. Roles could include workshops, publications, education both student and public, grants and sponsorships, programming, and roles in national agenda.

See more of this Session: Workshop On National Investment, Policy and Education in Simulation-Based Engineering and Science
See more of this Group/Topical: Topical 2: Simulation Based Engineering and Science

(365b) The Council on Competitiveness Initiative on US Manufacturing

Cynthia McIntyre, Council on Competitiveness, Washington, DC

Innovation has taken center stage in government and industry efforts worldwide to compete and prosper. Public and private sector leaders understand increasingly that cost and quality alone do not determine economic success. The lynchpin is innovation - the ability to develop new ideas and deploy them in the real world where they create economic and societal value.

Innovation creates high-margin business, sustains high-wage jobs and drives productivity, and it is the key to solving many of the most pressing challenges and addressing many of the greatest opportunities. The reality, however, is that this country is still at the beginning of understanding the dynamics of modern innovation and how the public and private sectors can enable and manage innovation most effectively.

See more of this Session: Workshop On National Investment, Policy and Education in Simulation-Based Engineering and Science
See more of this Group/Topical: Topical 2: Simulation Based Engineering and Science

(365c) Federal Agency View of Simulation Based Engineering and Science

Clark Cooper, NSF, Arlington, VA and Phillip Westmoreland, Chemical and Biomolecular Engineering, North Carolina State University, Raleigh, NC

US government agencies recognize that predictive simulation has become increasingly powerful for advancing scientific and technological goals of society. A key reason is high-performance computing that allows more complete treatment of physics and uncertainty. There is great optimism about the potentially dramatic increases in power due to hyperparallel multicore computers, including acquisition and use of increasingly massive amounts of data. At the same time, our challenge is to develop the computational tools and to prepare students and the workforce to take advantage of these changes.

See more of this Session: Workshop On National Investment, Policy and Education in Simulation-Based Engineering and Science
See more of this Group/Topical: Topical 2: Simulation Based Engineering and Science

(365d) Key Information and Data to Inform Discussions about AIChE Institute Roles

Jim Davis, Chemical and Biomolecular Engineering, UCLA, Los Angeles, CA

This presentation will provide information and data that will inform the discussion about AIChE's institute role in SBE&S. Included will be key findings from the High Performance Computing and Smart Manufacturing sessions and results of the program survey on SBE&S.

See more of this Session: Workshop On National Investment, Policy and Education in Simulation-Based Engineering and Science
See more of this Group/Topical: Topical 2: Simulation Based Engineering and Science

(365e) Facilitated Discussion On AIChE Institute Roles In Simulation Based Engineering and Science

Bond Calloway, Alternative Energy, Savannah River National Laboratory, Aiken, SC, **Jim Davis**, Chemical and Biomolecular Engineering, UCLA, Los Angeles, CA and **Phillip Westmoreland**, Chemical and Biomolecular Engineering, North Carolina State University, Raleigh, NC

This will be facilitated discussion with the speakers and the sessions participants on AIChE's potential roles with SBE&S and possible structures and approaches. The session is expected to produce a series of recommendations that will be reviewed, prioritized and developed with AIChE.

See more of this Session: Workshop On National Investment, Policy and Education in Simulation-Based Engineering and Science
See more of this Group/Topical: Topical 2: Simulation Based Engineering and Science

(366a) Open Forum: Investment Discussion

Larry Grogan, Grogan Advisory Services, Johnson City, TN

Grogan Advisory Services will answer your questions, discuss past, current, and future economic issues which may impact investment decisions. This session will be an open forum to discuss any investment topic.

See more of this Session: Open Forum: Investment Discussion
See more of this Group/Topical: Liaison Functions

Evolving Biological Engineering

Harvey Blanch, Department of Chemical Engineering, University of California, Berkeley, Berkeley, CA

Biochemical engineering evolved from unit operations related to antibiotic fermentations, through production of single cell protein, amino and organic acids, gasohol, and applications of immobilized enzymes. In the 1980s, revolutionary advances in life sciences spawned the growth of the biotechnology industry. New methods in molecular and cell biology have enabled engineers to develop sophisticated descriptions of cellular metabolism and bioproduct formation. Today, high-throughput and 'omics technologies underpin biological approaches to food, chemicals and pharmaceutics production, and the development of alternative energy sources. The subtleties of natural materials guide new generations of ceramics and polymers. With a look back at this rapid growth at the interface of biology and engineering, we will project where biological engineering will provide future impacts.

(398a) A Small Fraction of the USA Biomass Resource Would Meet 100% of National Transportation Fuel Needs

Percival Zhang and Weidong Huang, Virginia Polytechnic Institute and State University, Blacksburg, VA

Lignocellulosic biomass is a low-cost renewable chemical energy resource that would produce a large amount of transportation biofuels. We propose to calculate the energy conversion efficiency from biomass-to-kinetic energy (BTK) for driving so that we can avoid debates about the input uncertainties of life cycle analyses. Three different biomass-utilizing scenarios were compared: liquid cellulosic ethanol and the internal combustion engine; hydrogen, PEM fuel cells, and the electric motor; and electricity, battery, and the electric motor. The future sugar-hydrogen-electricity power train system called "sugar fuel cell vehicle (SFCV)" would not only have higher or similar BTK efficiencies than the electric vehicle but also have high energy storage densities plus benefits in operations, safety, infrastructure costs, environmental impacts, and so on. When SFCVs are widely implemented in 2030-2050, our analysis suggest that ~ 5% of the annual USA net biomass production or ~50% of the collected biomass would be sufficient to meet 100% of the national light-duty transportation fuel needs. Furthermore, we estimated that less than 1% of the global biomass resource would be sufficient to completely replace the gasoline consumption by using SFCV in 2050.

See more of this Session: Energy Efficiency, Challenges and Solutions
See more of this Group/Topical: Sustainable Engineering Forum

(398b) A New Principle for the Direct Application of Force to a Flame

Ludovico Cademartiri, Kyle J. M. Bishop, Aaron Mazzeo, Ryan C. Chiechi, Jabulani Barber, Charles Mace and George M. Whitesides, Chemistry and Chemical Biology, Harvard University, Cambridge, MA

In this talk we'll describe how electric fields can be designed to control combustion and flames. We'll propose the main mechanisms underlying this phenomenon and we will show how we can use it to suppress, stabilize, and manipulate flames, as well as prevent a flame from igniting a flammable gas mixture.

See more of this Session: Energy Efficiency, Challenges and Solutions
See more of this Group/Topical: Sustainable Engineering Forum

Radiative Properties of Coal and Biomass Slag and Deposits

Joseph Hoskisson, Austin Randall, Alex S. Romriell, Richard Reid, Larry Baxter

Abstract

When fuels are burned in boilers ash starts to build up on the interior walls of the boiler. This ash build-up decreases the amount of heat transferred from the fuel, through the boiler walls, and ultimately to the water side of the boiler, thereby decreasing the amount of steam generated by the boiler. The effect of this ash build up on the inside walls of the boiler decreases the efficiency of the boiler it also effects the size and design of new boilers when accounted for properly. The aim of this research is to characterize different ash deposits from a variety of fuel sources by using an FTIR spectrometer to measure the absorbance of each sample. Once the emittance of each sample is known, the effect that each sample has on heat transfer can be calculated. A few examples of the fuel samples that have been tested are Pebble River Basin Coal, Class C Bituminous Coal, Class F Sub-bituminous Coal, wood, and several other biomass fuels. These ash and slag samples have been gathered from various power-plants that use each of these fuels. Each sample is mixed with KBr (potassium bromide) powder to make a 2.5 wt% mixture of ash sample inside KBr. This mixture is then pressed into a thin circular wafer, 1.5cm in diameter and about 1mm thick. These wafers are placed in the FT-IR and tested against a KBr background to measure percent absorbance. Emittance is then calculated from the FT-IR data. Particle size is also measured and its effect on total emittance noted. Testing will be continued to strengthen the validity of the model.

Introduction

Ash and slag deposits from coal, biomass, and blended fuels affect heat transfer through boiler walls and at other locations in steam generating processes. Position in the process and fuel type affect the nature of deposit and therefore the heat transfer of the process. Total emittance of a material provides a useful tool in calculating heat transfer and thus, the cataloging of total emittances of varying types of deposits provides a great aid in modeling heat transfer in these processes and becomes significant in boiler sizing and design. The designs may depend on type of fuel used and the energy output desired. Operators may find, with boilers not designed by taking fuel types into account, that if they switch fuel types, then their boilers cannot meet output requirements. This can be an expensive realization.

FTIR spectroscopy allows the collection of spectral absorbance data of several ashes and slags. This, in turn, allows for the calculation of spectral emittance data. A Planck-weighted average of the spectral emittance yields a total emittance and represents an average of the emittance over significant wavelengths and over a hemispherical range of directions. This property of the deposit is useful in modeling heat transfer and required surface area in boiler driven processes.

Two types of FT-IR spectroscopy aided in data collection. They were a technique similar to Diffuse Reflectance Infrared Spectroscopy (DRIFTs) and standard transmission through pellets. In both techniques the samples were mixed into a matrix of Potassium Bromide (KBr) at generally a 2.5 wt% mixture.

Interestingly, as samples were ground to obtain more homogeneous mixing in the KBr pellets, particle size proved to have an unexpected effect on the spectral and total emittances: These properties increased with decreasing particle size for the two samples tested.

Experimental Method

FT-IR techniques measure the absorbance spectra of various deposits.

Diffuse reflectance

Traditional DRIFTS favors off-axis or integrating sphere optical systems to avoid Fresnel or specular reflections from particle surfaces. Since such specular components form an integral portion of radiative heat transfer on real surfaces, we used an on-axis optical approach, deliberately including them in the measurements. This is achieved through a set of on-axis mirrors coupled with a parabolic mirror as shown in Figure:

In diffuse reflectance sampling, the solid sample is mixed with a non-absorbing background in a 2-5 wt% concentration. Light reflects from the sample mixture and is directed to a detector. Given a broad-band light source and a Michelson interferometer, the detector records the Fourier transform of the spectral signature of the light. Since the optics, detectors, and several other components of the system have wavelength-dependent performance, a reference scan is needed. This reference scan is obtained using non-absorbing particles of the same size. The ratio of the absorbing sample to the non-absorbing sample provides a measure of the spectral emittance. The result is a plot of sample emittance over the mid-infrared region of the electromagnetic spectrum. These data used ash sample concentrations between 2 and 3 wt% in potassium bromide matrix. Particles also should be finely ground and isolated from atmospheric moisture as both the deposits and the KBr are hydroscopic.

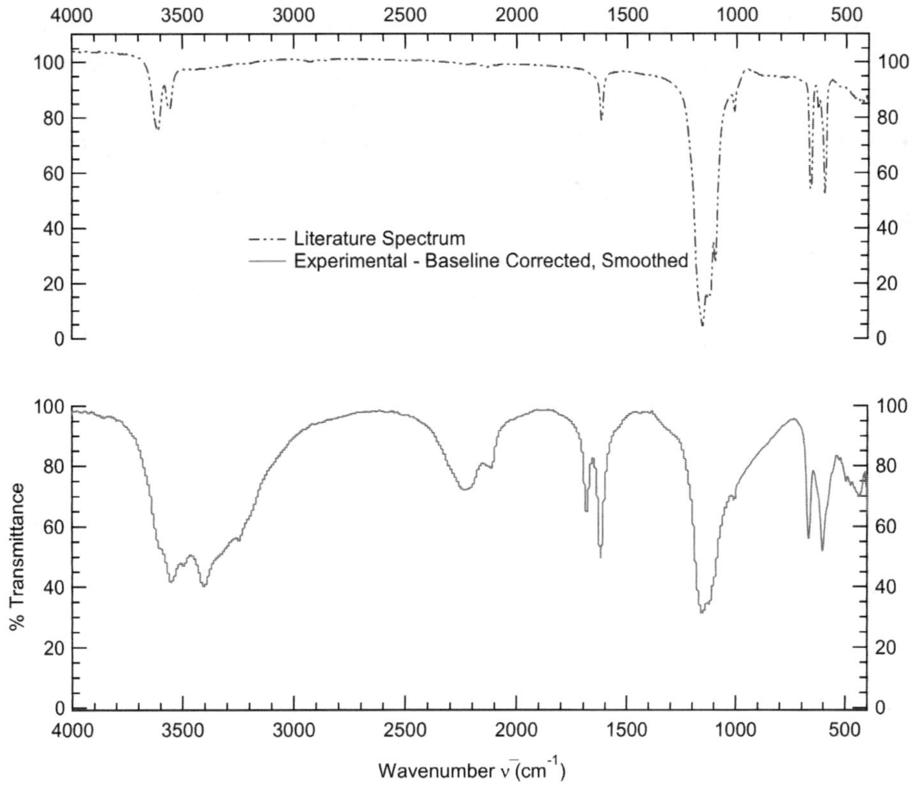

Figure 1 Ca_2SO_4 transmission spectra from the literature (Nyquist and Kagel 1971) compared to reflectance spectra using this procedure.

Comparisons of the experimental methods used here with literature spectra validate the procedures (Figure 1). Example results from our technique (bottom) and literature transmission data (Nyquist and Kagel 1971) for reagent-grade calcium sulfate appear in Figure 1. The peaks represent calcium sulfate in the bottom top plot at 3400, (2220, 2120,) 1620, 1680, 1120, 670, and 600 cm^{-1}. The peaks at 1130-1080 and 680-610 are especially characteristic of sulfates. The experimental data follow procedures similar to previous published work that cannot completely eliminate atmospheric water, CO, or CO_2 from the spectra (Baxter, Fletcher and Ottesen 1988; Baxter, Richards, Ottesen and Harb 1993).These are responsible for the broad peak (unresolved series of narrow peaks) near 3500 and 2200 cm^{-1} and some of the broadening of the peak near 1100 cm^{-1}. Other than these atmospheric interferences, the wavenumber accuracy of these experiments agrees well with the literature. The magnitude accuracy however varies as expected in these reflectance vs. transmission experiments. Reasons include multiple scattering events from the samples, sample-to-sample variation in particle density, number density, and surface smoothness, and chromatic combined with sample variation in Fresnel or specular content of the reflectance. The random sample-sample variations generally are small but the systematic variations require spectra to be renormalized after collection. Typically, experimental results are repeatable with an emittance of ± 0.1 without renormalization and by about ± 0.02 with renormalization.

This investigation attempts to quantify actual emittances, which differs from traditional DRIFTS, which is used to quantify optical properties, identify species and their concentrations, and other analytical chores. These differences influence experimental approaches. As one example, the sample particle size influences the measured results. In traditional DRIFTS experiments, sample particle sizes should be small enough that no more than 10% of the light entering any one particle would be absorbed even in the most intense absorption band. This generally translates into a particle size of less than 2 microns. The current investigation is less interested in generating optical properties, avoiding saturated absorbance or maintaining spectrally symmetric plots than in collecting practically relevant data. Therefore, no grinding or other method was used to reduce the primary particle sizes of our deposit samples nor were they uniform, resulting in larger primary particle sizes and size ranges, though the deposit primary particle size was still uniformly less than 40 microns and dominantly less than 20 microns for almost all samples. Among other effects, the potential variation in particle size makes the sample-to-sample variation in results large

Table 1 summarizes the samples measured in this project. The samples labeled *pxx* are slag from different locations in a gasifier – one from the post combustion chamber inlet, another from the post combustion chamber outlet, a third from further downstream, and a final one near the system exit. The next four samples come from biomass and biomass/coal blends – a 20% biomass in coal blend, a 10% biomass in Powder River Basin coal blend, a sawdust and bituminous coal blend, and a pure wood ash deposit. The final two samples are from a pure sub-bituminous coal (CC) deposit and a bituminous coal deposit (FF).

Table 1 Source and compositions of samples

Sample	Explanation	Concentration in KBr Matrix (wt%)
p17	Gasifier slag at post combustion chamber inlet	2.37
p18	Gasifier slag at post combustion chamber outlet	2.42
p19	Gasifier slag downstream	2.28
p20	Gasifier slag near system exit	2.48
20p	20% biomass in coal ash blend	2.49
PRB10%	10% biomass in Powder River Basin coal	2.35
Saw	Sawdust and bituminous coal blend	2.59
Wood	Wood deposit	2.40
CC	Sub-bituminous (PRB) coal deposit	2.60
FF	Bituminous coal deposit	2.57

Spectral to Total Emittance

Emittance depends strongly on wavelength since a material's ability to absorb radiation primarily through molecular vibrations and rotations depends on wavelength. By comparison, spectral emittances depend only slightly on temperature. Condensed-phase molecules interact strongly with one another, with the spectroscopic consequence of generally broad spectral features. The features are much broader than corresponding gas-phase features because both the energy levels and the time that molecules spend in any given vibrational state are highly uncertain due to strong near-neighbor interactions. The spectral locations of fundamental features depend very slightly on temperature, mainly through physical changes in atomic spacing and hence forces associated with thermal expansion. The molecular population at any given energy level also shifts with temperature, changing the relative intensity of individual peaks in a fully resolved spectrum. However, these temperature-related changes are small compared to the spectral changes, especially in condensed-phase materials. If, however, the material undergoes a phase or chemical change associated with temperature, the spectral emittance may change dramatically. In practical problems, the spectral emittances reported above rarely find application. A more general emittance that characterizes the entire wavelength region of interest is more typically required. This quantity is called a total emittance and in this paper is calculated from the measured spectral emittance as a Planck-weighted average.

The Planck function multiplied by the spectral emittance at each wavelength across the spectrum provides a weighting relative to the Radiative intensity at a given wavelength. Integrating this product over the spectrum and dividing by the area under the Planck function over the same wavelength region yields a Planck-weighted emittance that depends on temperature but not wavelength. The temperature dependence of this Planck-weighted emittance can be much higher than that of the spectral emittance because it includes the change of the Planck function shape (Figure 4), which is usually the dominant temperature effect, in addition to spectral temperature dependence. This investigation assumes spectral temperature dependence is small compared to the Planck shape effect. Quantitatively, Equation 4 provides the total or Planck-weighted emittance, where P is the Planck function expressed as a function of wavenumber

$$\varepsilon_T(T)=\frac{\int_{\bar{\nu}_{min}}^{\bar{\nu}_{max}} \varepsilon_\lambda(T,\bar{\nu})P(T,\bar{\nu})}{\int_{\bar{\nu}_{min}}^{\bar{\nu}_{max}} P(T,\bar{\nu})} \qquad 1$$

The spectral shape of the Planck function varies with temperature (Figure 2). As temperature increases, the spectral emittance values of greatest importance shift high wavelength (low wavenumber) to low wavelength (high wavenumber), or from right to left in most of the figures presented here.

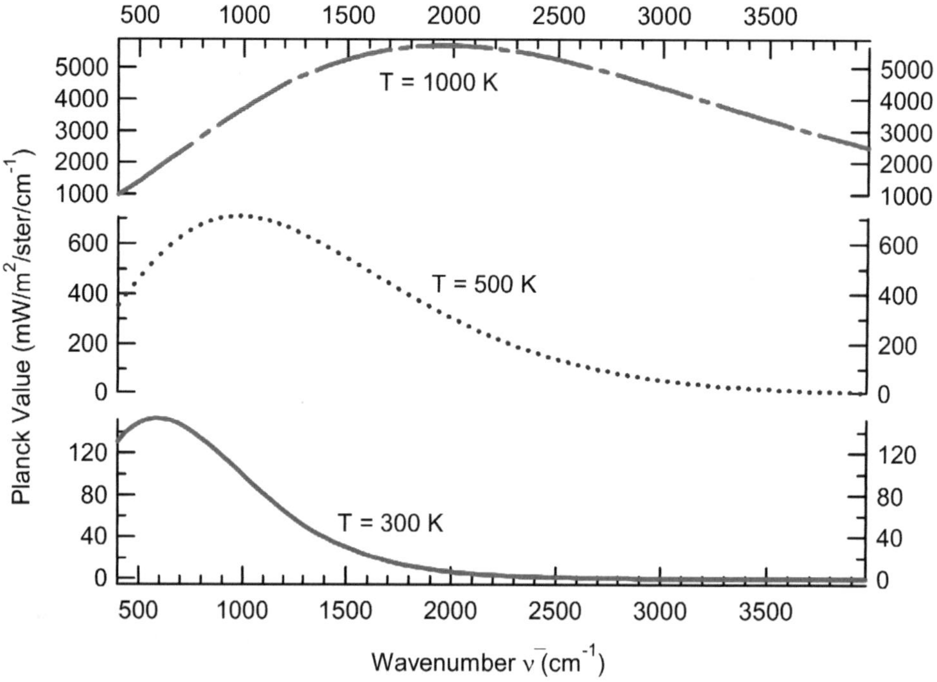

Figure 2 Spectral variation of the Planck function as a function of temperature.

Transmission

Samples tested using the transmission technique were the same as samples in the Diffuse Reflectance technique. Concentrations of sample-KBr matrices were however more consistent ranging from 2.49 – 2.51 weight% sample in KBr. Also the labels for bituminous coal changed from FF to CF.

KBr Pellet Preparation

In order to make transmittance measurements in an FT-IR spectrometer, a reliable method of creating quality pellets was needed. After a few variations and changes to the method, the process of creating pellets was perfected to the point where consistent data could be obtained.

Creating a usable background pellet was to be done initially. In order to make the background sample, KBr was placed in a circular die and the die was then placed in a hydraulic press. Forces around 12000 lbs were used to press the sample into a smooth round wafer. Then, similar pellets were made with sample added to the KBr and tested in the FT-IR spectrometer. After several variations of mass of the KBr, the force applied under the press and the step-by-step preparation of the KBr (i.e. grinding, placement of the uncompressed powder in the die and the speed of applying the force with the press), usable background pellets were finally produced. The pellets that first produced consistent, and therefore usable, data contained a clear circular section, generally in the center of the pellet. This clear section averaged a diameter of approximately 11 mm. The die itself is 19.05 mm (¾ in) in diameter, which meant that roughly half of the surface area on the pellet yielded a usable region for testing. See Figure 1 below.

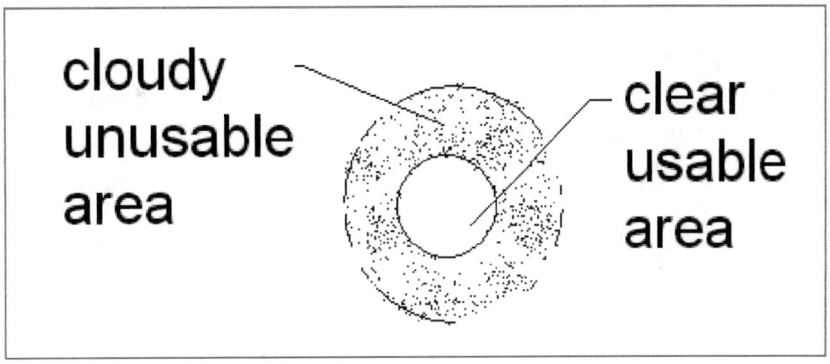

Figure 3 Original usable KBr background pellet.

The cloudy area, besides being unusable, was also very brittle. This property created difficulty when attempting to remove the pellets from the die and then secure them in the FT-IR spectrometer for testing. Often the edges would crumble and leave a trail of powder each time they were handled. Regardless of this challenge, and with the use of a custom pellet holder, consistent data was obtained and recorded.

Although the pellets, at this point, were usable, their crumbly nature provided motivation to further perfect their creation. After some research about how KBr pellets, or other pellets, had been successfully created in the past, it was determined that a larger mass of KBr was needed to create a thicker pellet. It was also determined that a greater force was needed to allow the entire pellet to be compressed sufficiently to eliminate the remaining cloudiness. The original background pellet was made using 0.4 grams of KBr and compressed with a force of 12,000 lbs. After various increments in mass and force, the best pellet (i.e. the pellet with the smallest cloudy area) was made using 0.65 grams of KBr with a force of 24,000 lbs, See Figure 2.

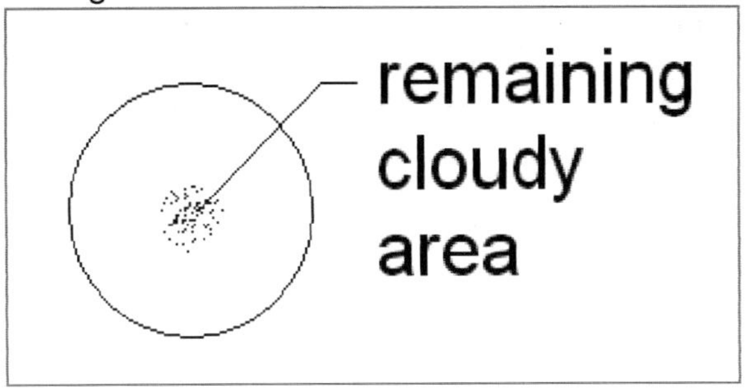

Figure 4 Thicker usable KBr background pellet.

It is assumed that an even greater force would eliminate all remaining cloudiness and yield an ideal pellet. Unfortunately, the press being used is limited to a maximum force of 24 lbs, which is why this particular quantity was chosen and tested.

These thicker pellets were then tested to ensure that they would also produce usable results. The thicker pellets yielded nearly identical results when compared to the original 0.4 gram background pellets. Even though usable pellets had already been produced, a higher quality design had now been obtained to ease the future pressing and testing of pellets containing the various samples.

Transmission Spectra

KBr pellets provide a background for each sample pellet. The KBr pellet placed orthogonal to the incident infrared beam absorbs what is absorbed the same as what is absorbed by the KBr matrix in which the sample is mixed in the sample pellet. This allows for distinguishing the KBr absorption patterns from the sample though it is mixed in with the KBr.

Spectra obtained for each sample through this technique allow for an average spectrum for each sample. Calculation of the total emittance of each sample employs the calculations described in the Spectral to Total Emittance section of this article above with the averaged transmission spectra.

Results and Discussion

Diffuse Reflectance - Spectral Emittance

The pilot-scale slag results appear in Figure 2. The peaks appear at fairly consistent wavenumbers and vary largely in height.

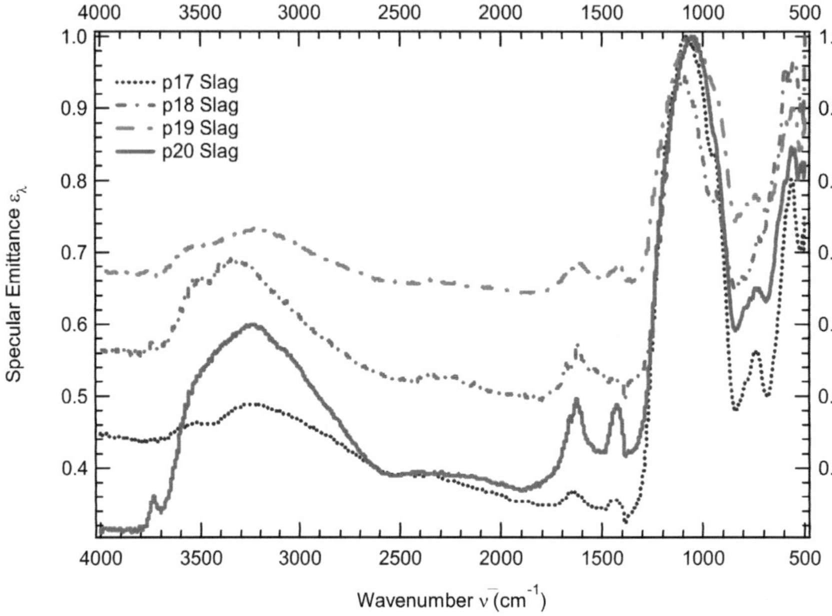

Figure 5 Normalized emittance spectra of slag samples.

Each slag sample shared a common peak around 1100 cm^{-1} and the spectra were normalized according to this peak by setting it to an emittance value of 1.0 and shifting the rest of the spectra accordingly. Without this normalization this peak and the rest of the spectra would vary in intensity and it was thought more descriptive of the samples to normalize the spectra to this peak which was expected to be all absorbing.

The various spectra collected for a sample were compared with each other. Comparison of the various spectra collected for a sample revealed a range of offsets between each run. Peaks appeared at consistent wavenumbers but varied along each point of the absorbance curve within a range of about 0.1 absorbance as seen in the figure below. This variation proves large compared to the entire range and normalizing each of the separate spectra for a sample to the same peak at 1100 cm^{-1} allowed for a more consistent data set before averaging the spectra and calculation of the total emittance with the resulting averaged spectrum. The figure below depicts the average of the spectra shown in the previous figure.

Diffuse Reflectance - Total Emittance

The total emittance of bituminous coal ash (CF) is greater across all temperatures than subbituminous coal ash (CC) as seen in the figure of Planck weighted total emittance of the two below:

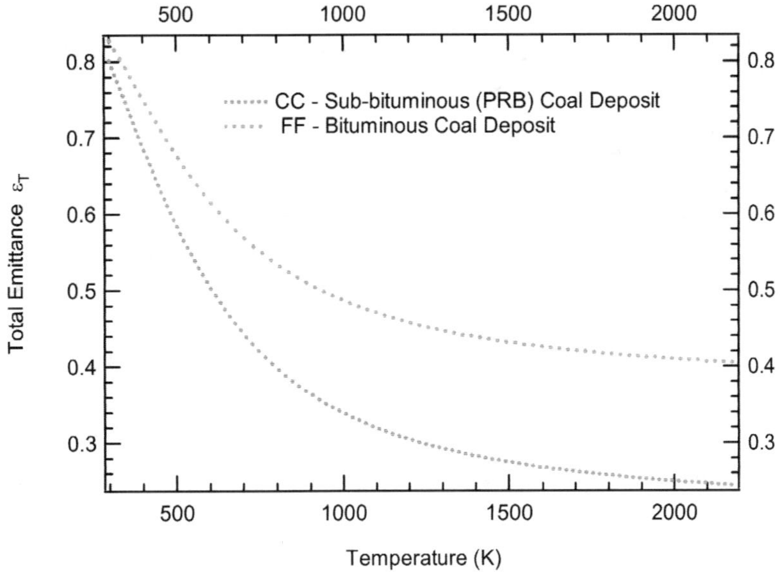

Figure 6 Sub bituminous and bituminous coal deposit total emittance from Diffuse Reflectance measurements.

Generally in these samples, total emittances slopes downward as temperature increases. The following figures include the temperature dependent of the Planck weighted total emittance of some of the ashes which also follow the downward trend as seen in the last figure:

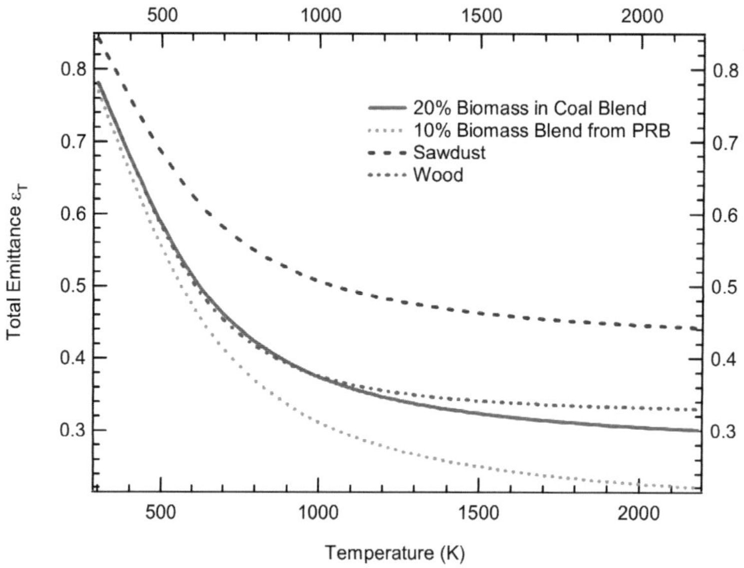

Figure 7 Ash total emittance from Diffuse Reflectance measurements.

Figure 8 shows total emittance as it relates to temperature for a control sample of slag. As predicted, total emittance decreases with increasing temperature. Emittances for slag samples generally vary between 0.4 and 0.7. This is one of the important practical, quantitative results of these measurements.

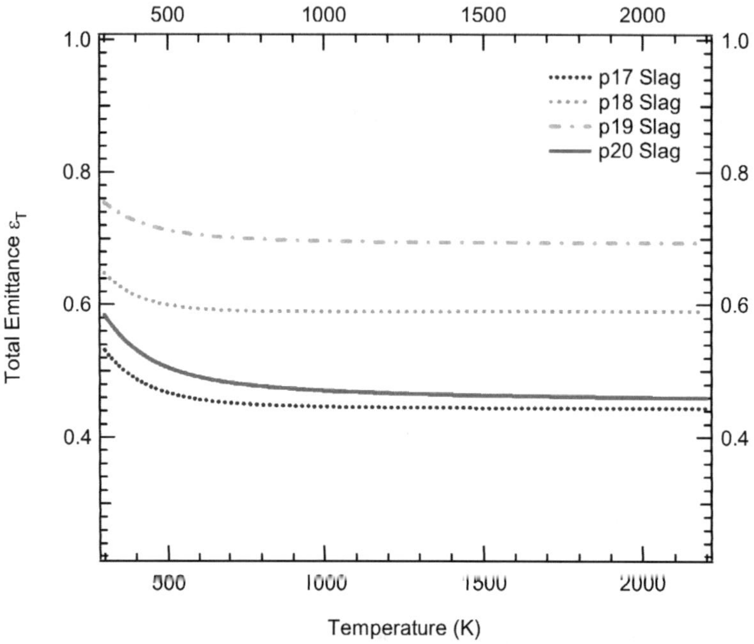

Figure 8 Variation of slag total emittances with temperature.

The p18 sample displayed physical trait differences in comparison to the others. Among these traits were its reddish color versus others black and charcoal colors, its larger original (before grinding for sample preparation) particle size, and its toughness which exceeded the other slags.

Transmission – Spectral Emittance

When research began initially, it was assumed that specular emittance could be calculated by equating it to absorbance. Further research determined that specular emittance relates to absorbance through the following equation obtained from Kirchhoff's Law (Griffiths, 2007):

$$T(\bar{\nu}) + R(\bar{\nu}) + \varepsilon(\bar{\nu}) = 1 \qquad 2$$

where $T(\bar{\nu})$ is transmittance, $R(\bar{\nu})$ is reflectance, and $\varepsilon(\bar{\nu})$ is spectral emittance. The emittance data obtained from the DRIFTS technique will be revised with this calculation of emittance before compared to the emittance data from the transmission technique. For this experiment's purposes, or because of the nature of the samples, reflectance can be set to zero and removed from the equation.

With this determined, the FT-IR spectrometer (Nexus 8700, Thermo Scientific) was set up to test samples by measuring the absorbance of various wavenumbers ranging from 4000 to 400 cm^{-1}; which is what was done in original measurements. Once the absorbance data was displayed to the screen in Omnic (Spectrometer software package), a "Transmittance %" calculation using this software was used. This calculation displays transmittance values as a function of wavenumber based-off data already measured in the form of absorbance. The transmittance data was then imported to Microsoft Excel where it was converted to spectral emittance with the following simplified equation:

$$\varepsilon(\bar{\nu}) = 1 - T(\bar{\nu}) \qquad 3$$

Calculations with all twelve samples of fly ash and slag and imported the data to Igor (plotting software) allowed for visual representation. Figure 9 represents the specular emittance for slag deposits.

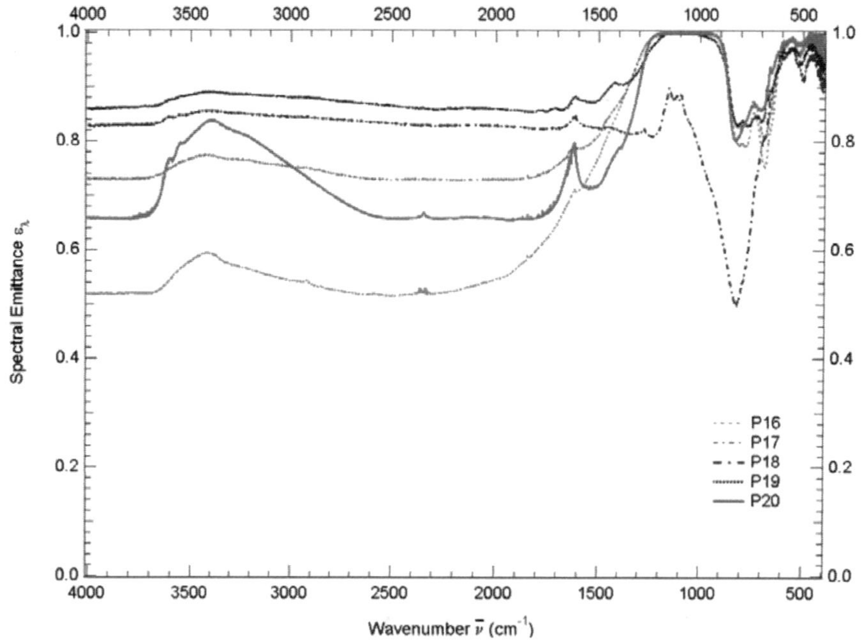

Figure 9 Slag specular emittance from transmission measurements.

Figure 10 represents specular emittance for bituminous and sub bituminous coal ash.

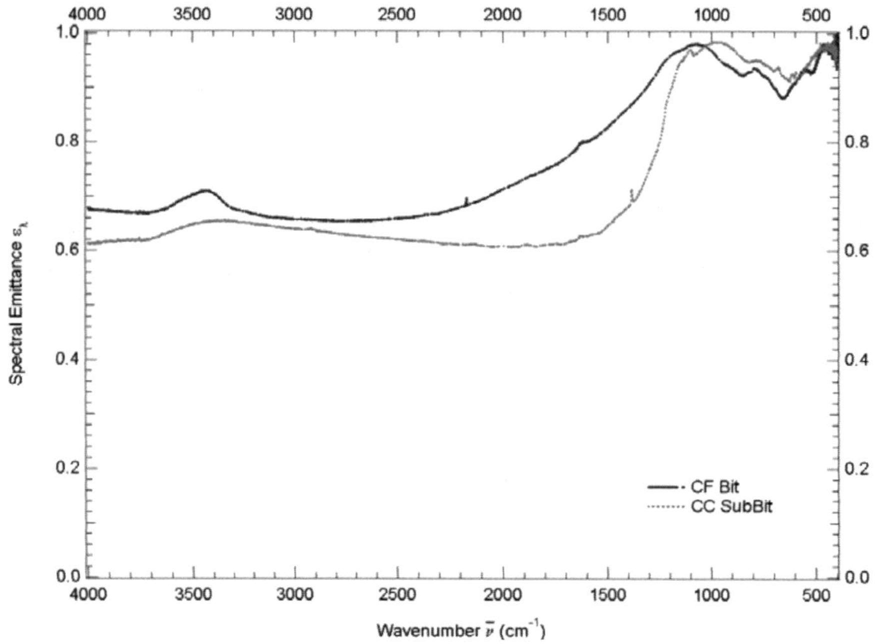

Figure 10 Sub bituminous (CC) and bituminous (CF) specular emittance from transmission measurements.

Ash samples from biomass and biomass coal blends are included in the following figure.

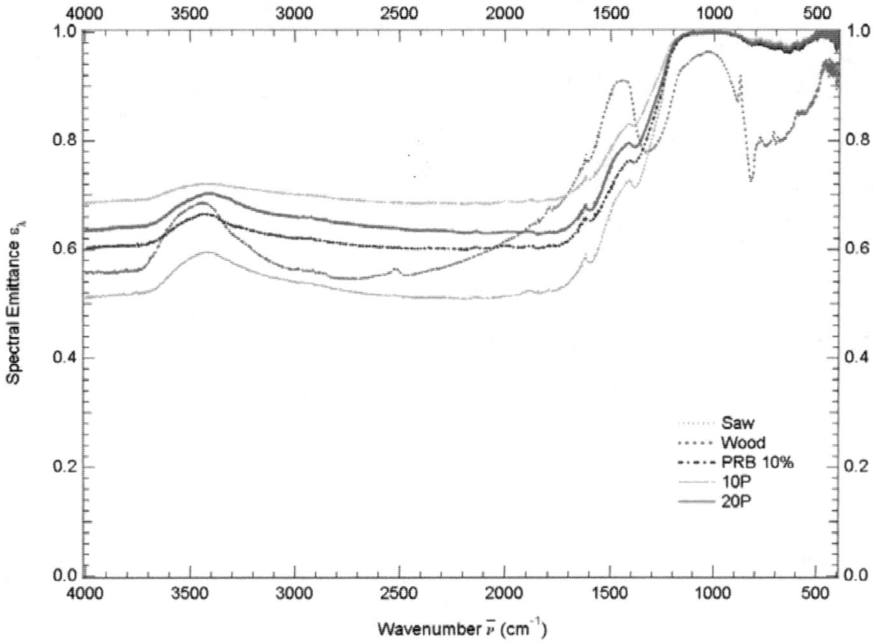

Figure 11 Ash specular emittance from Transmission Technique from transmission measurements.

Comparison between the specular emittances obtained from Diffuse Reflectance and transmission reveals significant differences between same sample spectra from each technique. This is explained because the specular emittance was calculated by equating specular absorbance to emittance in the Diffuse Reflectance technique and by Equation 3 in the transmission technique. In subsequent work, the first data will be analyzed using the latter method in order to compare the two spectra more adequately.

Transmission - Total Emittance

Total emittance is calculated here the same as in the Diffuse Reflectance technique, by employing the Planck weighted average of the specular emittance spectra in Equation 1.

The total emittances of bituminous and sub bituminous ash, other ash, and slag deposits are displayed in the graphs below respectively:

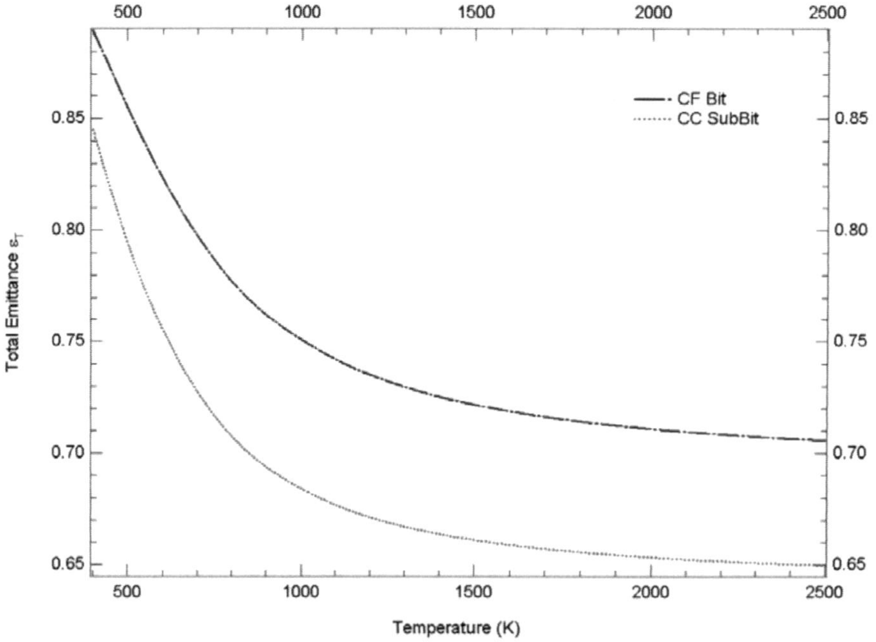

Figure 12 Sub bituminous and bituminous total emittance from transmission measurements.

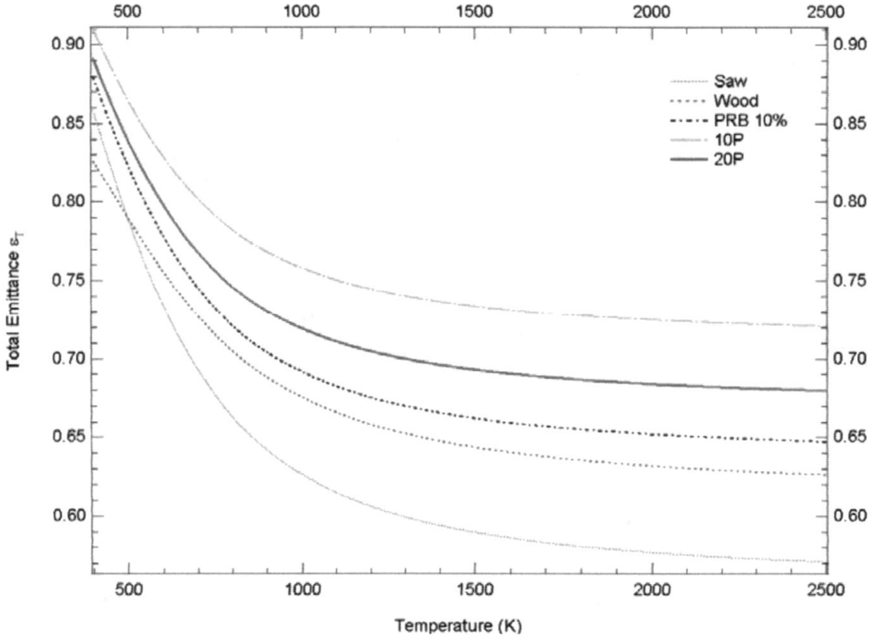

Figure 13 Ash total emittance from transmission measurements.

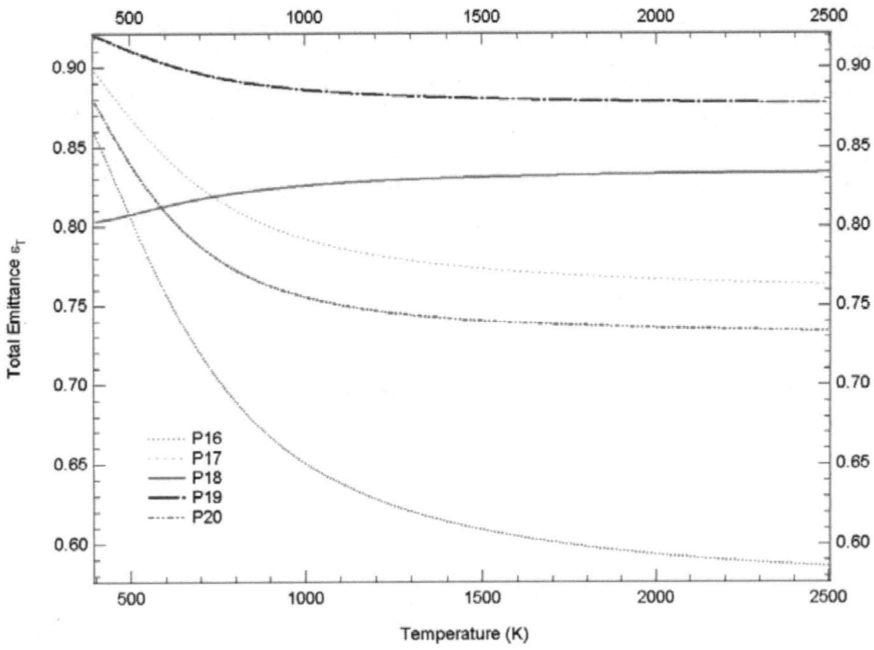

Figure 14 Slag total emittance from transmission measurements.

Note the irregularity of the P18 slag from the post combustion chamber outlet. This slag, reddish in hue, and much coarser in its pre-ground state is different than the other samples, which information is given to support its irregular behavior.

Particle size effect

To discover the affects of particle size on spectral emittance, an initial experiment was performed. A set of three pellets made from a finely-ground sample of ash was plotted against another set of three coarsely-ground pellets of the same ash; ensuring that the particle sizes of the two samples were significantly different. The results were surprising, as they demonstrated the opposite trend that was expected prior to performing the experiment. The results showed an increase in spectral emittance with a decrease in particle size. To validate this trend, the experiment was repeated with a different sample of ash, which showed a similar trend.

The current step is to perform the same experiment with multiple particle sizes under the assumption that further experiments will yield similar results, and thus, reveal a more detailed trend of the affect of particle size on spectral emittance. Currently, an efficient and accurate method of measuring particle size of the various samples has been under development and appears ready for testing purposes.

The samples of ash used in these experiments are of a dry and powdery nature, creating a difficulty in avoiding clustering of multiple particles in the analysis of individual particle size. The samples were viewed under a WILD M20/M21 microscope at 600X to verify separation of individual particles. Prior to placement under the microscope, the samples were mixed in distilled water, both with and without soap, to help avoid the clustering of multiple particles. Images of the prepared samples were then taken using a ten-megapixel digital camera, Nikon Coolpix. These images were then uploaded and processed using a program written in Matlab.

The Matlab program used for the analysis was written and designed to take a colored digital photo, convert it to a black & white image and calculate the diameter of each particle present in the image. All particle size calculations have and will use a set of 30-50 images for each particle size evaluation of each sample. Also, the final calculation of diameter of particle size is based-on a volumetric diameter.

During the execution of the program, the set of 30-50 images are read in one at a time and then distributed into a single matrix for calculation. Single particles are recognized by the program based-on pixel groupings in the black & white image. To prevent improper analysis due to clustering of particles, the program converts particles that appear to be too large, based-on the perimeter to minor-axis length ratio, into two or more particles. The program also eliminates any noise in the image by eliminating all particles, or pixel groupings, that don't coordinate with the appearance of the original image (i.e. all particles smaller than a specified diameter, or less than a specified number of pixels in a grouping are eliminated). The program displays each change made to the original image allowing the user to verify that final calculations are done accurately according to the appearance of the original image taken.

Although the program occasionally demonstrates small errors, these errors are considered negligible for the program's intended purpose. It is assumed that this program allows the user to determine that several samples of ash and slag differ in particle size significantly; thereby the several samples may be used in creating pellets, tested in an FT-IR spectrometer and plotted on a single graph for comparison. In other words, the primary purpose of the program is to provide the user an accurate and evenly spaced distribution of particle size for comparison. This program is not necessarily the most accurate method for simply measuring the size of particles contained in a sample.

Conclusion

At low temperatures, according to the transmission data, most total emittances fall between 0.8 and 1.0. As temperature increases slags generally maintain a higher total emittance than the ash deposits.

Interestingly, particle size has an effect on total emittance. For samples p18 and CC, decreasing particle size led to an increase in total emittance.

Sample p18 differs from the rest of the samples as its total emittance increases with increasing temperature.

Collection of emittance data of different ash types aids in boiler design and fuel type planning.

Works Cited

Coleman, P. B. (1993). *Practical Sampling Techniques for Infrared Analysis.* CRC Press, Inc.

Durig, J. R. (1990). *Applications of FT-IR Spectroscopy.* New York, NY: Elsevier Science Publishers.

Griffiths, P. R. (1986). *Fourier Transform Infrared Spectrometry.* John Wiley & Sons, Inc.

Griffiths, P. R. (2007). *Fourier Transform Infrared Spectrometry.* John Wiley & Sons, Inc.

Incropera, F. P. (2007). *Fundamentals of Heat and Mass Transfer (6E).* John Wiley & Sons, Inc.

The Sadtler Infrared Spectra Handbook of Minerals and Clays. (1982). Philadelphia, PA: 1982.

(398d) Chemical Looping Strategy and Its Novel, Exergetically Optimized Applications in Fossil Energy Conversions

Fanxing Li, Liang Zeng, Hyung Rae Kim and L.-S. Fan, Chemical and Biomolecular Engineering, The Ohio State University, Columbus, OH

Through the assistance of an oxygen carrier particle, the chemical looping gasification strategy indirectly converts fossil and biomass fuels into carbon free energy carriers such as hydrogen and/or electricity with 100% CO_2 capture. Besides serving as a stand alone process, the looping strategy can be integrated with other energy conversion systems to minimize the exergy loss of the overall process. In this paper, the developmental status of the Coal/Biomass Direct Chemical Looping (CDCL) gasification process, which is being tested at 25 kW_{th} scale at the Ohio State University for co-production of hydrogen and electricity, is briefly reviewed. Several novel, exergetically optimized CDCL configurations that convert coal and biomass into liquid fuel, electricity and/or hydrogen with ultra-high energy conversion efficiencies, are presented and analyzed. The analysis results indicate that, through the utilization of chemical looping strategy, the irreversibility of the fuel conversion process can be reduced, resulting in significantly improved energy and exergy conversion efficiencies. Life cycle analysis (LCA) indicates that, when biomass is used as the feedstock, the overall CO_2 emission from the chemical looping gasification process can be negative.

See more of this Session: Energy Efficiency, Challenges and Solutions
See more of this Group/Topical: Sustainable Engineering Forum

EXERGY ANALYSIS OF A SINGLE-STAGE MIXED REFRIGERANT LNG PROCESS

Danahe Marmolejo-Correa, Truls Gundersen;
Norwegian University of Science and Technology, Trondheim, Norway

Liquefied Natural Gas (LNG) is the fastest growing energy carrier in the world, with an average annual growth of 6,3% for the last decade [1]. There are many reasons for this trend, such as flexibility in the market, low environmental impact compared with oil and coal, and the fact that natural gas reserves and extraction grow faster than oil. In addition, LNG transport by ships is favorable to pipeline transport for larger distances and smaller volumes. A wide variety of processes have been developed for natural gas liquefaction in order to handle different gas compositions, different climates (i.e. cooling water and air temperatures) and different working environments (offshore vs. onshore).

LNG processes can be divided into three categories: (a) Pure component refrigerant cascade cycles; (b) Mixed refrigerant cycles and (c) Expander cycles. A general description of these processes is presented in Table 1.

This work is focused on the study of the exergy flow in a Mixed Refrigerant Process (MRP) for LNG production. A simplified version of the PRICO process which is commercialized by the Pritchard Company [2] is used as the main case study. The PRICO process consists of a single refrigerant cycle with a multi-component mixture as refrigerant. The composition of the refrigerant mixture should be similar to the LNG composition in order to decrease the thermodynamic irreversibility produced by heat transfer in the liquefaction process. The simplified MRP cycle shown in Figure 1 is characterized by two-stage compression with inter-cooling, one condenser, the main heat exchanger and one throttle valve.

Due to the large amounts of energy needed to operate such processes, an energy and exergy analysis should be carried out to evaluate its performance. In this work both, chemical and thermo-mechanical exergies are calculated for all process streams [3]. A common mistake found in the literature regarding chemical exergy, is that it is sometimes confused with fuel heating values (HHV, LHV).

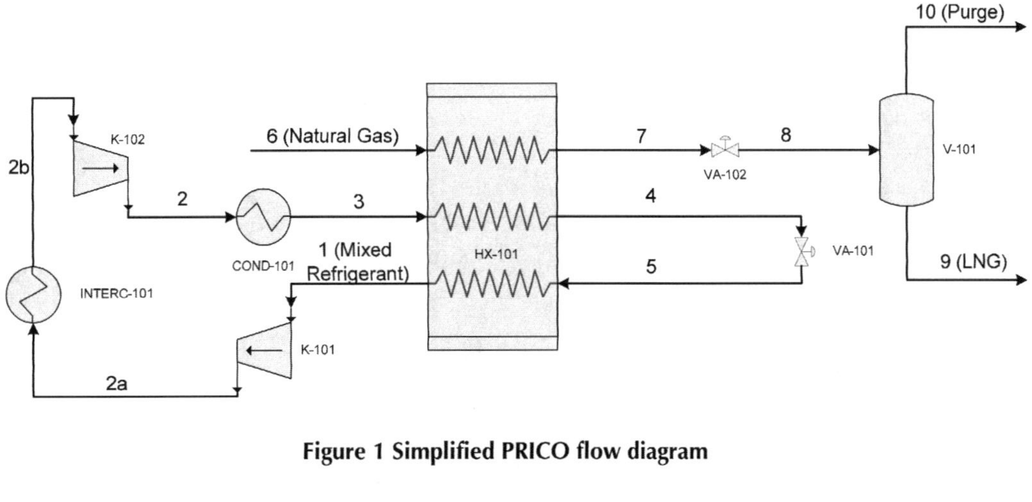

Figure 1 Simplified PRICO flow diagram

The chemical exergy is the amount of work that can be produced if a substance is brought to dead state with the environment. The heating value, on the other hand, is the heat that can be released in its combustion. From this point of view, any power plant should consider it to be more important to calculate the possible work obtained rather than the heat released produced. For this reason, the chemical exergy for mixtures (NG, Purge and LNG) is calculated. In this case the relative difference between both heating values (LHV and HHV) and the chemical exergy is around ±5%. In this analysis, both the individual (process unit) and the global process exergy efficiencies are calculated in addition to the generated exergy losses.

Table 1. An introductory description of LNG processes. [4-8]

Process	Representative Industrial Process / Licensor	General Characteristics
Cascade Refrigerant	• Classic cascade / Marathon and Philips • Optimized cascade /Conoco-Philips	• Single component refrigerant. • High process and control complexity • High energy efficiency • Plant size: >5 MTPA.
Mixed Refrigerant (MR)	• PRICO process /Black &Veatch • Propane Precooled / APCI • Single Refrigerant / APCI • C3MR and Dual-MR / Shell • Multifluid Cascade (MFC) / Linde-Statoil	• One or several multicomponent refrigerants • High energy efficiency • Large heat exchanger (small driving forces) • Medium to High process Complexity • Low operational flexibility • Highly dependent on inlet and site conditions • Plant size: Medium / small • C3MR is considered the LNG workhorse process with 0.5-6 MTPA per train
Turbine-based or Expansion Cycles	• Reversed Bryton cycle / TGE • Dual-Expander / ABB Lummus Global Inc. • AP-X[1] / APCI	• Low capital investment • High operational flexibility • Low energy efficiency • Plant size: < 1 MTPA • Industrial trend is to increase its use • Typically used for Floating Production Storage and Offloading (FPSO) plants

This work also aims to compare two of the most common definitions for exergy efficiency; the "classical" definition (Equation 1) and the one proposed by Tsatsaronis [9] in 1993 (Equation 2). These two definitions calculate two different efficiencies for the same process. Therefore, this discrepancy can lead to misunderstanding, mistrust and unfair comparisons in the analysis. Usually the efficiencies given by Equation 2 are larger than the ones given by Equation 1.

[1] AP-X is a combination of single refrigerant, multicomponent refrigerant and expansion cycles. Train capacities up to 8 MTPA are feasible.

$$\eta_{Ex} = \frac{\text{Exergy of the outputs}}{\text{Exergy of the inputs}} = \frac{\sum \dot{E}_{out}}{\sum \dot{E}_{in}} \qquad 1$$

$$\varepsilon = \frac{\text{Exergy of the Product}}{\text{Exergy of the Fuel}} = \frac{\dot{E}_P}{\dot{E}_F} = \frac{\dot{E}_{TF}}{\dot{E}_{CS}} \qquad 2$$

The exergy of the product represents the desired result produced by the system while, the *exergy of the fuel* stands for the sources expended to generate the product [9]. Equation 2 was developed for power plants, where the words *fuel* and *product* have a remarked importance. This might bring a bit of confusion when Equation 2 is applied to other industrial fields. For this reason, in this document the *exergy of the product* $\dot{E}_P$ and *exergy of the fuel* $\dot{E}_F$ will be named as *transferred exergy* $\dot{E}_{TF}$ and *consumed exergy* $\dot{E}_{CS}$ respectively.

Equation 2 also requires additional work to facilitate its use. It is necessary to find definitions for the transferred exergy and consumed exergy for each process unit. Some examples of these definitions are given in Table 2.

Table 2 Transferred and consumed exergy definitions for some process units

Component	Scheme	Transferred exergy rate $\dot{E}_{TF}$	Consumed exergy rate $\dot{E}_{CS}$
Compressor, Pump or Fan	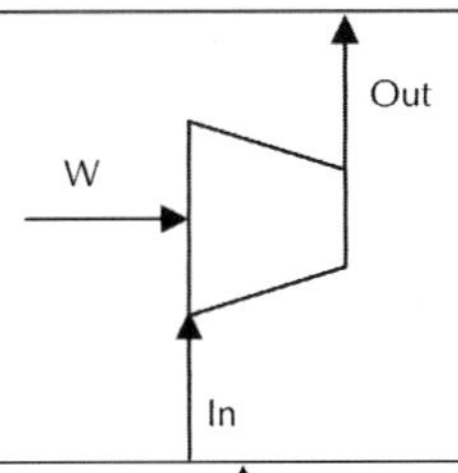	$\dot{E}_{out} - \dot{E}_{in}$	$\dot{W} = \dot{H}_{out} - \dot{H}_{in}$
Turbine or Expander	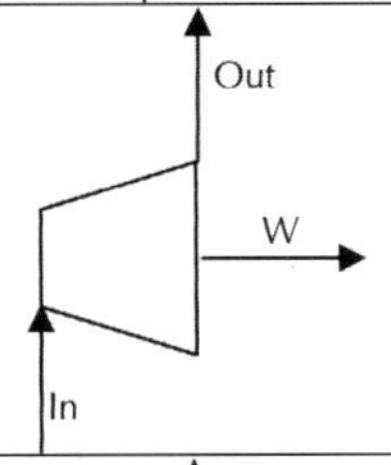	$\dot{W} = \dot{H}_{in} - \dot{H}_{out}$	$\dot{E}_{in} - \dot{E}_{out}$
Heater	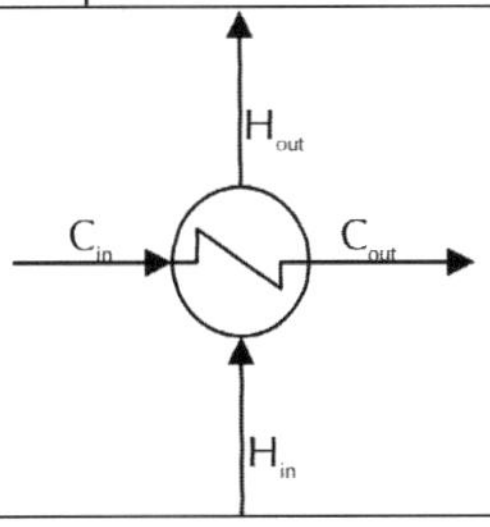	$\dot{E}_{C_{out}} - \dot{E}_{C_{in}}$	$\dot{E}_{H_{in}} - \dot{E}_{H_{out}}$
Sub ambient Expansion (Valve)	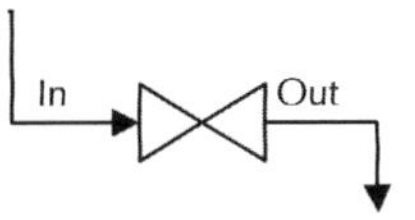	$\dot{E}^{TM}_{T_{out}} - \dot{E}^{TM}_{T_{in}}$	$\dot{E}^{TM}_{P_{in}} - \dot{E}^{TM}_{P_{out}}$

It is not always easy to find the correct definitions for all process units and sometimes it is necessary to do further exergy calculations and decompositions. For these cases, the thermo-mechanical exergy must be divided into its temperature-based and pressure-based components. A representative case for this division is the isenthalpic expansion below ambient temperature. In this case, the consumed exergy is the decrease in pressure-based exergy, while the transferred exergy is obtained by the increase in temperature-based exergy. The exergy decomposition used in this work was proposed by Aspelund et al. [10] in 2007 and it is represented by Equations 4 and 5. In this work the ambient conditions (T_0, p_0) are 298.15 K (25°C) and 1.013 bars (1.0 atm).

$$\dot{E}^{TM} = \dot{m}\left\{\left[h(T,p)-h_0(T_0,p_0)\right]-T_0\left[s(T,p)-s_0(T_0,p_0)\right]\right\} \qquad 3$$

$$\dot{E}_T^{TM} = \dot{m}\left\{\left[h(T,p)-h(T_0,p)\right]-T_0\left[s(T,p)-s(T_0,p)\right]\right\} \qquad 4$$

$$\dot{E}_p^{TM} = \dot{m}\left\{\left[h(T_0,p)-h_0(T_0,p_0)\right]-T_0\left[s(T_0,p)-s_0(T_0,p_0)\right]\right\} \qquad 5$$

$$\dot{E}^{TM} = \dot{E}_T^{TM} + \dot{E}_p^{TM} \qquad 6$$

It is worth noticing that these consumed and transferred exergy definitions are not general and they cannot be applied for all circumstances. Moreover, in some cases the transferred exergy is zero. One typical example is the isenthalpic expansion above ambient temperature. Here all the utilized exergy is lost (destroyed).

Therefore, the results obtained from the two equations for exergy efficiency are quite different. Using the "classical" efficiency (Equation 1), the global exergy efficiency is 50.1% and using Equation 2 it is 31.2%. The second global efficiency is calculated assuming the *transferred exergy* is the absolute exergy difference between the LNG + Purge streams (the sum of stream 9 and10) and the Natural Gas feed (stream 6). This is assuming that the purge has some value added. The *consumed exergy* is obtained by the total work supplied in the refrigerant cycle.

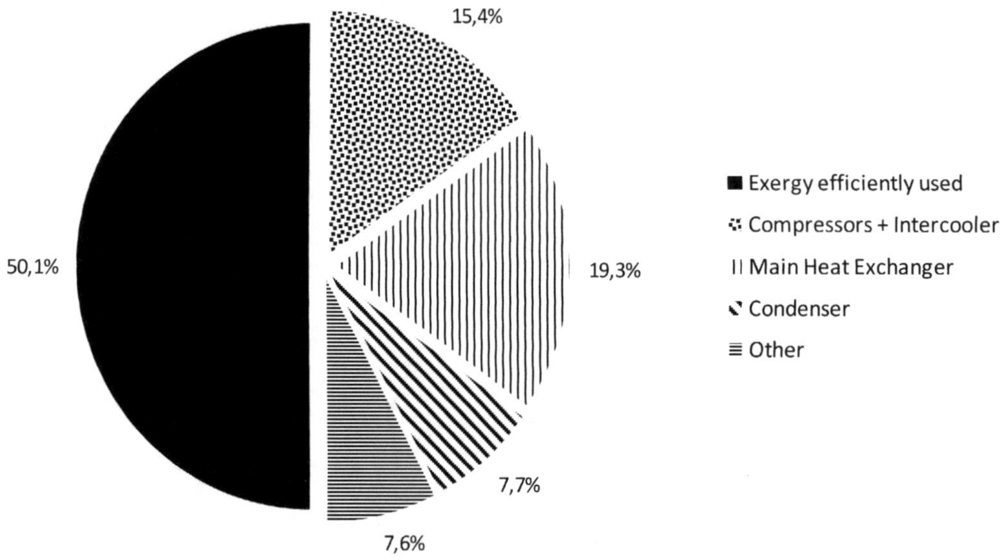

Figure 2 Global exergy efficiency (solid) and exergy losses (patterned) using Equation 1

Figure 2 shows the global exergy efficiency calculated with Equation 1 as well as, the exergy lost (destroyed) by the key process components. The main heat exchanger has the largest losses in the process. It contributes with the destruction of 19.3% of the total exergy input which represents 38.7% of the total losses (Figure 3). This destruction is generated by the temperature difference along the heat exchanger.

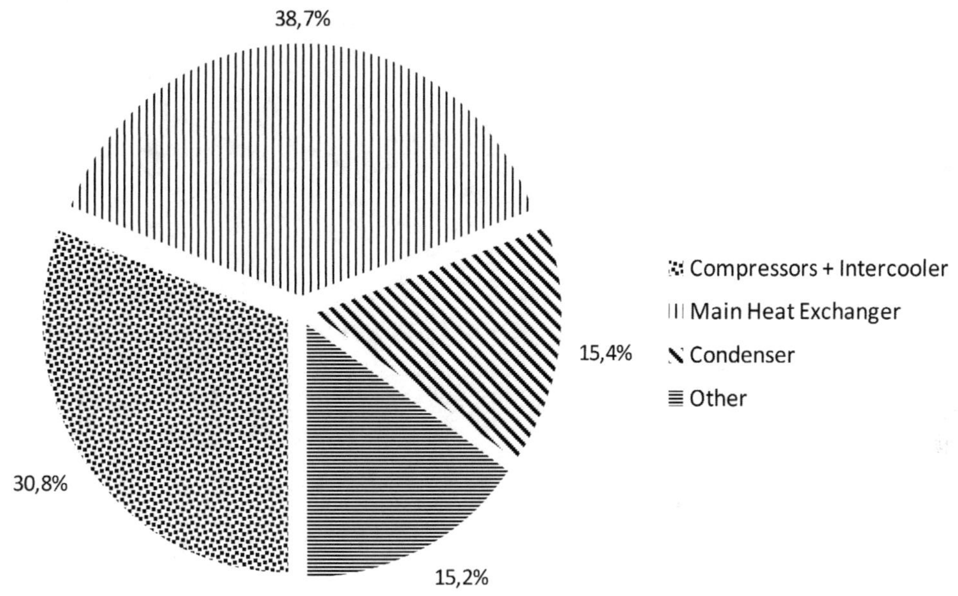

Figure 3. Exergy Lost distribution using Equation 1

The exergy destruction given by HX-101 might look large in comparison with other process unit destructions. However, the difference in temperature between the hot side and cold side of the heat exchanger is rather small. The minimum temperature difference in HX-101 is 2.5K. Figure 4 shows composite curves of HX-101. Therefore, even when the larger exergy destructions are in this unit, the improvement margin is small. In magnitude, the subsequent exergy destruction is given by the compression stage. The compressors and intercooler represents 30.8% of the exergy destroyed. Both compressors have a moderately high isentropic efficiency of 0.80 and a corresponding polytropic efficiency of 0.8126. Some authors report isentropic efficiencies from 0.72 to 0.78 for this type of processes [10, 11]. Hence, as with the main heat exchanger, the exergy saving opportunities are scarce.

On the other hand, even though the condenser and intercooler do not represent the larger losses, 15.4% and 6.5% respectively, these units can be improved. Both units use cooling water as cooling resource. The cooling water enters both heat exchangers at 288.15K (15°C) and exits at 305K (31.8°C). This temperature gap allows the heat exchangers to have a temperature difference correction factor (Ft) of 0.75 (shell &tube heat exchanger) which is relatively low. Having this physical restriction, the logarithmic mean temperature is approximately 20K producing large exergy destruction. Moreover, by using Equation 2 it is found that the efficiency of both units is zero. For example, analyzing the results for the condenser (COND-101) the cooling water exergy change is negative; it goes from a higher exergy level (288.15 K) to a lower exergy level (307K). Hence, this decrease in exergy should be recognized as the *consumed exergy* for Equation 2. However, the exergy change for the refrigerant is also negative. Stream 2 has a temperature of 361K and it is cooled down to 305K (stream 2). Consequently, the exergy change in the refrigerant stream is also exergy consumed and there is no exergy transformation. A possible improvement is to replace these exchangers by others that allow lower temperature approaches, such as plate-fin exchangers.

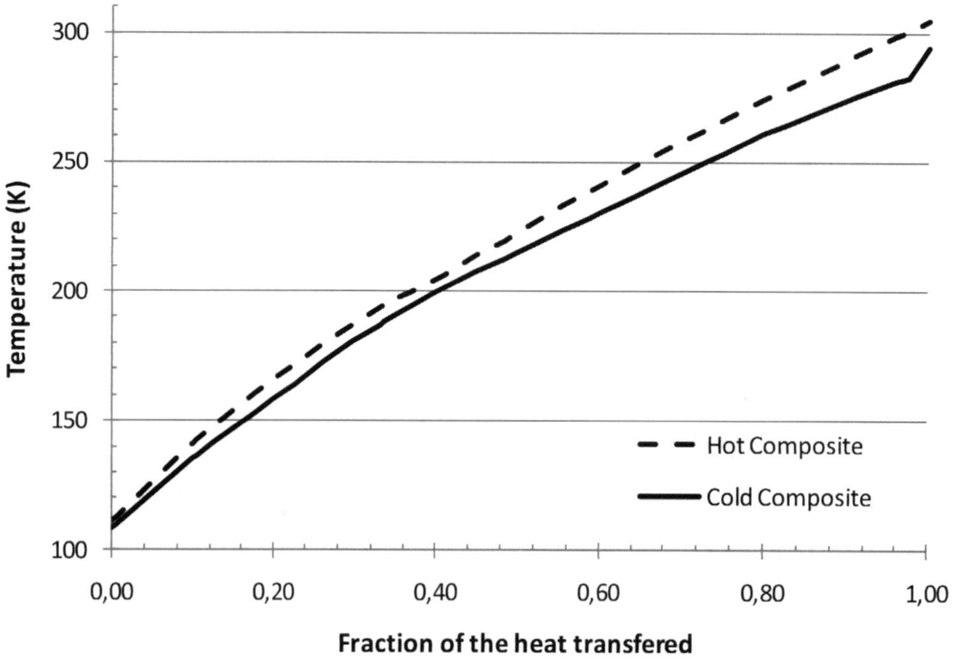

Figure 4. Main heat exchanger composite curves

Having illustrated this simple case study, the work aims to suggest a systematic procedure for the correct application of Exergy Analysis on energy intensive processes such as LNG plants. Besides, it is intended to show that by utilizing both exergy efficiency definitions mentioned above, it is easier to identify energy saving opportunities and disqualify false exergy analysis result interpretations more rapidly. In addition, a standardized exergy efficiency concept will be suggested.

References

[1] "British Petroleum: Statistical review of world energy," 2001-2010.

[2] L. K. Swenson, "Single mixed refrigerant, closed loop process for liquefying natural gas," U.S. Patent 4,033,735, 1977.

[3] T. J. Kotas, *The exergy method of thermal plant analysis*, Second ed. Malabar, Florida: Krieger Publishing, 1995.

[4] T. Shukry, "LNG Technology Selection," *Hydrocarbon Engineering,* pp. 71-74, February 2004.

[5] M. Barclay and N. Denton, "Selecting offshore LNG processes," *LNG journal,* October 2005.

[6] M. J. Roberts, *et al.*, "Technology Selection," *Hydrocarbon Engineering,* pp. 81-84, May 2004.

[7] K. D. Gerdsmeyer and W. H. Isalski, "On-board reliquefaction for LNG ships," in *Gas Processors Association Europe Conference*, London, 2005.

[8] J. H. Foglietta, "LNG production using dual independent expander refrigeration cycles," US Patent 6.412.302, 2002.

[9] G. Tsatsaronis, "Thermoeconomic analysis and optimization of energy systems," *Progress in Energy and Combustion Science,* vol. 19, pp. 227-257, 1993.

[10] A. Aspelund, *et al.*, "An Extended Pinch Analysis and Design procedure utilizing pressure based exergy for subambient cooling," *Applied Thermal Engineering,* vol. 27, pp. 2633-2649, 2007.

[11] G. Venkatarathnam, *Cryogenic Mixed Refrigerant Processes* New York, USA: Springer, 2008.

(398f) Energy Saving Design Methodology of Preheating Section for Gas Membrane Separation Based On Self-Heat Recuperation

Yasuki Kansha[1], Akira Kishimoto[1], Takafumi Kiuchi[2], Yoichi Ishibashi[3] and Atsushi Tsutsumi[1], (1) Institute of Industrial Science, The University of Tokyo, Tokyo, Japan, (2)Kitakyushu Environmental Technology Center, Nippon Steel Engineering Co., LTD., Kitakyushu-city, Japan, (3)Biomass Business Section, Business Development Center, Nippon Steel Engineering Co., LTD, Tokyo, Japan

In this research, a design methodology of a preheating section for gas membrane separation based on self-heat recuperation technology to reduce the energy consumption is developed. This design is consisted of the following three steps; 1) the heating load in the preheating section is evaluated, 2) the process is divided into the functions to balance the heating and cooling load, and 3) the process heat is recuperated and exchanged with the process self heat based on self-heat recuperation technology for providing an internal heat circulation to reduce the energy consumption.

In the process designed by the proposed methodology, the heat of the effluent streams from membrane separation unit is recuperated by compressors and exchanged with the heat of feed stream by using the self-heat recuperation technology. As a result, not only the latent heat but also the sensible heat of the process stream are circulated in the process and the process energy required can be drastically reduced, especially in the case that the phase change occurs during preheating section.

The simulation result demonstrates that the energy consumption of the process based on the proposed design methodology in an industrial application is less than 1/3 ofthat of the conventional counterpart.

See more of this Session: Energy Efficiency, Challenges and Solutions
See more of this Group/Topical: Sustainable Engineering Forum

(431a) The Implications of Bioenergy Derived From Forests for Net Greenhouse Gas Emissions

Werner A. Kurz, Pacific Forestry Centre, Canadian Forest Service, Victoria, BC, Canada

Greenhouse gas (GHG) accounting systems consider bioenergy from forests as carbon neutral because, under international greenhouse gas accounting rules, the carbon contained in wood is deemed emitted to the atmosphere at the time of transfer out of the forest ecosystem. This accounting simplification eliminates the need to track the fate of harvested carbon and to quantify the emissions from harvested wood when and where they occur. Life Cycle Analyses typically assume carbon neutrality of bioenergy with the justification that the regrowing forest will remove emitted carbon from the atmosphere. These simplifying assumptions fail to account for the timing of emissions and removals and the net impact on atmospheric GHG concentrations. To contribute to climate change mitigation bioenergy does not have to be carbon neutral: it merely has to produce fewer net GHG emissions than the fossil fuel alternative. Quantification of the timing of net GHG reductions from forest-derived bioenergy use is thus a prerequisite to the development of effective climate change mitigation portfolios. Net GHG emission estimates are obtained by accounting for carbon stock changes in forest ecosystems, the emissions from biomass use, and the avoided emissions from the displaced fossil fuels. Forest biomass typically causes higher emissions per unit of energy than fossil fuels. The excess emissions are reduced over time as regrowing forests remove C from the atmosphere. The type of forest biomass used, the conversion efficiency, the fossil fuel type that is displaced, and the rate of forest regrowth all affect the substitution benefits of bioenergy. Results indicate that the break-even point at which net emissions from bioenergy are equal to those from the displaced fossil fuel can be years or decades into the future. After that break even point, the net benefits of bioenergy use increase over time.

See more of this Session: Sustainable Biorefineries Plenary Session (Invited Papers)
See more of this Group/Topical: Sustainable Engineering Forum

(431b) Biochemical Refining of Lignocellulose for Biofuels: Status and Prospects

James D. McMillan, National Renewable Energy Laboratory, Golden, CO

A tremendous amount of work is underway at NREL and elsewhere to develop and demonstrate cost competitive process technologies to produce liquid fuels as well as other chemical and energy products from non-food lignocellulosic woody and herbaceous plant materials ("cellulosic biomass"). A variety of technologies have been shown to be technically feasible including those based on primarily biochemical or thermochemical conversion as well as hybrid approaches using a combination of biochemical and thermochemical processing. Many of these are under active development to prove economic viability.

This talk will focus on progress being made to develop and demonstrate cost-competitive biochemical refining technologies. Recent progress achieved at NREL to reduce the cost of converting cellulosic biomass to ethanol will be highlighted, as will new initiatives to convert biomass derived sugars to other advanced biofuels. The potential for process intensification to reduce capital equipment and process operating costs will also be illustrated. Remaining technical barriers and as-yet unrealized opportunities to stimulate biorefinery commercialization will be discussed. While process capital costs remain high and product titers relatively low for biochemical ethanol production from cellulosic feedstocks compared to production from grains (corn, wheat) or sugar juice (beet, cane), the gap between conventional and cellulosic ethanol production technology continues to close and NREL is on track to demonstrate cost-competitive cellulosic ethanol technology at the pilot scale in 2012.

See more of this Session: Sustainable Biorefineries Plenary Session (Invited Papers)
See more of this Group/Topical: Sustainable Engineering Forum

(431c) Sustainably Produced Sugars From Biomass: The Key to Commercialization

Sarah A., Teter, Novozymes, Inc., Davis, CA

Novozymes has launched the first commercially viable product for conversion of complex biomass to fermentable sugars. Continued enzyme improvement and integration of key process steps are the driving force enabling biorefineries to commercialize production of fuels and chemicals. On the enzyme front, we demonstrate increased catalytic activity, and improved tolerance to inhibition. We have developed enzymes that work well in a variety of processes and conditions on pretreated agricultural residues as well as woody based substrates. Our recent and continued success results from engineering improved specific activity of the enzymes present in our existing cellulase products, addition of accessory proteins that improve performance on some substrate, and introduction of superior proteins into high-yielding T. reesei production strains. In addition to developing new enzymes, we have addressed the interaction between pretreatment, enzymatic hydrolysis and fermentation in order to attain a holistic view of the entire biorefinery process and its associated costs. Novozymes works with industry leaders to integrate the technologies both upstream and downstream of enzymatic hydrolysis. This needed integration of technologies drives our open and broad approach to partnering, with a focus on innovative companies representing varied cellulose based substrates and cutting edge technologies. Novozymes' technologies enable companies to move aggressively from pilot scale testing to commercialization.

See more of this Session: Sustainable Biorefineries Plenary Session (Invited Papers)
See more of this Group/Topical: Sustainable Engineering Forum

(431d) Hydrocarbon Fuels From Renewable Biomass

David A. Glassner, Gevo, Inc., Englewood, CO

Existing gasoline, jet and diesel fuels are composed of mixtures of hydrocarbon molecules. Efforts to make hydrocarbon fuels, the so called 'drop-in fuels,' from renewable resources has escalated over the past few years. However, the process options available to make the fuels aren't well developed for policy makers and even the scientific and engineering community pursuing the fuels. The importance of drop-in fuels is their fit into gasoline, jet and diesel fuels without changes to the fuel or vehicle infrastructure. Meanwhile ethanol is by far the largest renewable fuel today and has provided an excellent start to sustainable, domestically produced renewable fuels. Additionally ethanol has provided an infrastructure, including technical expertise that is extremely valuable to support the development and commercialization of advanced biofuels. Longer term the desire is to continue to improve the sustainability profile for biofuels production and to convert plant matter into fuels and blend stocks that fit exactly, without special rules, into the existing infrastructure for crude oil derived transportation fuels and vehicles.

This contribution will focus on making drop-in hydrocarbon fuels and make the case that much of the technology required already exists. Additionally the ethanol industry infrastructure provides a platform to launch the drop-in fuels and blend stocks. There are multiple ways possible to produce hydrocarbon fuels from biomass a sampling of which will be described. The importance of conserving the energy content of the feed stock converted will be emphasized as this leads to the lowest cost, most efficient processes for production. Initially the technology for drop-in fuels is likely to be built on the backs of the existing ethanol infrastructure, migrating to cellulosic biomass as the technology and markets mature.

One route to drop-in hydrocarbons is the fermentative conversion of biomass to isobutanol followed by catalytic chemistry to yield a variety of chemicals, materials and fuels. Isobutanol from carbohydrates by fermentation has been enabled by pioneering biotechnology utilizing a novel five step pathway for the conversion of pyruvate to isobutanol. Additionally novel recovery technology has been developed providing for low cost purification of the isobutanol. Isobutanol is easily converted to isobutylene and a wide variety of 'drop-in' hydrocarbon chemicals and materials including isobutylene, para-xylene, isooctane, terephthalic acid, propylene, PET plastic, and polypropylene. The catalytic chemistry required has been previously developed and can be used to convert isobutanol to whole gasoline, jet and diesel fuels that meet all specifications for the fuels.

See more of this Session: Sustainable Biorefineries Plenary Session (Invited Papers)
See more of this Group/Topical: Sustainable Engineering Forum

(431e) Sustainable Biorefineries: Progress in the Making

Bob McDonald, Range Fuels, Broomfield, CO

At Range Fuels, our focus is on the production of low carbon biofuels, such as cellulosic ethanol, and clean renewable energy usingbiomass that cannot be used for food, and is sustainable, renewable, and in excess supply. In this presentation we will discuss the progress made by Range Fuels from being the first recipient of a DOE 932 Grant award, first to receive a biorefinery loan from the USDA, the first to build a thermochemical cellulosic ethanol facility and sell branded cellulosic ethanol in the USA.

See more of this Session: Sustainable Biorefineries Plenary Session (Invited Papers)
See more of this Group/Topical: Sustainable Engineering Forum

(436a) Chemical Engineering In a Complex World: Grand Challenges, Vast Opportunities

Julio M. Ottino, Chemical and Biological Engineering, Northwestern University, Evanston, IL

The Institute Lecture provides an opportunity to take a long view of chemical engineering – both where we have been and what opportunities lie ahead. By the mid-1970 to 1980s ChE had organized itself around an accepted set of principles, largely science-based, with mathematics and analysis providing the foundation. Beginning in the mid 1980s, new areas were incorporated, including materials, biology, and molecular-based elements. This broadening occurred even though ChE departments, when compared with areas such as mechanical engineering or electrical engineering, are typically smaller (at least in the US). Remarkably, often one or two people from ChE have risen to the very top levels of significant areas with relatively small footprints in ChE itself. I will illustrate, using part of my own work as a point of departure, how basic work can expand into new domains, and will present a broad view of ChE in the context of science and engineering. There are now tremendous opportunities for ChE and, in many respects, a current confluence of factors – energy, global health, environment, and many others arising from increased connectedness and complexity – make ChE more relevant than ever.

See more of this Session: 62nd AIChE Institute Lecture
See more of this Group/Topical: Liaison Functions

(461a) Hot Water and Dilute Acid Pretreatment of High and Low Specific Gravity Populus Deltoides Clones

Kris A Bunnell[1], **Ching-Shuan Lau**[1], Elizabeth M Martin[1], Matthew H. Pelkki[2], David W. Patterson[2], Edgar C Clausen[3], James A. Smith[1] and Danielle Julie Carrier[1], (1)Department of Biological and Agricultural Engineering, University of Arkansas, Fayetteville, AR, (2)School of Forest Resources, University of Arkansas-Monticello, Monticello, AR, (3)Ralph E. Martin Department of Chemical Engineering, University of Arkansas, Fayetteville, AR

Hybrid poplars, Populus sp. are hardwood feedstocks that grow as understory species in forest management areas that are logged for softwood. Because of the mixed cultures found in forested environments, this study involves sampling not only wood and bark from hybrid poplar, but also mixed cultures containing low and high specific gravity specimens. The objective of this research was to use hot water and dilute acid pretreatment to obtain extracts that contain high percentages of hemicellulose sugars. Dilute sulfuric acid (0.1%) or water was added to 7.5 g of wood and bark chips that were ground to a particle size of approximately 4 mm. These samples were placed into thick-walled stainless steel tubes and placed into a fluidized sand bath at temperatures between 130-240 °C and run for 20-100 min. Filtered solutions from the sulfuric acid treatment were neutralized with calcium carbonate. Extracts from both dilute acid and water treatments were filtered with 0.2 μ m syringe filter and analyzed by HPLC, using either a carbohydrate method or an organic acid method. Results from this study suggest that hot water and dilute acid pretreatments can help in optimizing the extraction of hemicellulose sugars from hybrid poplar bark and wood.

See more of this Session: Developments in the Pretreatment of Lignocellulosics for Bioconversion II
See more of this Group/Topical: Sustainable Engineering Forum

(461b) Fungal Pretreatments for Lignocellulosic Bioconversion

Richard L. Giles[1], Emily R. Galloway[1], Gloria D. Elliott[2], Barry Edwards[3] and Matthew W. Parrow[1], (1)Department of Biology, University of North Carolina at Charlotte, Charlotte, NC, (2)Mechanical Engineering and Engineering Science, University of North Carolina at Charlotte, Charlotte, NC, (3) Department of Utilities and Engineering, Catawba County, NC Government, Newton, NC

Lignin and complex polysaccharides limit the efficiency of chemical and enzymatic conversion of lignocellulosic biomass into ethanol. Processing technologies that address these problems are needed to advance lignocellulosic biofuel production. In the current research, whole-organism wood decay fungi were utilized as a pretreatment process for improved lignocellulosic bioconversion. Corn stover, Miscanthus, switchgrass, kenaf, and tulip poplar were selectively decayed with white rot (Ceriporiopsis subvermispora) and/or brown rot (Postia placenta) fungal treatments. Both white-rot and brown-rot fungi were successfully grown alone or in succession on all biomass types tested. Significant lignin removal and minimal polysaccharide loss were observed in some treatments, resulting in biomass with significantly improved chemical and mechanical properties for downstream conversion into ethanol. Glucose analysis determined that select fungal pretreatments effectively depolymerized lignin and increased the efficiency of enzymatic polysaccharide hydrolysis. Fungal bioconversion treatments determined at the laboratory scale to be best-suited for lignocellulosic ethanol production were applied on site at the Catawba County, NC Regional EcoComplex in a small scale pilot study.

See more of this Session: Developments in the Pretreatment of Lignocellulosics for Bioconversion II
See more of this Group/Topical: Sustainable Engineering Forum

(461c) Pretreatment and Fractionation of Corn Stover Using Zinc Chloride and Ammonium Hydroxide

Chang Geun Yoo, Agricultural and Biosystems Engineering, Iowa State University, Ames, IA, Tae Hyun Kim, Department of Agricultural and Biosystems Engineering, Iowa State University, Ames, IA and Monlin Kuo, Department of Natural Resources Ecology and Management, Iowa State University, Ames, IA

A novel pretreatment and fractionation method of corn stover using zinc chloride (ZnCl2) and ammonium hydroxide (NH4OH) was developed to achieve the complete utilization of biomass and improve the economics of the biorefinery process. Zinc chloride, one of the strongest swelling chemicals, is effective reagent for hemicellulose hydrolysis while ammonium hydroxide was proven to be effective for lignin removal. In this study, two different modes of reactions were investigated: (1) single-stage treatment using zinc chloride and (2) two-stage treatment using zinc chloride in the first stage and ammonium hydroxide in the second stage. We found that zinc chloride is highly selective for hemicellulose hydrolysis (70-80% of xylan), but it retained most of lignin and cellulose with the solids. In the optional second stage, ammonia was added to solubilize lignins and remaining cellulose after treatment was highly digestible by commercial cellulase enzymes. Both processes enhanced enzyme digestibility of remaining cellulose (80-90%) and effectively fractionated biomass into glucan-rich solid, xylooligomer-rich stream, and soluble lignin stream. Zinc chloride and ammonia can be recovered and reused in this process. The optimal process conditions for effective pretreatment and fractionation were explored. Various effects on the compositional changes, enzyme digestibility, fermentability, and other technical aspects related to development of process will be presented.

See more of this Session: Developments in the Pretreatment of Lignocellulosics for Bioconversion II
See more of this Group/Topical: Sustainable Engineering Forum

(461d) Update On the Redistribution of Lignin Caused by Dilute Acid Pretreatment of Biomass and Its Effect On the Enzymatic Digestion of Cellulose

David K. Johnson[1], Bryon S. Donohoe[1], Rui Katahira[2], Melvin P. Tucker[2], Todd B. Vinzant[1] and Michael E. Himmel[1], (1)Biosciences Center, National Renewable Energy Laboratory, Golden, CO, (2) National Bioenergy Center, National Renewable Energy Laboratory, Golden, CO

The purpose of this research was to advance our understanding of the effect of pretreatment on the lignin component in biomass so that engineering research can be directed more efficiently. Research conducted at NREL over the past few years has shown that lignin undergoes a phase transition during thermochemical pretreatments conducted above its glass transition temperature. The lignin coalesces within the plant cell wall and appears as microscopic droplets on the surfaces of cells. In recent years, we have studied the redistribution of lignin that occurs during dilute acid pretreatment using small lab scale reactors operated under conditions that were not strictly process relevant. Typically, samples were pretreated in flow reactors with high liquid to solid ratios, with no acid, and for relatively long periods of time. However, microscopic analysis showed that changes were occurring both within the cell wall and on cell surfaces. It was clear that pretreatment could cause significant changes in the distribution of lignin. More recently we have studied the changes in lignin distribution that could occur under more process relevant conditions. Dilute acid pretreatments have been performed in a steam gun reactor at high solids loadings (43%), although at somewhat lower temperatures (150 and 170oC) and longer times (5-60 min) than normally used in our pilot scale reactors. Our studies have shown that lignin still undergoes a significant change in distribution and that coalescence and droplet formation still occur under these conditions. A method for selectively isolating these lignin droplets from the surface of pretreated cell walls using extraction with aqueous dioxane has been successfully employed. Characterization of the extracts by 2D NMR (HSQC) showed they contained carbohydrates and that the carbohydrates were covalently bound to the lignin in at least 2 types of lignin-carbohydrate complexes. Chemical analysis of the extracts showed the extracts contained only low levels of carbohydrates (2-6%) with the highest levels observed in the extracts from pretreatments performed at the lowest severity conditions. There was also evidence that the carbohydrates were at least oligomeric in size (i.e., not monomers or dimers). Based on this evidence, it appears likely that the frequency of the lignin-carbohydrate linkages found in the lignin extracts were low at about one cross-linkage per 80 to 100 lignin units. The effect of lignin redistribution on the digestibility of the pretreated solids was also tested, using the solids remaining after extracting the lignin droplets from the cell wall surfaces It was clear that removal of the droplets produced an increase in the digestibility of the pretreated corn stover. Whereas the digestibility of the cellulose was strongly correlated with the level of xylan removal, droplet removal increased cellulose conversion by about 15%. The improved digestibility could be due to a decrease in non-specific binding of the enzymes to the lignin in the droplets, or it could be because the droplets no longer blocked access to cellulose in the pretreated solids.

See more of this Session: Developments in the Pretreatment of Lignocellulosics for Bioconversion II
See more of this Group/Topical: Sustainable Engineering Forum

INTEGRATED DYNAMIC MODEL OF THE ALKALINE DELIGNIFICATION PROCESS OF LIGNOCELLULOSIC BIOMASS

John M. Fuertez[1], Angela A. Ruiz[1*], Hernan D. Alvarez[2] and Alejandro Molina[1], (1) Bioprocesses and Reactive Flows, (2)Research Group on Dynamic Processes-KALMAN, School of Processes and Energy, National University of Colombia, Medellin campus, Medellin, Colombia

ABSTRACT

Although in the public literature there are several studies that describe models of alkaline delignification, they were originally developed for the paper industry, and do not include the effects of important operating variables such as temperature, hydroxide-ion concentration, solid to liquid weight ratio, particle size, biomass composition (hemicellulose, lignin fraction) and mixing. This lack of detailed models of the pretreatment stages prompted the current study that describes a model which includes the variables listed above and provides an important tool to predict the degree of lignin removal in lignocellulosic materials such as sugarcane bagasse (*Saccharum officinarum L*). The model considers kinetic expressions available in the literature. The kinetic parameters were determined by fitting the model to experimental data obtained for that purpose in our lab. The experimental matrix considered eighteen, 24-h isothermal experiments in which bulk and residual delignification stages were observed to occur in parallel. Carbohydrate removal and hydroxide consumption were related to lignin removal by effective stoichiometric coefficients that were calculated by fitting the experimental data. A mixing compartment network model that represented mixing inside the reactor was included into a temporal superstructure based on the similarity between plug flow reactors and ideal batch reactors to model a non-ideally mixed batch reactor. The kinetic model was validated with data obtained in this study.

Keywords: mathematical model, alkaline and kraft delignification, pretreatment, lignocellulosic biomass, sugarcane bagasse.

[*] Corresponding author. Tel.: +0544255300.
E-mail address: aaruiz@ unal.edu.co

INTRODUCTION

Extensive research could decrease the cost of biomass pretreatment processes [1]. While most pretreatment methods are either physical or chemical, some incorporate both effects [1, 2, 3]. Alkali pretreatment technologies, such as lime pretreatment, are similar to the Kraft paper pulping technology. The major effect of alkaline pretreatment is the removal of lignin from biomass, thus improving the reactivity of the remaining polysaccharides [4].

Analogous to alkaline delignification in the paper industry, biomass pretreatment can be simplified as three parallel processes of lignin removal, all occurring at the same time, but with one dominating during different stages of the processes. The first process is an *initial stage* which takes place rapidly. Small parts of lignin molecules are dissolved by breaking of phenolic α-0-4 links which are easy to break and some β-0-4 links. The second process called *main* or *bulk stage*, involves breaking of the β-aryl-ether (β-0-4) links. The reaction rate is selective and carbohydrates dissolution is relatively low. Finally, in the *residual stage,* carbohydrates degradation is much slower than in the main stage and some carbon-carbon links of lignin brake [5, 6, 7, 8].

The existence of two or three stages for carbohydrates breakdown has caused some controversy [9]. However, reference [10] and [11] observed three parallel stages during Kraft deslignificación of softwood, as well as [12] has considered that the reaction rate of carbohydrates follows a linear function of the delignification rate.

Although there are a lot of alkaline and kraft delignification models [10, 13, 14, 15, 9, 16, 17], they are generally used to describe the delignification process of the paper industry. Nevertheless, it seems possible to propose a model that considers effects such as particle size, solid-liquid ratio, temperature and mixing intensity. The objective of this paper is to present such a model for the alkaline delignification of biomass, as pretreatment to produce ethanol from sugarcane bagasse (*Saccharum officinarum L*).

EXPERIMENTAL SECTION

Materials

Sugarcane bagasse (*Saccharum officinarum L*) is widely used for ethanol production [18, 19, 20, 5, 21, 22] and was the material selected in this study.

Raw sugarcane bagasse was manually cut to 2 cm long pieces and dried at 45ºC for 48 hours in a convection furnace, milled, sieved, and stored in sealed containers at 4°C until it was used. The stored sugarcane bagasse had the follow composition: 5.79% humidity, 13.34% lignin, 42.88% cellulose, and 27.38% hemicellulose, all dry weight percentages. Calcium hydroxide was used to carry out the delignification process.

During the delignification reaction, samples were taken every four hours, until a 24 hour test was completed.

The resulting mixture was neutralized with 1.8N hydrochloric acid to determine hydroxide consumption and avoid interferences during lignin and hemicellulose determination. Finally, the samples were washed and dried during 48 hours at 45 °C in a drying convection drying oven.

Analysis Methods

The lignin and hemicellulose content of the solid phase was determined by the VanSoest method [23]. Sieving and optical microscopy were used to estimate the average particle diameter. The hemicellulose content was calculated according to (1).

$$C_s = \frac{Hemicellulose(\%)}{100}\, Y\, M_s \qquad (1)$$

All terms are explained in the nomenclature section. The remaining lignin fraction (L) was calculated as:

$$L = \frac{L_s}{L_{so}} \qquad (2)$$

Experimental Design

Experimental conditions were defining according to [18], [24], [6], [25], [4], [26] and are summarized in Table 1. The experimental design consisted of two groups of isothermal tests. The first group (from F1 to F9) had a high liquid to solid ratio (R=50L/kg) and guaranteed almost constant chemical concentration during the delignification process. Experiments for the second group (from G1 to G9) were carried out at a low liquid to solid ratio (R=12L/kg) to identify changes in the hydroxide concentration. The liquor volume was 4.5L and 0.09kg or 0.375kg of biomass were added for liquid to solid ratios of 50L/kg and 12L/kg respectively.

Experiments were carried out in a cramping bracket, 316 stainless steel reactor with a 0.5HP stirring system. The reactor had two baffles. The following are the most important geometric characteristics of the reactor and stirring system: tank diameter (D_T): 0.205m, tank height (L_T): 0.205m, baffle width (W_B): 0.015m, baffle height (L_B): 0.13m, 30° inclined agitator, impeller diameter (D_A): 0.17m, agitator height from the reactor bottom (L_A): 0.035m, blade width (Wa): 0.03m.

MODEL

Kinetics of Alkaline Delignification

The three-stage parallel model has been extensively used [5, 6, 22] to describe the alkaline delignification of material used in the ethanol production. According to this model and references [15, 19, 27], that included the particle size, temperature and reagents concentration by means of a single first order kinetic equation, the equation (3) is presented to describe the lignin dynamics:

$$-\frac{dL_t}{dt} = \frac{A_t}{\delta^{w_t}}\, e^{\frac{E_t}{R_c T}} \left[C_{OH^-}\right]^{a_t} L_t \qquad (3)$$

In this expression, the hydroxide concentration follows a power law and the temperature dependence is given by an Arrhenius expression. The variable corresponding to the average particle diameter (δ) is introduced in the kinetic equation as well, in order to consider that effect in the chemical reaction.

Kinetics of the Carbohydrate Reactions

This study considers a constant (g_1) that determines the relation between hemicellulose and lignin kinetics for low temperatures (<100ºC) (see equation (4)). This approximation reduces the number of parameters and is required because the process stoichiometry is unknown.

$$\frac{dC_s}{dt} = g_1 \, Lo \frac{dL}{dt} \qquad (4)$$

Kinetics of the Hydroxide Reaction

In the initial stage most hydroxide is consumed in the hydrolysis of acetyl groups that are associated with the hemicellulose, in the neutralization of acid products and in possible reactions of carbohydrate degradation. In the main stage, it is consumed in the neutralization of products formed during carbohydrates breakdown and lignin-phenolates formation. In the residual stage, hydroxide consumption is mainly due to the neutralization of products from the breakdown of carbohydrates. The change in the hydroxide concentration can be expressed as [16, 28]:

$$\frac{dC_{OH^-}}{dt} = \left(q_1 \frac{dC_s}{dt} + q_2 \, Lo \frac{dL}{dt} \right) \frac{1}{V}$$

(5)

In (5), q_1 and q_2 are empirical constants that are required because the real stoichiometry is unknown.

Mixing Effect

To consider mixing effects we followed a temporal superstructure, similar to that established by [29] and summarized bellow and that considers the Reynolds number for agitated tanks with suspended solids (6), the apparent viscosity for pseudoplastic fluids (7).

$$Re = \frac{N D_a^2 \rho}{\mu_a} \qquad (6)$$

$$\mu_a = K(\gamma)^{n-1} \qquad (7)$$

where the shear rate (γ) is calculated as:

$$\gamma = kN \qquad (8)$$

for laminar regime (Re <10), according to Metzner and Otto equation [30], and

$$\gamma = \left(\frac{4 \rho N_p D_a^2}{\pi H_L D_T^2 K} \right)^{1/(n+1)} N^{3/(n+1)} \qquad (9)$$

for pseudoplastic fluids in turbulent flow [31].

Time Intervals

[29] calculated the length of time intervals or mixing time (t_m) as four times the circulation time, which is determined with the pumping capacity for one-phase systems. However, with solid materials in suspension, it is possible to use the follow equation [32, 33]:

Table 1. Summary of experimental conditions

	No.	T (°C)	C (kg $Ca(OH)_2$/kg biomass)	Average particle diameter (mm)	Stirring rate N* (rpm)
Group 1 R= 50L/kg	F1	40	0.3	0.455	100
	F2	40	0.1	0.455	100
	F3	60	0.3	0.455	100
	F4	70	0.3	0.455	100
	F5	40	0.5	0.455	100
	F6	40	0.3	0.220	100
	F7	40	0.3	1.065	100
	F8	40	0.3	0.455	50
	F9	40	0.3	0.455	200
Group 2 R= 12L/kg	G1	40	0.3	0.455	100
	G2	40	0.1	0.455	100
	G3	60	0.3	0.455	100
	G4	70	0.3	0.455	100
	G5	40	0.5	0.455	100
	G6	40	0.3	0.220	100
	G7	40	0.3	1.065	100
	G8	40	0.3	0.455	50
	G9	40	0.3	0.455	200

$$t_m = \frac{3{,}5}{k_m}\left(\frac{R_s}{5}\right)^{\beta} \phi^{\alpha} \quad (10)$$

where km is giving by (11).

$$\frac{k_m}{N} = d\left(\frac{D_a}{D_T}\right)^{b}\left(\frac{D_T}{H_L}\right)^{0.5} \quad (11)$$

Equation (10) is valid for 97% uniformity, solid to liquid ratio of $5 kg/m^3$ and particle size larger than 0.1 mm. The form factor was estimated by equation (12) which is valid for diameters from 0.1mm to 2.8mm [34].

$$\phi = 0{,}124 d_p^{1{,}333} \quad (12)$$

Temperature program

In the experiments, the temperature was varied according to the temperature program in (13) and (14) that represents a temperature ramp followed by a constant temperature period:

$$T(t) = \begin{cases} To + et & t \le \dfrac{T_p - To}{e} \quad (13) \\ Tp & t \ge \dfrac{T_p - To}{e} \quad (14) \end{cases}$$

PARAMETRIC ESTIMATION

The parametric determination of the kinetic parameters in the model was carried out with Matlab®. The experimental data of lignin kinetic were adjusted initially using the Matlab® **nlinfit** function, designed for nonlinear models, in order to estimate the kinetic reaction constant (k_i) and the initial values of the degradable lignin fractions for the main stage (Lo_1) since the bulk (main) and residual delignification stages were observed. A second step involved linearization of the lignin kinetic equation (3) and the Matlab® **regress** function was used. For the parametric estimation of the hemicellulose kinetics, hydroxide consumption and mixing model parameters, the Matlab® **sim_anneal_MR** function was used for nonlinear systems with multiple responses. The confidential intervals (CI) were calculated by each Matlab® function.

[6] reported activation energies of 50.15kJ/mol and 54.21kJ/mol for the main and residual stages in the delignification of corn straw with calcium hydroxide under oxidative conditions. Similarly, [35] and [36] reported that Ea varies between 120kJ/mol and 130kJ/mol for the main stage and between 110 kJ/mol and 119kJ/mol for residual stage when the raw material was wood. The activation energy values found in this study, presented in Table 2, are lower than those in [35] and [36]. Given the highly empirical character of the model presented in this study, one needs to be careful when giving any physical interpretation to the kinetic values obtained after regression. Nevertheless, the low values of E_1 and E_2 could imply that sugarcane bagasse is more easily removed than corn straw or wood. References [5] and [22] reported values of activation energy for the alkaline delignification of sugarcane bagasse that varied between 4.56kJ/mol and 42.04kJ/mol depending on pretreatment, that agree with those in Table 2.

The (a_i) exponential factor values for alkaline pretreatment with sugarcane bagasse has not been reported in refereed literature, however references [37] and [38] reported values similar to those in Table 2 for the Kraft delignification of *Eucalyptus globulus*; poplar white birch, maple sugar and pine fiber.

The effective coefficients values (q_i) show the low hydroxide consumption. As expected, a significant amount of hydroxide is consumed by the hemicelluloses, possibly in the hydrolysis of acetyl groups, acid neutralization and polymer degradation [5, 28].

Degradable Lignin Fraction (Lo_i)

The degradable lignin fraction of the main stage (Lo_1) was adjusted to equation (15) using the Matlab® **regress** function. The fraction of residual stage (Lo_2) was calculated as 1 - Lo_1. The experiments showed that Lo_1 increases with temperature and hydroxide concentration and decreases when the average particle diameter (δ) increases.

$$La_1 = 0.0849 + 0.00239\, x T^{\circ} + 0.0406\, x \ln\left(Ca(OH)_2\right) - 0.109 \ln \delta \qquad (15)$$

Table 2. Parameter values obtained after regression of the experimental data.

Parameter	**Estimate**	**CI**
E_1 (kJ/mol)	15.11	±0.084
E_2 (kJ/mol)	25.81	±1.309
A_1 (mm^{wi} / min (mol $L^{-1})^{ai}$)	0.728	±0.258
A_2 (mm^{wi} / min (mol $L^{-1})^{ai}$)	0.646	±0.235
a_1	0.372	±0.061
a_2	0.539	±0.006
w_1	0.178	±0.021
w_2	0.330	±0.029
q_1 (mol hydroxide/kg hemicellulose)	4.500	±0.002
q_2 (mol hydroxide /kg Lignin)	3.005	±0.001

Effective Coefficient (g_1)

The effective coefficient (g_1) was adjusted to equation (16). This expression shows that by increasing temperature, hydroxide concentration and material size there is an increase in the hemicellulose dissolution. This indicates that hemicellulose has a sensitive structure during the alkaline pretreatment and agrees with previous observations [5, 8, 11, 39].

$$g_1 = 5.745 x T^{0.00023}\, Ca(OH)_2^{\,0.074}\, x\, \delta^{0.839} \qquad (16)$$

Mixing Model Parameters

Mixing was approximated by three interconnected CSTRs that considered three mixing regions: impeller section (V_a), main section (V_b) and baffle section (V_c). The volumes for each section were defined according to [29], leaving free a parameter (*Flow fraction*, λ) that divides the flow into a temporary structure (see Figure 1). The estimated value of (λ) was 0,730±0,015m^3min^{-1}/m^3min^{-1}.

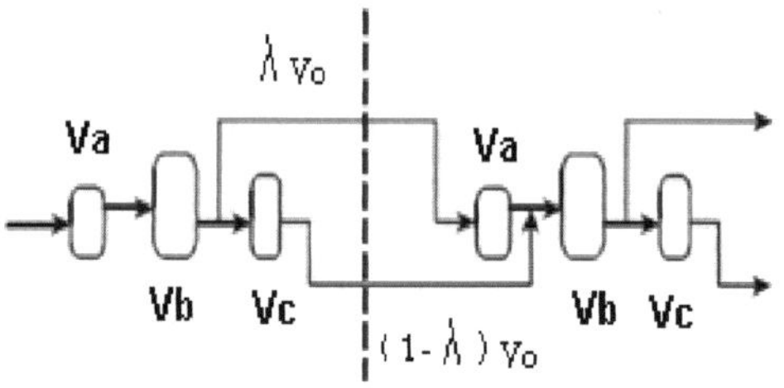

Figure 1. Connection of mixing compartments. Adapted from [29].

Table 3 describes the different parameters used to simulate mixing.

Table 3. Input data for simulation of mixing in the reactor.

Parameter	Value	Reference
ρ (kg/m^3)	1005	This work
Np	2.5	[40]
n	0.88	[41]
K (kg s$^{(n-2)}$ /m)	0.009	[41]
β	1.400	This work
α	0.010	This work
b	1.150	[32]
d	0.102	[32]
e (°K/min)	1.0782	This work

Table 4. Time intervals, temporal superstructure

N* (rpm)	δ (mm)	t^*_m (min)
50L/kg		
50	0.455	4.683
	0.220	2.318
100	0.455	2.342
	1.065	2.367
200	0.455	1.171
12L/kg		
50	0.455	34.48
	0.220	17.09
100	0.455	17.24
	1.065	17.39
200	0.455	8.621

RESULTS

The values of each time constant for the temporary structure in the mixing model at different rates of stirring, liquid-solid relations and average chips diameter after 2000 minutes of pretreatment are shown in Table 4. The time interval was higher when the solids amount was higher (12L/kg liquid to solid ratio), and increased by increasing the average particle diameter (δ) but decreased by increasing the stirring rate.

Table 5 presents estimated values of apparent viscosity and flow regime for the stirring levels evaluated in this work. The results are similar to those reported for [42] and [43], but slightly lower probably because of carbohydrates losses, a different particle size used in this study and other the undetermined stirring effects.

Table 5. Estimates values of apparent viscosity and flow regime

N* (rpm)	μ_a* (mPa.s)	Re
50	5.219	4637
100	4.570	10591
200	4.002	24189

The statistical characterization of a comparison between predicted and experimental values showed that the residuals varied between -0.06 and 0.04, following a Gaussian distribution with a small variance. The residual calculation was made as the difference between hemicellulose, lignin and hydroxide consumption fractions for each prediction and the experimental data. The error was calculated as the sum of squared residual.

MODEL VALIDATION

In order to validate the model predictions a test with 0.3kg of sugarcane bagasse for 48 hours independent from experimental design used in the parametric estimation was carried out. The conditions were: T^*=57ºC, 0.24kg $Ca(OH)_2$/kg biomass, R=15L/kg biomass, δ=0.77mm, and N^*=150rpm.

Figure 2 shows good adjustment between prediction and experiments when the following values were calculated t^*_m = 8.475min, # intervals = 413, Re = 17.17, μ_a^* = 4.229mPa.s.

a.

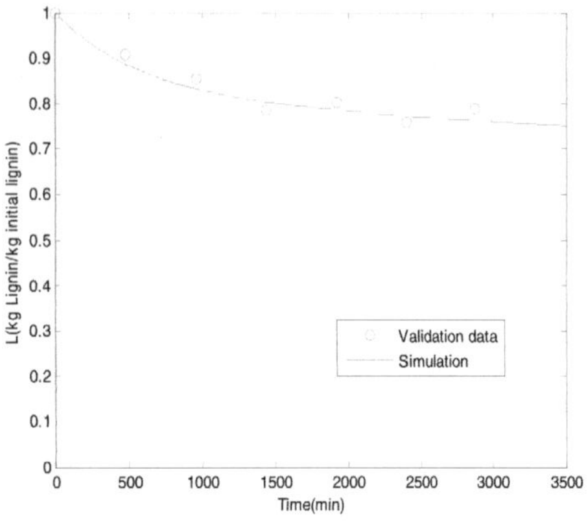

b.

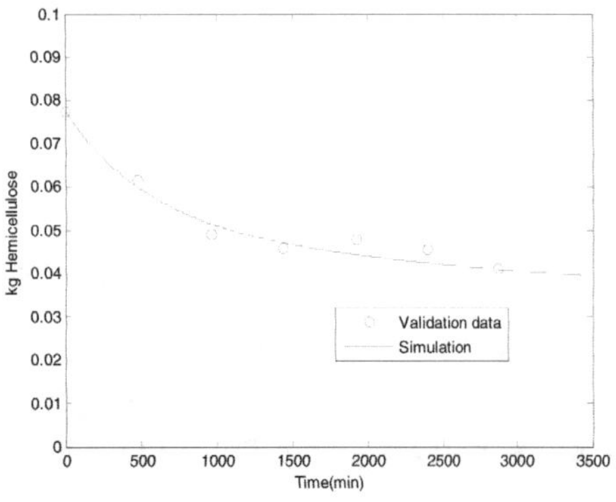

c.

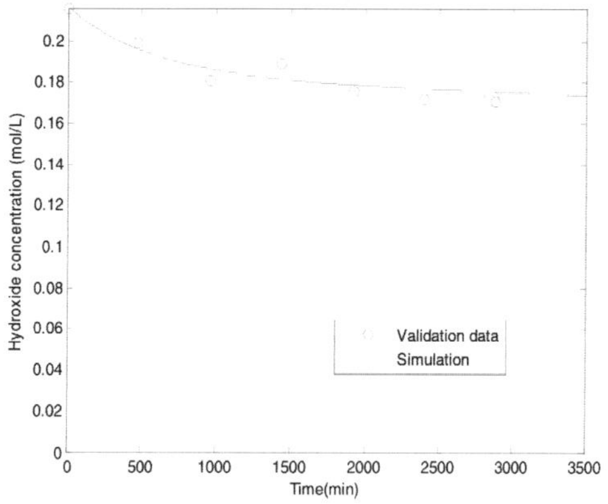

Figure 2. Comparison of model predictions (lines) with experimental data (points) for the variation of the concentration of a. Lignin, b. Hemicellulose and c. hydroxide with time.

Predictions with the model were compared as well with data from [5] with relative good success. A detailed comparison is available in [44].

CONCLUSIONS

The alkaline deslignificación process may be adequately modeled considering the effect of temperature, particle mean diameter, reagent concentration, liquid to solid ratio and stirring effects using the model proposed in this paper.

The sugarcane bagasse delignification happens faster as temperature and hydroxide concentration increase, while the material size and liquid to solid ratio decreases, in an environment with adequate mixing.

The mixing model composed of batch reactors formed with a temporary superstructure of mixing compartments correctly describes the flow within the reactor and predict the lignin, hemicellulose and hydroxide kinetics during alkaline pretreatment.

NOMENCLATURE

μ_a	Apparent viscosity. [Pa. s]
μ_a*	Apparent viscosity. [mPa. s]
γ	Shear rate. [s^{-1}]
K	Consistence index. [kg $s^{(n-2)}$ /m]
n	Flow behavior index
k	Constant that depend on the system geometry
N	Stirring rate. [rps]
N*	Stirring rate. [rpm]
Re	Reynolds number
Np	Power number
D_T	Reactor diameter. [m]
V_T	Reactor tank volume. [m]
L_T	Height reactor tank. [m]
W_B	Width baffles. [m]
W_a	Width blades. [m]
L_B	Height baffles. [m]
L_A	Impeller height respect the reactor bottom. [m]
H_L	Height liquid. [m]
ρ	Suspension density. [kg/m^3]
D_a	Impeller diameter. [m]
M	Total mass. [kg]
M_s	Biomass mass. [kg]
T	Absolute temperature. [°K]
T*	Temperature. [°C]
To	Initial temperature. [°K]
Tp	Isothermal operating temperature. [°K]
e	Heating rate [°K/min]
L_i	Lignin fraction. [kg Lignin/kg initial lignin]
Lo	Initial mass of lignin. [kg]
Ls	kg remaining lignin at the time t/kg dry matter.
Lso	kg initial lignin/kg dry matter.
Loi	Degradable lignin fraction, stage i. [kg Lignin/kg initial lignin]
Y	Dry matter fraction. [kg dry matter/kg biomass mass]
g_1	Effective coefficient. [kg Hemicellulose/kg Lignin]
q_1	Effective coefficient. [mol hydroxide/kg hemicellulose]
q_2	Effective coefficient. [mol hydroxide /kg Lignin]
V	Volume of liquor. [L]
R_s	Solid-liquid relation. [kg/m^3]
R_c	Universal gas constant. [J/mol °K]
θ	Form factor
β	Correction factor associated with the solid concentration
α	Correction factor associated with the form factor
d	Constant equation
b	Constant equation
d_p	Average particle diameter. [mm]
C_s	Hemicellulose matter. [kg remanent hemicellulose]
C_{OH^-}	Hydroxide concentration. [mol/L]
t	Time. [min]
t_m	Mixing time. [s]
t_m*	Mixing time. [min]
A_i	Preexponential factor. [mm^{wi} / min (mol $L^{-1})^{ai}$]
δ	Chips average diameter . [mm]
Ei	Activation energy, stage i. [J/mol]
Ei*	Activation energy, stage i [kJ/mol]
a_i	Associate constant, hydroxide concentration in the stage i
w_i	Associate constant, particle size in the stage i
ki	Kinetic constant, stage i. [min^{-1}]
v_o	Volumetric flow. [m^3/min]
λ	Flow fraction. [m^3 min^{-1}/m^3 min^{-1}]
V_a	Volume of the impeller section. [m^3]
V_b	Volume of the main section. [m^3]
V_c	Volume of the baffles section. [m^3]

REFERENCES

[1] Chandel AK, Es C, Rudravaram R, Narasu ML, Rao LV, Ravindra P. Economics and environmental impact of bioethanol production technologies: an appraisal. **Biotechnol Molec Biol Rev** 2007; 2:14–32.

[2]McMillan J.D. Pretreatment of lignocellulosic biomass. Enzymatic Conversion of Biomass for Fuels Production, ACS Symposium Series, vol. 566. ACS, Washington, 1994, pp. 292–324.

[3] McMillan JD. Bioethanol production: status and prospects. **Renew Energy** 1997; 10:295–302.

[4] Mosier N, Wyman C, Dale B, Elander R, Lee Y, Holtzapple M, Ladisch M. Features of promising technologies for pretreatment of lignocellulosic biomass. **Bioresource Technology** 2005; 96: 673–686.

[5] Granda C. Sugarcane juice extraction and preservation, and long-term lime pretreatment of bagasse. Texas A&M University. Tesis doctoral. Ingenieria Química. 2004, pp 369.

[6] Kim S, Holtzapple M. Delignification kinetics of corn stover in lime pretreatment. **Bioresource Technology** 2006; 97:778-785.

[7]Mocchiutti P. Fundamentos de la producción de pastas celulósicas - Tema II – 1: Pulpado Químico - Pulpado Alcalino. 2007.

[8] Sierra R, Granda C, Holtzapple M.T. Lime Pretreatment. Jonathan R. Mielenz (ed.), Biofuels: Methods and Protocols, **Methods in Molecular Biology**, vol. 581. 2009. 115-124.

[9] Gustafson R, Sleicher C, Mckean W. Theoretical Model of the Kraft Pulping Process, Ind. Eng. Chem. **Process Design and Development** 1983; 22(1): 87-96.

[10] Andersson N, Wilson D, Germard U. An improved kinetic model structure for softwood kraft cooking. **Nordic Pulp and Paper Research Journal**, 2003; 18 (2): 200-209.

[11] Johansson D. Carbohydrate degradation and dissolution during Kraft cooking. Licentiate thesis. Karlstad University Studies 2008:12. 62p.

[12] Gustavsson G. On the interrelation between kraft cooking conditions and pulp composition. Royal Institute of Technology, Department of Fibre and Polymer Technology Division of Wood Chemistry and Pulp Technology, Stockholm. Thesis Doctoral. 2006

[13] Bogrena J, Brelida H, Theliandera H. Evaluating reaction kinetic models using well-defined kraft delignification data. **Technical Articles** 2008: 1-11.

[14] Castro J, Baptista C, Carvalho M, Egas A, Simao J. Heterogeneous studies in pulping of wood: Modelling mass transfer of alkali. **Chemical Engineering Journal** 2008; 139:615-621.

[15] Dang V, Nguyen K. Characterisation of the heterogeneous alkaline pulping kinetics of hemp woody core. **Bioresource Technology** 2006; 97: 1353-1359.

[16] Giudici R, Won S. Kinetic model for kraft pulping of hardwood. Industrial and Engineering Chemistry Research. **Ind. Eng. Chem. Res** 1996; 35: 856-863.

[17] Yang L, Liu S. Kinetic model for Kraft pulping process. **Eng. Chem. Res** 2005; *44:*7078–7085.

[18] Chang, V.S., Nagwani, M., Holtzapple, M.T., Lime pretreatment of crop residues bagasse and wheat straw. **Applied Biochemistry and Biotechnology**. 1998, 74, 135–159.

[19] Dang V, Nguyen K.. A Universal kinetic model for characterisation of the effect of chip thickness on Kraft pulping. Short Communication. **Bioresource Technology** 2008; 99: 1486-1490.

[20]Dawson L, Boophaty R. Cellulosic ethanol production from sugarcane bagasse without enzymatic saccharification. **BioResources** 3(2), 2008, pp 452-460.

[21] Hernandez, J. Comparative hydrolysis and fermentation of sugarcane and agave bagasse. Centro de Investigación en Biotecnologia Aplicada del Instituto Politécnico Nacional (CIBA-IPN), México. 2009.

[22] Sabatier J, Peniche C, Fernández N. Soda pulping of bagasse: Delignification phases and kinetics. **Holzforschung**. Vol 47. 1993, 313-317.

[23] Mora I. Nutrición animal. Primera Edición. Ed Universidad Estatal a Distancia San José. Costa Rica. 1991. P.120

[24] Chang, V.S., Holtzapple, M.T. Lime pretreatment of switchgrass. **Applied Biochemistry and Biotechnology**. 63-65; 1997. 3-19.

[25] Karr, W. E.; Holtzapple, T. Using lime pretreatment to facilitate the enzymatic hydrolysis of corn stover. **Biomass Bioenergy**, 2000, 18, 189–199.

[26] Playne, M.J. Increased digestibility of bagasse by pretreatment with alkalis and steam explosion. **Biotechnology and Bioengineering**, 1984, 26 (5), 426–433.

[27] Dang V, Nguyen K. A universal kinetic equation for characterising the fractal nature of delignification of lignocellulosic materials. **Cellulose** 2007; 14:153-160.

[28] Santos A, Rodríguez F, Gilarranz M, Moreno D, García F. Kinetic modeling of kraft delignification of eucalyptus globules. **Ind. Eng. Chem. Res.,** 1997, *36* (10), pp 4114–4125.

[29]Zhang, J., Smith, R. Design and optimization of batch and semi-batch reactors. **Chem. Eng. Sci**. 2004; 59: 459 –478.

[30] Bhole M, Ford C, Bennington CPJ. Characterization of axial flow impellers in pulp fibre suspensions. 13th European Conference on Mixing London, 14-17 April 2009. Pages 1-8.

[31]Sánchez J, Rodríguez E, Casas J, Fernández J, Chisti Y. Shear rate in stirred tank and bubble column bioreactors. **Chemical Engineering Journal** 124, 2006, p 1–5

[32] Kuznamié N, Ljubieié B. Suspension of floating solids with up-pumping pitched blade impellers; mixing time and power characteristics. **Chemical Engineering Journal** 84. 2001. 325–333.

[33]Kuznamié N, Zanetié R, Akrap M. Impact of floating suspended solids on the homogenisation of the liquid phase in dual-impeller agitated vessel. **Chemical Engineering and Processing** 47. 2008. 663–669.

[34] Roca G, Sanchez C, Gómez E; Barbosa L. Caracterización del bagazo de la caña de azúcar. Parte I: características físicas. An. 6. Enc. Energ. Meio Rural 2006.

[35] Chiang VL, Yu J. Isothermal reaction kinetics of kraft delignification of Douglasfir. **Journal of Wood Chemistry and Technology** 10(3): 1990, 293-310.

[36] Dolk M, Yan J, McCarthy. Kinetics of delignification of western hemlock in flow through reactors under alkaline conditions. **Holzforschung**, 1989, 43(2): 91-98

[37] Pérez J, Gilarranz M, Rodríguez F, Oliet M, García J. Estudio sobre la cinética de la deslignificación en la fase residual de las cocciones kraft. Congreso Iberoamericano de investigación en celulosa y papel. 2000. 1-9p.

[38] Sixta H, Rutkowska E. Modeling of eucalyptus globulus kraft pulping. Lenzing AG Process Innovation 2006; 1-7.

[39] Taherzadeh M J, Karimi K. Pretreatment of Lignocellulosic Wastes to Improve Ethanol and Biogas Production: A Review. **Int. J. Mol. Sci**. 2008, 9, 1621-1651.

[40]Martín A. Agitation and mixing. Suspension of solid particles. Chapter 9. 2003. 7p

[41] Karim K, Thoma G, Al-Dahhan. Gas-lift digester configuration effects on mixing effectiveness. Water research 41. 2007, 3051 – 3060.

[42] Goncalves A, Costa S, Esposito E. Panus trigrinus strains used in delignification of sugarcane bagasse prior to kraft pulping. **Applied Biochesmistry and Biotechnology**. Vol 98-100, 2002. 373-382.

[43] Ruzene DS, Goncalves A, Teixeira J, Pessoa M. Carboxymethylcellulose obtain by etanol/water organosolv process under acid conditions. **Applied biochemistry and Biothecnology** vol. 136-140. 2007.

[44] Fuertez J. Modelo dinámico integrado de base fenomenológica del proceso de deslignificación por precipitación alcalina de material lignocelulósico. Tesis de Maestría, Universidad Nacional de Colombia Sede Medellín. 2010. P200.

(461f) Biofuel Production: Performance of New Generation of Cellulase Enzymes in Combination with Esterases On Ionic Liquid Pretreated Lignocellulosic Biomass

Thehazhnan (Thihal) K. Ponnaiyan and Sasidhar Varanasi, Chemical & Environmental Engineering, University of Toledo, Toledo, OH

Successful commercialization of lignocellulosic biofuels mandates the cost-effective release of fermentable sugars from the biomass substrate. A process based on enzymatic hydrolysis and fermentation is a promising route in converting the biomass to ethanol with high yields and low production cost. However, the components of biomass (cellulose, hemicellulose, and lignin) are intricately bound together, thereby requiring suitable pretreatment of the substrate prior to hydrolysis. Pretreatment and enzymatic hydrolysis are critical steps, and most of the current pretreatment methods do not effectively disrupt the crystallinity of cellulose, which is a major barrier for efficient enzymatic hydrolysis. Ionic liquids (ILs) being non volatile, non-flammable, recyclable and designer friendly, are gaining wide recognition as green solvents, and their unique solvating properties make them ideal for pretreating lignocellulosic substrates like corn stover and poplar. Strategic addition of enzyme mixtures can produce enhanced enzymatic hydrolysis yields of glucan to glucose and xylan to xylose. In our earlier studies, we reported the use of complex cellulase enzyme mixtures in the hydrolysis step. Covalent lignin-carbohydrate linkages exist in lignocellulose from wood and herbaceous plants. The addition of esterases can help in breaking the ester linkages that bind lignin to carbohydrates. In this study, we address the following: (1) the pretreatment of lignocellulosic biomass using different ionic liquids, and (2) a comparative assessment of the efficiency of old generation and new generation of cellulase enzymes, and the effect of addition of esterases on hydrolysis yields.

See more of this Session: Developments in the Pretreatment of Lignocellulosics for Bioconversion II
See more of this Group/Topical: Sustainable Engineering Forum

(474a) Engineering a Fungal Conversion System for the Novel Production of Alkanes

John Blazeck, Department of Chemical Engineering, University of Texas at Austin, Austin, TX and Hal Alper, Chemical Engineering, The University of Texas at Austin, Austin, TX

Bio-derived alkanes are attractive second generation biofuels that circumvent some of the disadvantages associated with traditional biofuels such as ethanol. Engineering alkane production in yeasts relies upon the metabolic engineering of fatty acid metabolism. An attractive host for this task is Yarrowia lipolytica, an oleaginous yeast capable of growing on lignocellulosic materials and adept at producing high levels of fatty acids. Here, we report the novel heterologous production of pentane in a fungal system by engineering the yeast Yarrowia lipolytica. To accomplish this, a heterologous pentane-production pathway was synthetically assembled using plant-derived enzymes. Pentane production was optimized using environmental and genetic techniques including increasing precursor availability and employing promoter engineering and codon optimization for high enzyme expression levels. Here, we will present the results of this pathway optimization and discuss strategies for improved pentane (and other alkane) production in this strain of yeast.

See more of this Session: Microbial Engineering of Microorganisms for Biofuels and Chemicals II
See more of this Group/Topical: Sustainable Engineering Forum

(474b) Engineering Production of Fatty Acid Derived Products in Escherichia Coli through Functional Genomics and High Throughput Screening

Brian F. Pfleger, Chemical and Biological Engineering, University of Wisconsin Madison, Madison, WI

Biological hydrocarbons offer many advantages over cellulosic ethanol including higher energy density, lower water solubility, and compatibility with current engines. Biodiesel, the methyl or ethyl esters of fatty acids, is currently the second leading biofuel to ethanol. Similar to corn-based ethanol, the production of biodiesel is limited by the availability of oil crops. Metabolic engineering offers the tools to link cellulosic technologies with the production of more desirable fuels. One potential platform for the synthesis of diesel-like hydrocarbons, is the production of lipids in bacteria such as *Escherichia coli* for subsequent reduction to hydrocarbon fuels. We have successfully engineered E. coli to overproduce C12 fatty acids by overexpressing a plant thioesterase in a β-oxidation mutant, Lennen et al., *Biotechnology and Bioengineering*, 2010. In these initial studies, we identified the thioesterase catalyzed cleavage of acyl chains from the corresponding aycl-ACP as the key step in the biosynthesis of free fatty acids. If the gene encoding the thioesterase is overexpressed beyond an optimal level, we observe a dramatic decrease in cellular fitness and fatty acid titer. Here we present recent progress using functional genomics and high throughput screening to elucidate the molecular link between thioesterase overexpression and reduced fitness.

See more of this Session: Microbial Engineering of Microorganisms for Biofuels and Chemicals II
See more of this Group/Topical: Sustainable Engineering Forum

(474c) Metabolic Engineering of E.Coli for Increased Production of Novel Fatty Acids

Paul Handke, Sean A. Lynch, Joseph R. Warner and Ryan T. Gill, Chemical and Biological Engineering, University of Colorado at Boulder, Boulder, CO

Renewable transportation fuels produced by microorganisms must meet various criteria to be considered viable alternatives to petroleum-based fuels. Chemical properties, infrastructure compatibility and sensible production routes are essential considerations. Microbial fatty acid derivatives are promising biofuel candidates as they have numerous desirable properties and are readily produced by Escherichia coli. The goal of this work is to produce specific monounsaturated free fatty acids for subsequent catalytic processing into alkane and alcohol fuels.

Unsaturated fatty acid biosynthesis depends on the function of the fabA gene, responsible for introducing a double bond to the growing fatty acyl chain. Deletion of this essential gene and simultaneous expression of a plant desaturase, capable of introducing specific double bonds to saturated acyl chains, created a selection system for novel unsaturated fatty acids. The production of these novel fatty acids was further improved through the over-expression of the acetyl-CoA carboxylase complex which catalyzes the first dedicated step in fatty acid biosynthesis, and expression of E.coli native and non-native thioesterases which alleviate feedback inhibition and further increase total fatty acid production.

These rational engineering approaches exhaust current knowledge of improving fatty acid production in E.coli. For this reason a novel genome-wide tool developed in the Gill Group was employed to identify further targets that can improve fatty acid production. The TRackable Multiplex Recombineering (TReMR) method uses strong promoters and ribosomal binding sites to increase gene expression or deletes existing ribosomal binding sites to decrease expression. These "up" and "down" chromosomal insertions were introduced in E.coli ΔfabA and growth selections were performed. Mutants were identified using microarray analysis and evaluated based on their contribution to improved growth in E. coli ΔfabA.

See more of this Session: Microbial Engineering of Microorganisms for Biofuels and Chemicals II
See more of this Group/Topical: Sustainable Engineering Forum

(474d) Engineering of Oleaginous Escherichia Coli for Hydrocarbon Production

Fengming Lin[1], Yu Chen[2], Ying-jin Yuan[1] and Xiaoxia (Nina) Lin[2], (1)Department of Pharmaceutical Engineering, School of Chemical Engineering and Technology, Tianjin University, Tianjin, China, (2) Department of Chemical Engineering, University of Michigan, Ann Arbor, MI

Fatty acid derived hydrocarbons, such as long-chain alkane, alkene and biodiesel, are promising biofuels. Previous works have demonstrated the feasibility of their production in E. coli. However, the yield and efficiency remains low, which prevents economical scale-up to industrial processes. The goal of this project is to design and construct efficient E. coli strains to produce and accumulate fatty acids in the form of wax ester and triacylglycerol, similar to oleaginous species. This approach can potentially enable very high level of hydrocarbon production as it overcomes the toxicity issue associated with long-chain hydrocarbons by forming carbon-rich self-organized lipid bodies.

Our genetic modifications of E. coli focus on both the fatty acid biosynthesis/conversion pathways and the central carbon metabolic network that provides key precursors. We employed metabolic and regulatory network modeling to design optimal strains for fatty acid production and specific reactions involved in the central carbon metabolism have been suggested for knockout or reactivation. We also modified the fatty acid pathway substantially and our designed manipulations include knockout of acyl-CoA dehydrogenase (fadE) to block β-oxidation, cytosolic expression of an E. coli native tioestesrase (TE), and overexpression of four subuits of acetyl-CoA carboxylase (ACC). In order to convert the over-produced fatty acids to wax ester and triacylglycerol, heterologous expression of Arabidopsis fatty acyl-CoA reductase and bacterium wax ester synthase/acyl-CoA-diacylglycerol acyltransferase (WS/DGAT) will also be incorporated. The resulted E. coli strains are expected to produce and accumulate abundant fatty acids, which can be further converted to hydrocarbon either biologically or chemically.

See more of this Session: Microbial Engineering of Microorganisms for Biofuels and Chemicals II
See more of this Group/Topical: Sustainable Engineering Forum

(474e) Increasing Product Tolerance through Metabolic Engineering: Short-Chain Fatty Acids

Liam A. Royce[1], Matthew Stebbins[1], Maria Rodriguez-Moya[2], Erin Boggess[3], Julie Dickerson[3], Ramon Gonzalez[2] and Laura Jarboe[1], (1)Chemical and Biological Engineering, Iowa State University, Ames, IA, (2)Departments of Chemical & Biomolecular Engineering and Bioengineering, Rice University, Houston, TX, (3)Electrical and Computer Engineering, Iowa State University, Ames, IA

As we aim to increase the number of fuels and chemicals produced by biocatalysts in an economically competitive process, biocatalyst inhibition by these products becomes an ever-increasing problem. Here we discuss our strategy to increase the tolerance of *Escherichia coli* to short-chain fatty acids. Our approach is two-pronged: analysis of the inhibition response were used to understand the mechanism of fatty-acid mediated growth inhibition in the wild-type strain, while directed evolution and reverse engineering of a fatty-acid tolerant mutant identifies additional mechanisms of fatty acid tolerance. Analysis methods include transcriptome analysis, lipid profile characterization, membrane fluidity, and intracellular pH measurements.

See more of this Session: Microbial Engineering of Microorganisms for Biofuels and Chemicals II
See more of this Group/Topical: Sustainable Engineering Forum

(474f) Microbial Production of High Value-Added Chiral Hydroxyacids

Hsien-Chung Tseng, Catey L. Harwell and Kristala L. J. Prather, Chemical Engineering, Massachusetts Institute of Technology, Cambridge, MA

Hydroxyacids have the potential to serve as useful chiral building block molecules for a diverse range of products, including polyhydroxyalkanoates-based biodegradable plastics and optically-active fine chemicals. Through metabolic engineering efforts, we have constructed unique pathways for the biosynthesis of both enantiomers of 3-hydroxybutyrate (3HB) in E. coli, where the chirality was controlled by utilizing two enzymes of opposite stereoselectivity. We have further profiled several enzyme homologs to promote production, and analyzed 3HB chiralities, in vitro enzyme activities, and intracellular cofactor levels to fully characterize the 3HB pathway. This work led to strains of E. coli that produce enantiopure (R)-3HB and (S)-3HB with titers of both in excess of 2.5 g/L in shake flask cultures within two days. Furthermore, the successfully constructed chiral 3HB pathway provides a feasible production platform for synthesis of other versatile short-chain hydroxyacids. Recent progress has been made on exploiting biosynthetic potential of the 3HB pathway on synthesis of chiral 3-hydroxyvalerate (3HV) with titers of both up to 2.0 g/L as well as 3-hydroxyhexanoate (3HH) and their isomers by either supplying a second carbon source or harnessing the power of natural metabolic pathways. In addition, multiple genetic alternations are being investigated to boost the ratios of higher hydroxyacids to 3HB. This presentation will summarize our engineered microbial system for production of the above-mentioned chiral hydroxyacids.

See more of this Session: Microbial Engineering of Microorganisms for Biofuels and Chemicals II
See more of this Group/Topical: Sustainable Engineering Forum

(532a) Comparisons of SPORL and Dilute Acid Pretreatments for Sugar and Ethanol Productions From Aspen

J.Y. Zhu, USDA Forest Service, Forest Products Laboratory, Madison, WI, Shen Tian, Capital Normal University, Beijing, China, Wenyuan Zhu, South China University of Technology, Guangzhou, China, Roland Gleisner, Forest Products Laboratory, US Forest Service, Madison, WI and Xuejun Pan, Biological Systems Engineering, University of Wisconisn, Madison, Madison, WI

This study reports comparative evaluations of sugar and ethanol production between sulfite (SPORL) and dilute acid (DA) pretreatments from a native aspen. All aqueous pretreatments were carried out in a laboratory wood pulping digester using wood chips at 170oC with a liquid to oven dry (od) wood ratio (L/W) of 3 at two levels of acid charge on wood of 0.56 and 1.11%. Pretreatment duration ranged from 0 to 30 min with an increment of 10 min. The only difference between corresponding SPORL and DA pretreatment was sodium bisulfite charge with 0 for DA and 1.5 or 3.0% on od wood for SPORL. Both SPORL and DA pretreatments were effective in removing hemicelluloses. SPORL removed about the double amount of lignin (20-40%) than DA did (10-20%). All substrates produced by both pretreatments (except DA with pretreatment duration of 0) had good enzymatic digestibility of over 80%. However, SPORL pretreatment produced higher enzymatic digestibility than its corresponding DA pretreatment for all the experiments conducted. As a result, SPORL produced higher glucose yield from enzymatic hydrolysis, EHGY, total sugar yield, and ethanol yield from simultaneous saccharification and fermentation (SSF) of cellulosic substrate, than its corresponding DA pretreatment did. The enzyme loading in SSF were 10 FPU cellulase/cellullose and 15 CBU beta-glucosidase/g cellulose for SSF runs. The maximum EHGY, total sugar, and ethanol yields were 442 kg/ton, 590 kg/ton, and 238 L/ton, respectively, for SPORL, vs 408 kg/ton, 543 kg/ton, and 214 L/ton, respectively, for the corresponding DA pretreatment. Both SPORL and DA pretreatment significantly reduced energy consumption for wood chip size-reduction, however SPORL was more effective than its corresponding DA pretreatment. SPORL, with the lower energy input and higher sugar and ethanol yield, produced higher sugar and ethanol production energy efficiencies than its corresponding DA pretreatment. The maximal sugar and ethanol production energy efficiency was 395 kg/GJ and 273%, respectively, for SPORL, vs 339 kg/GJ and 213% for DA. Although SPORL produced more acetic acid than the corresponding DA pretreatment, the effectiveness of SPORL pretreatment affords to use a mild pretreatment condition (such as short pretreatment duration) than DA to achieve similar sugar yield. This can help to reduce the production of other fermentation inhibitors, such as furfural and HMF, favorable to the fermentation of the pretreatment hydrolysate that mainly consists of hemicelluloses sugars.

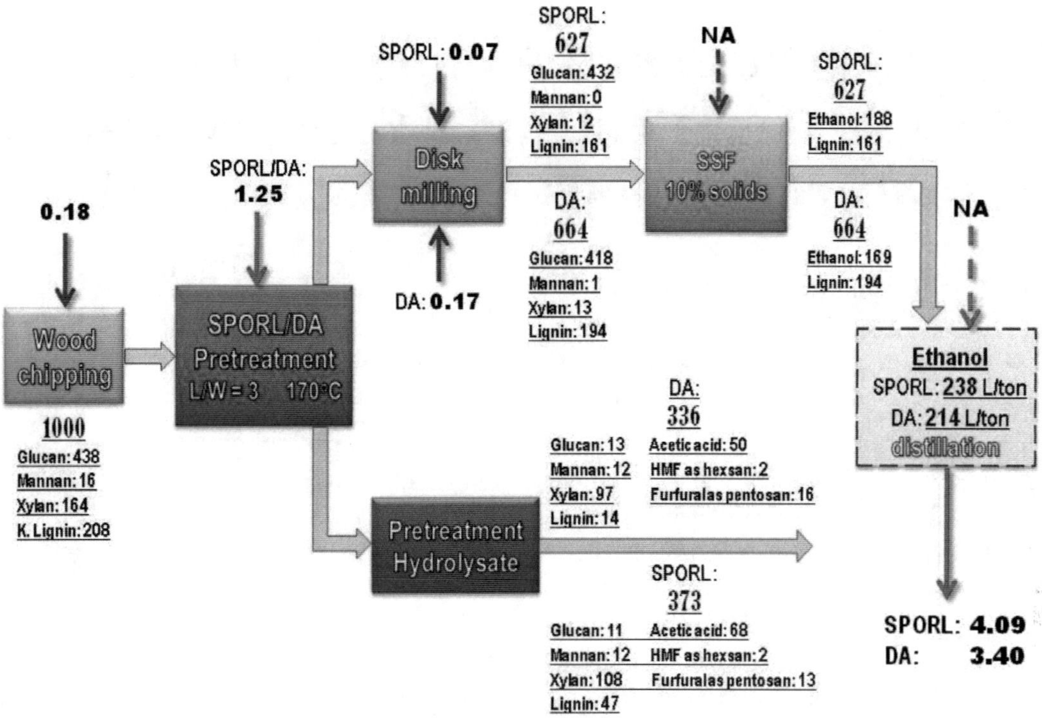

Unless indicated, energy data (bold font) in GJ/ton wood; mass data (underlined) in kg

See more of this Session: Developments in the Pretreatment of Lignocellulosics for Bioconversion III

See more of this Group/Topical: Sustainable Engineering Forum

(532b) An Ionic Liquid Pretreatment Process That Includes IL Recycling and a Characterization of Pretreatment Fractions From Corn Stover

Dean Dibble[1], Chenlin Li[2], Lan Sun[3], Peter Benke[2], Anthe George[3] and Seema Singh[2], (1)Joint BioEnergy Institute, Sandia National Laboratory, Emeryville, CA, (2)Joint BioEnergy Institute, Emeryville, CA, (3)Deconstruction Division, Joint BioEnergy Institute, Emeryville, CA

Ionic liquid pretreatment has been shown to be an effective means of delignification and decrystallization of biomass and result in enhanced enzymatic saccharification rates. An additional advantage of this technology is the ability to fractionate biomass components into separate product streams. In this talk we describe a new method of recovering and fractionating solutes, including lignin and soluble carbohydrate oligomers, from the ionic liquid [C2mim][OAc] after the pretreatment of corn stover. This new method avoids the formation of gel-phase intermediates during biomass precipitation, can result in very low residual IL content in processed biomass, and simplifies the recovery of IL and residual IL solutes without requiring the addition of difficult to recover materials such as salts or acids. We also present our extensive characterization of biomass fractions from corn stover by 2-D NMR, wet chemical analysis, enzymatic saccharification rate, and provide details, including mass flow and closure, of a closed-loop ionic liquid recovery and recycling process.

See more of this Session: Developments in the Pretreatment of Lignocellulosics for Bioconversion III
See more of this Group/Topical: Sustainable Engineering Forum

(532c) Hemicellulose-Lignin Oligomers Dissolution Fundamentals in Hot Water Flowthrough Pretreatment of Cellulosic Biomass

Bin Yang, Center for Bioproducts & Bioenergy, Washington State University Tri-Cities, Richland, WA and Lishi Yan, Center for Bioproducts and Bioenergy, Washington State University, Richland

Hot water flowthrough pretreatment removes hemicellulose as well as lignin from the biomass providing a unique opportunity to examine the dynamic deconstruction pattern of various components of biomass, including lignin, hemicellose and their complex etc. Efficient hot water flowthrough pretreatment ensures high hemicellulose and lignin removal very high yield in sequential enzymatic hydrolysis that are vital to the economic viability of biomass conversion. More importantly, it is believed that dissolvement of higher molecular weight hemicellulose-lignin oligomers could play an important role in explaining why the performance of flowthrough systems differs from conventional batch approaches. In the presentation we will show results from a number of studies using flowthrough pretreatment of Miscanthus with hot water under varying temperature (160-220°C), time (0-120min), and flowrates (1-50mL/min) to understand the influence on lignin and hemicellulose removal into the hydrolyzate on further processing of the resulting pretreated biomass material. The dynamic characteristics of hemicellulose, lignin and their complex in the hydrolyzate during pretreatment, including compositions, structural features and solubility will be discussed. Finally we will present a model using kinetic data comparing conventional kinetic behavior versus that by dissolution and solubility limits with reaction and mass transfer effects. The model was further verified using an analog of a lignin-hemicellulose complex. Our model was found to be useful for providing insights into the thermochemical pretreatment mechanism.

See more of this Session: Developments in the Pretreatment of Lignocellulosics for Bioconversion III
See more of this Group/Topical: Sustainable Engineering Forum

(532d) Pretreatment of Herbaceous Biomass by Sodium Carbonate in Batch and Percolation Reactors

Urvi D. Kothari, Chemical Engineering Department, Auburn University, Auburn University, AL and Y. Y. Lee, Chemical Engineering Department, Auburn University, Auburn, AL

Alkyl-aryl linkages in lignin are readily cleaved in alkaline conditions and the reaction is independent of [OH-] concentration, as long as pH is maintained in the alkaline region. For this reason, alkaline reagents generally show much higher selectively for reaction with lignin than carbohydrates in biomass. With this understanding, we have explored newer and cheaper reagents for alkaline pretreatment of biomass. Sodium Carbonate is among such candidates as it is 4-6 times cheaper than sodium hydroxide and also easier to recover than NaOH. In this study, sodium carbonate was used for pretreatment of Switchgrass and Corn stover in two different reactor systems. One was Soaking in Aqueous Alkali (SAAL) and the other was Sodium Carbonate Percolation (SCP). Both processes were effective in delignifying biomass. The SAAL process removed 65-70% lignin from Switchgrass while SCP process removed 70-75% lignin. With these treatments, the enzymatic glucan digestibility of 70% was obtained with Switchgrass, and 85% with Corn stover. The SCP process required higher temperature, but much less treatment time (20 min or less) than SAAL. Corn Stover was pretreated more effectively than Switchgrass by both processes, giving 10-15% higher digestibility than the latter. Optimum process conditions were determined considering the enzymatic digestibility and retention of carbohydrates. Effects of pretreatment conditions were further investigated over a broad range of temperature, time of treatment and solid/liquid ratio. TEM and SEM images were taken to examine the effects of pretreatment on both substrates. Increase in crystallinity index by XRD indicated that amorphous parts of biomass were lost during pretreatment, while there was no change in cellulose crystalline structure. FTIR reaffirmed the delignification of biomass after pretreatment.

See more of this Session: Developments in the Pretreatment of Lignocellulosics for Bioconversion III
See more of this Group/Topical: Sustainable Engineering Forum

(532e) Ionic Liquids and Biomass: Untangling Biomass Dissolution and Pretreatment Efficacy

Xinglian Geng, E. Kate Brown and Wesley A. Henderson, Department of Chemical & Biomolecular Engineering, North Carolina State University, Raleigh, NC

The most frequently used ionic liquids (ILs) for biomass pretreatment are 1-ethyl-3-methylimidazolium acetate ([C2mim][Ac]) and 1-butyl-3-methylimidazolium chloride ([C4mim][Cl]). These two ILs induce different structural and compositional changes of biomass components, however, resulting in varied enzyme digestibility of the pretreated material. To more fully understand this, four ILs – [C2mim][Ac], 1-ethyl-3-methylimidazolium chloride ([C2mim][Cl]), 1-butyl-3-methylimidazolium acetate ([C4mim][Ac]) and ([C4mim][Cl]) – have been used to explore how the ion structure affects the dissolution and fractionation process of biomass components. In particular, the carbohydrate yield after pretreatment, separation/utilization of non-carbohydrate components and recycleability of the ILs are critical factors influencing the practical application of ILs for biomass pretreatment. This research is crucial for defining an effective dissolution and fractionation process for corn stover and other biomass feedstocks based upon IL properties, cellulose dissolution ability, fractionation efficiency of biomass components, and enzyme digestibility of the pretreated materials.

See more of this Session: Developments in the Pretreatment of Lignocellulosics for Bioconversion III
See more of this Group/Topical: Sustainable Engineering Forum

(532f) Enhanced Xylan Conversion of Process Relevant Dilute Sulfuric Acid Pretreated Corn Stover

Joseph Shekiro III, Erik M. Kuhn, Nicholas J. Nagle, Richard T. Elander, Melvin P. Tucker, David A. Sievers and Xiaowen Chen, National Bioenergy Center, National Renewable Energy Lab, Golden, CO

Dilute sulfuric acid pretreatment is well established as an effective pathway for the conversion of lignocellulosic biomass to fermentable sugars using a variety of reactor configurations. However, a high solids continuous reactor is strongly favored economically. While such a reactor configuration poses a number of challenges, the National Renewable Energy Lab's (NREL) horizontal screw pretreatment reactor has historically achieved nearly 80 percent conversion of xylan to monomeric xylose on a normalized basis using corn stover as feedstock. Further enhancement of conversion to monomeric xylose is driven by two approaches: improvement of xylan solubilization with a reduction in degradation products and development of processes to better convert xylo-oligomers to xylose.

A series of statistically designed experiments in a high solids batch reactor, the NREL Steam Digester, attempted to maximize xylose yields (monomeric and oligomeric) and minimize furfural production through optimization of reaction conditions (acid loading, temperature and residence time). Results indicated that improved xylan conversion yields could be achieved at temperatures slightly higher than the conditions used historically for the pilot-scale continuous reactor (158 °C, 5 min, ~1% H2SO4). The findings of these bench-scale experiments were used to refine the reaction space for pilot-scale operation.

Preliminary results indicate that enhanced xylo-oligomer conversion is achieved through a hybrid secondary mild thermochemical-enzymatic process compared to either stage alone. To refine the secondary mild thermochemcial xylo-oligomer conversion process, statistically designed experiments investigating reaction time, temperature, catalyst and co-catalyst addition were executed on the bench scale. Enzymatic conversion of xylo-oligomers achieved monomer yields as high as 90% when preceeded by the mild thermochemical conversion step. This set of enzymatic hydrolysis experiments was performed at dilutions mimicking late stage fermentation. While enzyme activity was low in undiluted samples, likely due to end-product inhibition, further experiments were conducted to evaluate enzyme activity throughout fermentation. Because the amount of solubilized xylo-oligomers is low in comparison to cellulose, the cost contribution of additional enzymes for xylo-oligomer conversion is relatively insignificant to the final fuel selling price.

The xylan conversion results of these integrated strategies using the NREL 200 kg/day process-relevant, pilot scale dilute sulfuric acid continuous pretreatment reactor will be presented.

See more of this Session: Developments in the Pretreatment of Lignocellulosics for Bioconversion III

(566a) Investing In 2011-Encore

Larry Grogan, Grogan Advisory Services, Johnson City, TN

Market woes are just the beginning to managing your portfolio and goals. Now, we have new taxes going into effect which will impact your income potential. So, what can you do with your portfolios to manage the new tax environment? We will look at specific investments, as well as, investment strategies to minimize the new tax environment.

See more of this Session: Investing In 2011-Encore
See more of this Group/Topical: Liaison Functions

(582a) Designing Reforming Catalysts to Condition Biomass Derived Syngas

Kim Magrini[1], Whitney Jablonski[2], Yves Parent[2] and Matthew M. Yung[2], (1)National Renewable Energy Laboratory, Golden, CO, (2)National Bioenergy Center, National Renewable Energy Laboratory, Golden, CO

One approach to the efficient thermochemical conversion of biomass to produce liquid fuels using indirect gasification utilizes a catalytic steam reforming process to remove and reform tars to produce clean syngas which additionally maximizes the use of the biomass carbon and hydrogen content. The catalysts used in the steam reforming reactor can comprise a significant portion of the overall process cost to produce biofuels. In order to reduce the catalyst cost and improve the overall fuel synthesis economics, highly active and stable reforming catalysts are required. To achieve this goal, we are using rational catalyst design to understand, design, and synthesize robust, tailor-made catalysts capable of handling biomass-derived impurities such as tars, polyaromatic hydrocarbons (PAHs), and H2S. Emerging reforming catalysts designed by industrial collaborators are also assessed for reforming performance. Results of catalyst design and reforming performance at laboratory and pilot scales are discussed.

See more of this Session: Catalytic Conversion of Renewable Resources to Synthesis Gases and Pyrolysis Oils
See more of this Group/Topical: Sustainable Engineering Forum

(582b) A DFT Study of Desulfurization and Tar Cracking of Gasifier Effluents Over Ceria-Based Rare-Earth Oxides

Adam D. Mayernick, Department of Chemical Engineering, Pennsylvania State University, University Park, PA, Michael Janik, Department of Chemical Engineering, Pennsylvania State University-University Park, University Park, PA and Kerry M. Dooley, Dept. of Chemical Engineering, Louisiana State University, Baton Rouge, LA

Cleanup of biomass gasifier effluents involves both the removal of sulfur-containing compounds such as H_2S and the cracking of tars. Mixed rare earth oxides of ceria containing lanthanides or transition metal additives are effective H_2S sorbents and tar cracking catalysts. We report density functional theory (DFT $+U$) results used to evaluate the H_2S adsorption capacity and tar cracking activity of pure ceria and ceria doped with a series of lanthanides and group VIIB-VIIIB transition metals. We use ab initio thermodynamics to construct phase diagrams for the surface layers of these rare earth oxides as well their corresponding sulfides and oxysulfides as a function of H_2S and O_2 partial pressures. Our results illustrate the relationship between oxide reducibility and H_2S sorbent capacity, and quantify changes in the surface activity for sulfur adsorption as a result of doping. We also evaluate the effects of transition metal additives on C-H and C-C bond activation, and discuss our results in the context of designing improved rare earth oxide catalysts for simultaneous desulfurization and tar cracking of gasifier effluents.

See more of this Session: Catalytic Conversion of Renewable Resources to Synthesis Gases and Pyrolysis Oils
See more of this Group/Topical: Sustainable Engineering Forum

(582c) Development of a Catalytic Reactor for Conditioning of Biomass Generated Producer Gas and Evaluation of Selected Commercial Catalysts

Ajay Kumar, Hasan K. Atiyeh, Danielle Bellmer and Raymond L. Huhnke, Biosystems and Agricultural Engineering, Oklahoma State University, Stillwater, OK

Gasification of biomass is one of the promising conversion technologies to produce fuels, chemicals and power from renewable resources. Removal of tar is one of the most challenging barriers for commercial-scale use of producer gas generated from gasification of biomass. Tar can be reduced to an acceptable level by cracking in presence of catalysts at high-temperature. This method can also increase carbon monoxide and hydrogen compositions which are low in biomass generated producer gas. In this project, we have designed and developed a catalytic reactor specifically for evaluating commercially-available catalysts for conditioning producer gas which is supplied from a lab-scale gasifier with a biomass feedrate of 3-7 kg/h. In our prior study, selected commercially-available catalysts were evaluated in decomposing toluene as a model tar compound. In this project, the catalysts and reaction conditions will be further evaluated using producer gas with tar generated by gasification of switchgrass as the feedstock. Results are expected to show the effects on selected catalysts, shape of catalysts, and reactor temperature on reducing tar and improving producer gas composition.

See more of this Session: Catalytic Conversion of Renewable Resources to Synthesis Gases and Pyrolysis Oils
See more of this Group/Topical: Sustainable Engineering Forum

(582d) Comparison and Characterization of Bimetallic, Ni-Based Catalysts for Conditioning of Biomass-Derived Syngas

Matthew M. Yung[1], Kimberly A. Magrini-Bair[2], Whitney S. Jablonski[2] and Jessica Olstad[2], (1) National Bioenergy Center, National Renewable Energy Laboratory, Golden, CO, (2)National Renewable Energy Laboratory, Golden, CO

Due to the decreasing availability of low-cost petroleum, efforts to develop technology for transportation fuel production from alternative and renewable sources are being actively researched. Biomass is one resource that has the potential for significant fuel production based on feedstock availability. Thermochemical conversion of biomass to syngas has the potential to be used as a cost-effective and renewable source of transportation fuels, displacing petroleum consumption for this purpose. In order to efficiently utilize the carbon in the biomass and protect downstream processing equipment, the light hydrocarbons and tars in the gas stream following gasification should be converted into additional synthesis gas through steam reforming. To this end, we have designed and evaluated several Ni-based tar reforming catalysts for the conditioning of biomass derived syngas. Catalysts of varying composition are evaluated through reaction experiments for hydrocarbon steam reforming in the presence of hydrogen sulfide, using a model syngas that is representative of the composition that would be created through biomass gasification. Catalyst characterization through several techniques (including XRD, TPR, SEM/EDS, and laser Raman spectroscopy) are also discussed.

See more of this Session: Catalytic Conversion of Renewable Resources to Synthesis Gases and Pyrolysis Oils
See more of this Group/Topical: Sustainable Engineering Forum

(582e) Autothermal Oxidative Pyrolysis of Alternative Hydrocarbon Feedstocks

Christine M. Balonek, Joshua L. Colby and Lanny D. Schmidt, Department of Chemical Engineering and Materials Science, University of Minnesota, Minneapolis, MN

Traditional pyrolysis units are heated externally, requiring large energy inputs, and produce gases, solid chars, and complex liquid mixtures. We have developed and demonstrated an oxidative pyrolysis reactor able to run autothermally with negligible solid char formation. The heat necessary for pyrolysis is generated internally, eliminating the heat transfer issues associated with external heating. This process can produce both synthesis gas and hydrocarbon products from alternative feedstock models of biomass and plastic wastes. Solid cellulose and polyethylene (PE) particles were fed individually with air to a Rh-Ce/γ-Al_2O_3 catalytic fixed-bed reactor. Catalyst temperature and product species were studied as a function of the inlet C/O ratio, the ratio of carbon moles in the solid feed to oxygen moles in the O_2 feed. At all C/O ratios reported, the solid feed was completely converted with negligible coke formation at millisecond time scales. Cellulose and PE pyrolysis produced large amounts of synthesis gas at low C/O ratios (0.5 to 1.0). Cellulose pyrolysis produced up to 20 % carbon selectivity to C_{3+} hydrocarbons at C/O of 1.3 (autothermal operation was unsustainable at higher C/O ratios). At similar conditions, PE pyrolysis had 50 % selectivity to C_{3+}, and selectivity continued to increase with increasing C/O ratio to 90 % at C/O = 3.4. This high initial yield of C_{3+} hydrocarbons from cellulose and PE pyrolysis demonstrates the potential for autothermal oxidative pyrolysis to utilize alternative hydrocarbon feedstocks in a fast, clean, and efficient way.

See more of this Session: Catalytic Conversion of Renewable Resources to Synthesis Gases and Pyrolysis Oils
See more of this Group/Topical: Sustainable Engineering Forum

Production of Bio-oil and Hydrogen from Sawdust via Pyrolysis and Catalytic Steam Reforming

Ebrahim Salehi, Jalal Abedi, Tom Harding, and Fakhry Seyedeyn Azad
Chemical and Petroleum Engineering, University of Calgary, Calgary, AB T2N 1N4, Canada
E-mail: jabedi@ucalgary.ca; Tel: 403-220-5594

ABSTRACT
Fixed-bed and fluidized-bed lab-scale reactors were developed to produce bio-oil using pyrolysis of biomass including sawdust and agricultural wastes such as wheat straw. The role of pyrolysis temperature, nitrogen gas flow rate, and condensation temperature on product yield was investigated. The chemical characteristics of the pyrolysis bio-oil and char obtained under the optimal operational conditions were identified by elemental and calorific analyses. The bio-oil was analyzed using chromatographic and spectroscopic techniques. Pyrolysis temperature had the most significant effect on the bio-oil yield. In addition, an experimental system was designed and constructed to make hydrogen from produced bio-oil in a catalytic steam reforming micro-reactor. Gas chromatographic technique was used to determine the concentration of hydrogen and the other components in produced flue gas from the catalytic micro-reactor. The effect on the hydrogen yield of Ni-based catalysts with different Ni loading was investigated.

Keywords: sawdust; fast pyrolysis; hydrogen; bio-oil, catalyst

1. Introduction
Concerns related to the depletion of fossil fuel reserves as well as environmental issues like global warming make hydrogen, in combination with fuel cells, an attractive alternative energy source. The conventional method of hydrogen production is steam reforming of fossil fuels which results in emission of green-house gasses. Renewable sources, such as biomass, may be used as alternative feedstocks for hydrogen production. Biomass for producing energy has a special place among all other renewable energy sources, and it is estimated to contribute 10-14% to the world's total energy supply [1-7].

To produce hydrogen from biomass, two main technologies are steam gasification and catalytic steam reforming of pyrolysis oil (bio-oil) [8]. The second method has an advantage over steam gasification. In this technique, the first step is to apply fast pyrolysis to bulky and low density biomass and thus convert it into an easy-to-transport intermediate called bio-oil which has a high volumetric energy density. The next step is to steam reform the produced bio-oil into a combustible gas mixture including hydrogen. In addition, bio-oil can be used as a feedstock in conventional petroleum refineries, as well as for the production of high value chemicals [5].

Pyrolysis of biomass is a process in which biomass is subjected to heat in an oxygen-free environment resulting in its conversion to char, liquid, and noncondensable gases. The steam reforming of the bio-oil can be simplified as the catalytic steam reforming of the oxygenated organic compounds (CnHmOk) [9-11], followed by water–gas shift reaction (WGS):

$$C_nH_mO_k + (n-k)H_2O \leftrightarrow nCO + (n+m/2-k)H_2$$

$$nCO + nH_2O \leftrightarrow nCO_2 + nH_2$$

Therefore, the overall process can be represented as follows:

$$C_nH_mO_k + (2n-k)\ H2O \leftrightarrow nCO_2 + (2n+m/2-k)H_2$$

This study aimed to develop first a fixed-bed, and then a fluidized-bed pyrolysis system to produce bio-oil from sawdust. Bio-oil was then steam reformed in another developed experimental apparatus using various Ni-based catalysts made in-house.

2. Experimental

2.1. Feedstock
The feedstock used for pyrolysis was sawdust provided by a waste wood recycling company. The sawdust was sieved into fractions of particle diameter < 590 μm. Before the experiments, the sample to be used was dried overnight at 100°C in an oven. A commercial bio-oil supplied by BTG Company, The

Netherlands, was used in catalytic steam reforming tests. Table 1 shows the elemental analyses and the calorific value of the biomass and bio-oil samples.

2.2. Fixed-bed pyrolysis

The fixed-bed pyrolysis system developed consisted of a stainless steel tubular reactor (1.9 cm I.D., 2.2 cm O.D., and 50 cm length) which was placed in a tubular electrical furnace. In a typical experiment, a sample of 10 g of sawdust was placed in the reactor and nitrogen was used as the carrier gas. After heating the reactor, the biomass particles decomposed to char, non-condensable gas and volatile materials. The gas and the volatile materials passed through two sequential condensers maintained at -10°C by a bath circulator. The schematic diagram of the fixed-bed pyrolysis system may be found in another publication [1].

Table 1. Characteristics of feedstocks

Properties	Bio-oil	Sawdust
Elemental analysis (wt%)[a]		
Carbon	45.83	47.37
Hydrogen	7.15	6.03
Nitrogen	0.87	2.40
Oxygen[b]	46.15	44.2
Heating value[c] (MJ/kg)	17.50	19.05
Water content wt (%)	23.10	-
Density (kg/m3)	1255.00	-

[a] dry basis
[b] by difference
[c] using bomb calorimeter

As discussed in our previous work [1], it was recommended that future tests be performed under fast pyrolysis conditions in a fluidized-bed system. Changing the pyrolyzer from the fixed-bed to a fluidized-bed system allows all feedstock particles to be heated faster and more homogenously which results in higher bio-oil yield and different quality.

2.3. Fluidized-bed pyrolysis

Figure 1 shows a picture of the developed fluidized-bed fast pyrolysis system. This system was designed to operate at temperature between 400-500°C, atmospheric pressure, and with a 100 g/hr biomass feed rate. The system comprised mass flow controllers to control and measure fluidizing gas (nitrogen), fluidized-bed reactor and plenum, biomass feed system, cyclone char collection system for char separation from gas stream, vapor condensation system for bio-oil recovery, instrumentation for data acquisition and control, and heating system to provide heat for the system.

Figure 1. Developed fluidized-bed pyrolysis system

The reactor and the cyclone were sized to provide residence time of less than 1 second for gas phase before reaching the condenser. At the end, the product gas is cooled in a series of water-glycol glass condensers to recover bio-oil from gas steam.

2.4. Catalytic steam reforming

Because at the point of starting steam reforming experiments there was not enough produced bio-oil from the pyrolysis system, a commercial bio-oil was used. Figure 2 shows a picture of the developed catalytic steam reforming system. The heart of the system is the catalytic micro-reactor which is a stainless steel fixed-bed rector. Two syringe pumps, one for water injection and the other one for bio-oil injection, provide feeds to the reactor on top of the catalyst bed. Nitrogen is used as the carrier gas. Before running an experiment, catalyst is reduced in the medium of nitrogen and hydrogen (50:50 vol%) at the temperature of 850 °C.

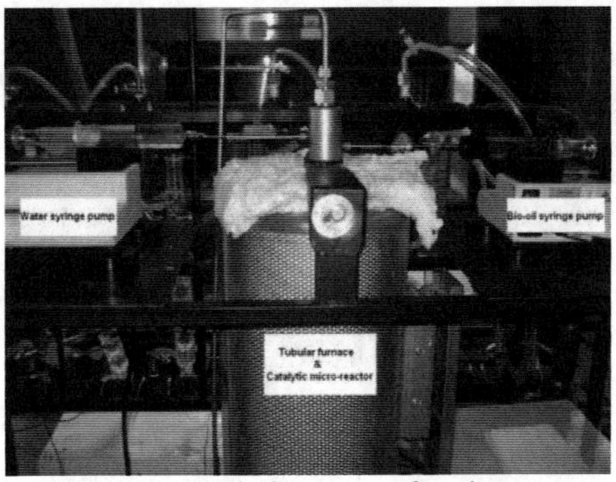

Figure 2. Catalytic steam reforming system

Ni-based catalysts which are very common for the steam reforming of natural gas were prepared in our lab in four different Ni loading (5.6%, 10.7%, 14.1%, and 18%) on alumina. Down stream of the catalytic steam reforming reactor, the product gases, after passing a glass condenser to recover excess steam, were analyzed by an on-line micro-GC capable of detecting H_2, N_2, CO, CO_2, CH_4, C_2H_6, C_3H_8, and C_4H_{10}.

3. Results and Discussion

In fixed-bed pyrolysis experiments, three operational parameters including reaction temperature, condensation temperature, and nitrogen gas flow rate were selected to investigate their effects on pyrolysis product yields. Maximum bio-oil yield of 45% could be achieved. GC-MS analysis was performed allowing detection of ten components in the produced bio-oil. Results indicated that bio-oil is a complex mixture of organic compounds containing a wide variety of aromatics and oxygenated compounds, such as carboxylic acids, phenols, and aldehydes [1].

Regarding the fluidized-bed pyrolysis system, more than 30 experiments have been done to date. Maximum bio-oil yield of ≈50% has been achieved. The char and gas yields were15% and 35%, respectively. During these initial tests challenges involving the char separation system and bio-oil recovery system have been gradually overcome. Now the main objective is to determine the optimal operational conditions to obtain the highest bio-oil yield and to investigate the effects of operational parameters on the bio-oil quality.

Catalytic steam reforming test results have shown yield of 73% for hydrogen on 14% Ni/Alumina. Compared to the rest of the Ni-based catalysts, the 14% Ni loading provided higher hydrogen yield while the temperature was set at 950 °C and steam to carbon ratio was 5. The main problem regarding catalytic steam reforming of bio-oil is coke deposition on the catalyst. In the next step, the plan is to investigate the effect of temperature and steam to carbon ratio on the hydrogen yield, and also to prepare noble metal catalysts for the bio-oil steam reforming to improve the hydrogen yield.

4. Conclusion

In this study, fast pyrolysis experiments of sawdust were carried out in fixed-bed and fluidized-bed pyrolysis systems. Bio-oil yields of 45% were achieved in the fixed-bed system and a 5% increase in yield to 50% was obtained in the fluidized-bed system. For the pyrolysis phase, the next step is to investigate the effect of operational parameters like biomass particle size, pyrolysis temperature, and vapor residence time on bio-oil yield and quality. Results of catalytic steam reforming show the hydrogen yield of 73%. Improving Ni-based catalysts by adding other metals like Mg, and preparing noble metal catalysts have potential to help prevent or at least decrease coke deposition.

Acknowledgement

Financial support from the Natural Sciences and Engineering Research Council of Canada (NSERC), the Agricultural Funding Consortium of Alberta, Alberta Innovates Bio Solutions, and the Institute for Sustainable Energy, Environment and Economy (ISEEE) at the University of Calgary are gratefully acknowledged.

References

[1] E. Salehi, J. Abedi, T. Harding, Energy & Fuels, 23 (2009) 3767–3772.
[2] N. Ozbay, A. E. Putun, E. Putun, Int. J. Energy Res. 30 (2006) 501.
[3] A. E. Putun, B.B. Uzun, E. Apaydin, E. Putun, Fuel Process. Techno. 87 (2005) 25.
[4] S.Sensoz, D.Angın, Bioresour. Technol. 99 (2008) 5492.
[5] H.J. Park, J.I. Dong, J.K. Jeon, Y.K. Park, K.S. Yoo, S.S. Kim, J. Kim, S. Kim, Chem. Eng. J., 143 (2008) 124–132.
[6] G. Chen, J. Andries, H. Spliethoff, Energ. Convers. Manage. 44 (2003) 2289.
[7] M. Krzesinska, M.B. Pilawa, S.Pusz, Biom. Bioenerg. 30 (2006) 166.
[8] A.C. Basagiannis, X.E. Verykios, Inter. J. Hydrogen Ener. 32 (2007) 3343 – 3355.
[9] A. C. Basagiannis, X. E. Verykios, Catalysis Today 127 (2007) 256–264.
[10] Z. Wang, Y. Pan, T. Dong, X. Zhu, T. Kan, App. Catal. A: General 320 (2007) 24–34.
[11] C. Rioche, S. Kulkarni, F. C. Meunier, J. P. Breen, R. Burch, App. Catal. B: Environmental 61 (2005) 130–139.

(617a) Calculation of Sugar Yields From High Solid Hydrolysis of Lignocellulosic Biomass

Yongming Zhu[1], Marco Malten[2], Mads Torry-Smith[1] and Hui Xu[1], (1)Novozymes North America, Franklinton, NC, (2)Novozymes A/S, Bagsvaerd, Denmark

Accurate determination of sugar yields is important for the evaluation of an enzymatic hydrolysis process. Calculation of sugar yields at high solids is complicated primarily by the varying liquid volume and liquid density owing to solid solubilization and sugar liberation. Ignoring these factors can lead to significant errors in yield calculation. In this paper, a mathematical method was developed for yield calculation in high solids hydrolysis. The information needed in the calculation includes the composition of the substrate, initial solid loading, initial liquid density, and the sugar concentrations before and after hydrolysis. All of these parameters are measurable with conventional procedures. The method was validated experimentally for a range of different substrates and solid loading up to 30 wt%. Compared to the other methods found in the literature, this method is easy to use and represents good accuracy.

See more of this Session: Reaction Kinetics and Transport Fundamentals for Biomass Conversion I
See more of this Group/Topical: Sustainable Engineering Forum

(617b) Diffusion-Adsorption Measurements of Solutes in Cellulose Fiber Beds Using Magnetic Resonance Imaging

David M. Lavenson, Chemical Engineering & Materials Science, University of Calfiornia Davis, Davis, CA, Emilio J. Tozzi, Chemical Engineering and Materials Science, University of Calfiornia Davis, Davis, CA, Michael J. McCarthy, Food Science & Technology, University of Calfiornia Davis, Davis, CA and Robert L. Powell, Chemical Engineering and Material Science, University of California-Davis, Davis, CA

Effective mixing and transport of cellulosic suspensions for enzymatic conversion to glucose is an important biofuels process bottleneck. Most recent literature studies have highlighted the lack of efficient mixing strategies available for combining substrate and enzymes at the initial stages of high solids enzymatic hydrolysis. Design of new energy efficient mixers requires quantitative information about rheology and diffusion in concentrated slurries. We investigated diffusion and adsorption of cellulose suspensions using magnetic resonance imaging. MRI allows for spatially resolved measurements of concentration distributions in highly concentrated suspensions. Paramagnetic tracers of various molecular sizes were used as diffusing solutes in suspensions of various fiber types. Effective diffusion coefficients are compared with predictions of a simple diffusion-adsorption model which accounts for the system porosity, tortuosity, and adsorption.

See more of this Session: Reaction Kinetics and Transport Fundamentals for Biomass Conversion I
See more of this Group/Topical: Sustainable Engineering Forum

Breaking the Barrier to Obtain High Yield on Enzymatic Hydrolysis Using Low Enzyme Loadings

Yoshiki Ueno, Makoto Ikeo, Naohisa Inoue, and Daisuke Taneda
Technology Development Department, JGC Corporation
Higashiibaraki-gun, Ibaraki Pref., Japan

Abstract

JGC Corp. has been developing a method of enzymatic hydrolysis of cellulosic biomass through an approach focused on development of the process. During this effort, we found that by hydrolyzing cellulose without agitation, the enzyme loading can be reduced without decreasing the glucose yield. Moreover, the application of this hydrolysis method allowed more than a 10 fold reduction in the enzyme loading. The studies indicated that there were changes in the characteristics of cellulase, such as adsorption, desorption, and inactivation, when this method was applied. These changes may contribute to the more efficient hydrolysis under the low enzyme loading, and provide insights on further development of the process.

Introduction

Ethanol is one of the biofuels that has been attracting attention as a substitute for fossil fuels. Up to the present, most ethanol has been produced from sucrose and starch by fermentation ; using feedstocks such as sugarcane, beet, and corn. On the other hand, those feedstocks are also used as food. In order to meet an expected increase in the demand for both energy and food in the future, it is necessary to find alternative feedstocks for the ethanol production that will not adversely affect world food resources. To solve this issue, lots of fundamental research and demonstration projects, whose objectives are to develop technologies to produce ethanol from cellulosic biomass such as wood, corn stover, and wheat straw, have been carried out all over the world [1] - [3].

The main issue when utilizing cellulosic biomass is the conversion of cellulose, the main component of cellulosic biomass, to glucose. Enzymatic hydrolysis is one of the technologies that enable this conversion. The cost of the enzymes is a critical issue for the commercialization of the enzymatic hydrolysis technology, because a large amount of enzyme (cellulase) is required to obtain a high glucose yield, and commercial cellulase is expensive.

JGC Corp. has developed a method of enzymatic hydrolysis that enables the reduction of the enzyme costs through an approach focused on the process development. We have found a method that achieves a high glucose yield using low enzyme loadings, which will be introduced in this report. The method can be scaled-up easily, and does not depend on the particular enzyme used.

Materials and Methods

Substrates

Filter paper (No.2, Toyo Roshi Kaisha, Tokyo, Japan) and pulp made from broad-leaved trees were used as substrates. Each substrate was shredded by a paper shredder. The size of the shredded substrate was approximately 15mm-length, 2mm-width, and 1mm-thickness.

Enzymes

A commercial *Tricoderma reesei* cellulase (Cellulase SS), purchased from Nagese ChemteX (Osaka, Japan), was used in this study. This Cellulase SS contained 120 mg of protein per ml as measured by the Bio-Rad protein assay (Bio-Rad Laboratories, Hercules, CA), using Bovine Gamma Globulin (BGG) as a standard protein. The hydrolytic activity of this enzyme was 158 filter paper units (FPU) per gram of protein as measured by the procedure described in the literature [4].

Enzymatic Hydrolysis

Enzymatic hydrolysis experiments were performed in 500ml Erenmeyer flasks placed in a 50 deg-C incubator. The agitation of the flask was controlled using a rotary shaker, and two conditions, 115 rpm (mixing condition) and 0 rpm (static condition) were tested. The original enzyme reagent was diluted in 50mM sodium-acetate buffer (pH 5.0) containing 0.02 w/v% sodium azide to inhibit contamination. The substrate concentration was 10 w/v%, which means that 100ml of the diluted enzyme was used against 10 g of substrate, in all of the experiments.

Analytical Method

HPLC (Shimadzu, Kyoto, Japan) was used for the analysis of monosaccharides. Samples taken during the experiment period were centrifuged under conditions of 13,000rpm and 15 minutes. Then the supernatants were filtrated by disposable filters (Millipore), the pore size of which was 0.45 micro-meters. The filtrated supernatants were used for the HPLC analysis.

Results and Discussion

Hydrolysis of cellulose under the mixing and static conditions

The influence of the enzyme loading on the sugar yield was investigated with respect to the mixing condition. Figure 1 shows the change in the glucose concentration over time. With the exception of enzyme loadings of 158 FPU/g cellulose and 79 FPU/g cellulose, the lower enzyme loadings resulted in lower glucose concentrations. This relationship is difficult to explain from the perspective of conventional catalytic reaction. In a chemical reaction, the catalyst loading generally doesn't affect the final yield of product, although it affects the reaction rate. However, in the case of the enzymatic hydrolysis of cellulose, influences of the enzyme loading on both the yield and the reaction rate have been reported repeatedly, although enzymes work as catalysts [5] - [9]. This phenomenon is a major issue when commercializing the enzymatic hydrolysis technology, because reduction of the enzyme loadings is necessary to reduce the enzyme cost. This barrier has already been identified, but no one has been able to find a solution to it.

However, we found that high glucose yield was achieved at the low enzyme loading when the enzymatic hydrolysis was conducted without agitation. Figure 2A shows the experimental results under both the mixing and the static conditions. The final glucose concentrations obtained from both conditions were the same for an enzyme loading of 79 FPU/g cellulose. However, when the enzyme loadings were reduced to 7.9 FPU/g cellulose, the final glucose concentration obtained under the static condition became higher than that obtained under the mixing condition. Moreover, the final glucose concentration reached the same level as that achieved by hydrolysis with the enzyme loading of 79 FPU/g cellulose. This phenomenon will be called "The Static Effect" from here on.

The results of The Static Effect were also observed in the cases where pulp was used as a substrate (Figure 2B), and where different commercial enzymes were used. (Data are not shown.) In Figure 2B, the sugar concentration indicates the total value of glucose and xylose since the hydrolyzate produced from pulp contains both of these monosaccharides.

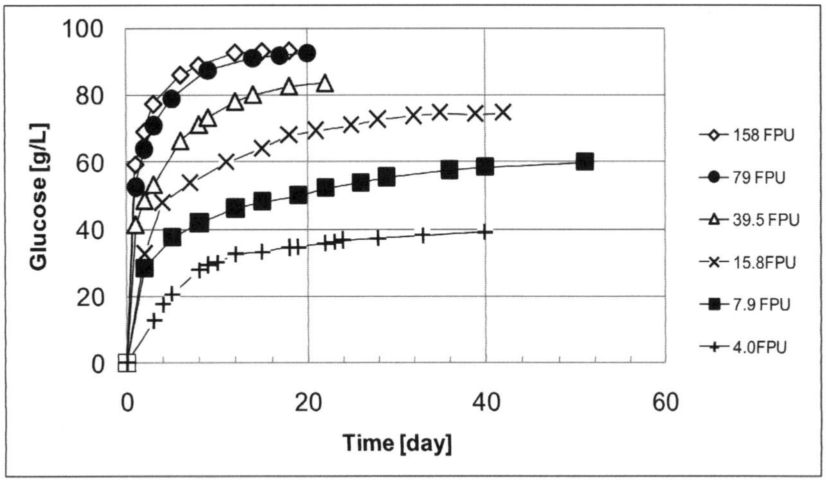

Figure 1. Effect of enzyme loadings (FPU/g cellulose) on enzymatic hydrolysis of cellulose under mixing (115rpm) condition

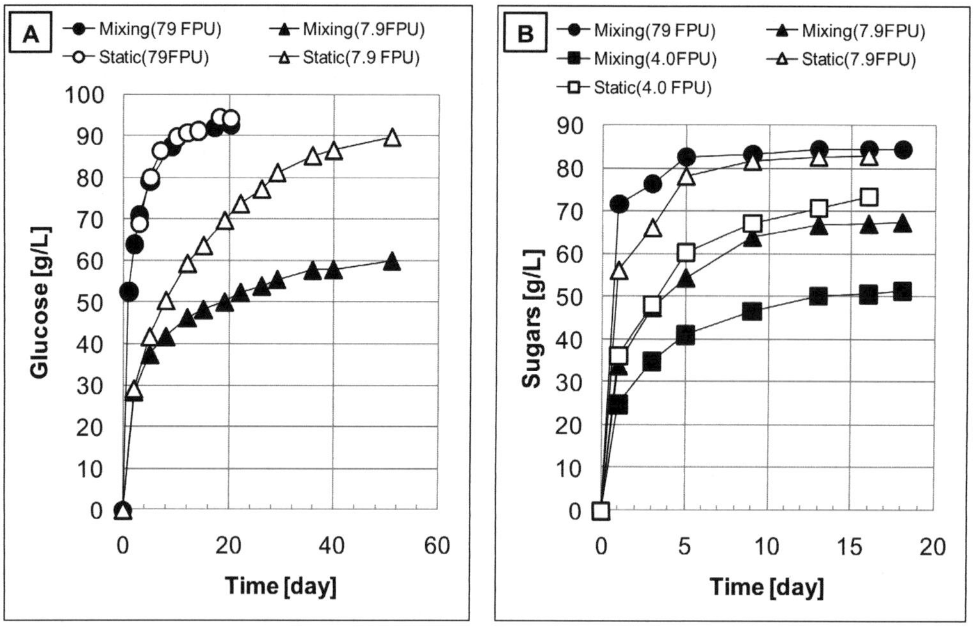

Figure 2. Comparison of enzymatic hydrolysis of cellulose at several enzyme loadings (FPU/g cellulose) under mixing (115rpm) and static condition : (A) filter paper, (B) pulp

Behavior of cellulase under the static condition

Figure 3 shows the changes in the enzyme concentrations when the loading condition was 79 FPU/g cellulose of enzyme. Although the final glucose concentrations of the two experiments were the same (Figure 2), the trends in the enzyme concentrations in the hydrolyzate were different. During the hydrolysis under the mixing condition, the enzyme concentration decreased to 25% of the initial concentration by the adsorption of enzyme on the substrate at the beginning of the reaction, and it didn't change for the succeeding few days. At the end of the reaction, the enzyme concentration in the hydrolyzate was about a half of the initial concentration. However, under the static condition, the enzyme concentration decreased only to 40% of the initial concentration at the beginning, and by the end of the reaction, it had reached a level of up to 75% of the initial concentration. These results indicated that the hydrolysis of cellulose was effectively conducted under the static condition so that a small amount of enzyme absorbed on substrates and much more free enzyme existed in the hydrolyzate compared with the case of the mixing condition. In addition, microscopic observation revealed an increase in the presence of small substrate particles under the mixing condition. The difference in the final enzyme concentrations might have been caused by the adsorption of the enzyme onto these small substrate particles immediately after desorption from substrates under the mixing condition.

One hypothesis is that this adsorption on to the small particles is non-productive, because the final glucose concentrations of the two experiments were the same, and the decrease in this adsorption under the static condition helps to give rise The Static Effect. The influence of the non-productive adsorption will be more significant under low enzyme loading condition, since there is a smaller amount of the enzyme in the solution to start with.

Another possible cause of The Static Effect is the decrease in the inactivation of the enzyme. The effect of agitation on enzymes has been studied, and it was reported that the shear stress can damage the enzyme and decreases the enzyme activity [10] [11].

When the solution containing cellulase was put into the incubator, precipitation was observed. The precipitate was assumed to be the denatured enzyme. The change in the amount of precipitate over time was investigated under the mixing and the static conditions (Figure 4). The amount of the precipitate in the mixing condition was larger than that in the case of the static condition. This result corresponds with the hypothesis that The Static Effect is caused by reduced inactivation of the cellulase. When the influence of the enzyme concentration on the precipitation was investigated, a larger portion of the initial enzyme precipitated at lower enzyme concentrations. The results of these studies indicate that by putting the enzymatic hydrolysis under the static condition, the amount of inactivation of enzyme through non-productive adsorption and denaturing is reduced. The effect of these reductions becomes more significant under the low enzyme loading, and this is assumed to bring about The Static Effect.

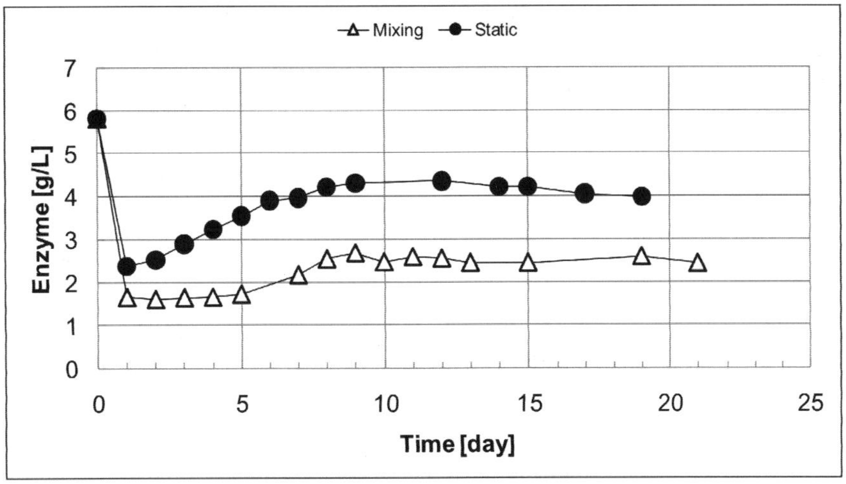

Figure 3. Profiles of enzyme concentration in the hydrolyzate under mixing (115rpm) And static condition : substrate ; filter paper, enzyme loading ; 79 FPU/g cellulose

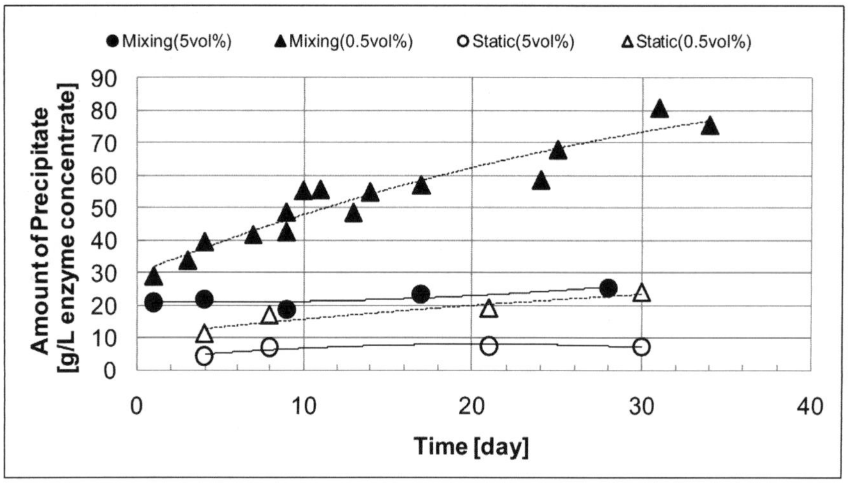

Figure 4. Influence of enzyme concentration and mixing condition (115rpm) on amount of precipitate : substrate was not added

Conclusion

It was found that the static condition is preferable for the enzymatic hydrolysis. Through The Static Effect, hydrolyzate with higher glucose concentration can be obtained with lower enzyme loadings. The study indicates the application of the static method allowed a more than 10 fold reduction in the enzyme loading in the hydrolysis of cellulosic biomass, while maintaining the same level of the glucose yield. This new method is suitable for commercial scale plants, because it obviates the need to mix high solid-content slurry, it reduces the energy consumption required, it uses a simpler saccharification tank, and it lowers the enzyme loading. Moreover, the knowledge acquired from further studies of this method will be applied for the development of microorganisms that will produce a more powerful cellulase.

Experiments using other cellulosic biomass such as rice straw and bagasse are being conducted currently to investigate the preferable pretreatment method.

Acknowledgment

The authors gratefully acknowledge the funding of this work by the New Energy and Industrial Technology Development Organization (NEDO).

References

1. Z. Xiao, X. Zhang, D. J. Gregg, and J. N. Saddler (2004), "Effects of sugar inhibition on cellulases and β-glucosidase during enzymatic hydrolysis of softwood substrates", Appl. Biochem. Biotechnol., Vol. 115, pp.1115-1126.
2. T. I. Georgieva, X. Hou, T. Hilstrom, and B. K. Ahring (2008), "Enzymatic Hydrolysis and Ethanol Fermentation of High Dry Matter Wet-Exploded Wheat Straw at Low Enzyme Loading", Appl. Biochem. and Biotechnol., Vol. 148, pp.35-44.
3. J. S. Tolan (2002), "Iogen's process for producing ethanol from cellulosic biomass", Clean Techn. Environ. Policy, Vol. 3, pp.339-345.
4. T.K. Ghose (1987), "Measurement of cellulase activities", Pure Appl. Chem., Vol. 59, pp. 257-268.
5. W. Sattler, H. Esterbauer, O. Glatter, and W. Steiner (1989), "The effect of enzyme concentration on the rate of the hydrolysis of cellulose", Biotechnol. Bioeng., Vol. 33, pp. 1221-1234.
6. Y. Lu, B. Yang, D. Gregg, J. N. Sadler, and S. D. Mansfield (2002), "Cellulase adsorption and an evaluation of enzyme recycle during hydrolysis of steam-exploded softwood residues", Appl. Biochem. and Biotechnol., Vol. 98-100, pp.641-654.
7. F. Teymouri, L. L. Perez, H. Alizadeh, and B. E. Dale (2005), "Optimization of the ammonia fiber explosion (AFEX) treatment parameters for enzymatic hydrolysis of corn stover", Bioresour. Technol., Vol. 96, pp.2014-2018
8. M. Chena, L. Xia, and P. Xue (2007), "Enzymatic hydrolysis of corncob and ethanol production from cellulosic hydrolysate", Int. Biodeterior. Biodegrad., Vol.59, pp. 85-89
9. C. Tengborg, M. Galbe, and G. Zacchi (2001), "Influence of enzyme loading and physical parameters on the enzymatic hydrolysis of steam-pretreated softwood", Biotechnol. Prog., Vol. 17, pp. 1100-117.
10. S. I. Mussatto, G. Dragone, M. Fernandes, A. M. F. Milagres, and I. C. Roberto (2008)," The effect of agitation speed, enzyme loading and substrate concentration on enzymatic hydrolysis of cellulose from brewer's spent grain", Cellulose, Vol. 15, pp.711-721.
11. S. Mukataka, M. Tada, J. Takahashi (1983), "Effects of Agitation on Enzymatic Hydrolysis of Cellulose in a Stirred-Tank Reactor", J. Ferment. Technol., Vol.61, pp.615-621.

(617d) A Dynamic Model for Cellulosic Biomass Hydrolysis: Validation of Hydrolysis and Product Inhibition Mechanism

Jonas Eeclo[1], Chien-Tai Tsai[2], Ricardo Morales-Rodríguez[1], Krist V. Gernaey[3], Anne S. Meyer[4] and **Gürkan Sin**[1], (1)CAPEC, Department of Chemical and Biochemical Engineering, Technical University of Denmark, 2800 Kgs. Lyngby, Denmark, (2)BioEng, Department of Chemical and Biochemical Engineering, Technical University of Denmark, 2800 Kgs. Lyngby, Denmark, (3)PROCESS, Department of Chemical and Biochemical Engineering, Technical University of Denmark, 2800 Kgs. Lyngby, Denmark, (4)Department of Chemical and Biochemical Engineering, Technical University of Denmark (DTU), 2800 Kgs. Lyngby, Denmark

Production of bioethanol from agricultural waste such as lignocellulosic biomass forms a candidate alternative renewable energy resource. The process of converting biomass to ethanol is complex with many unit operations including the physical pre-treatment and hydrolysis of lignocellulosic material to free simple sugars. One of the challenges in the process is the underlying enzymatic hydrolysis mechanism. The kinetics of multiple enzymes on insoluble substrate is a challenging phenomenon to describe. Although a number of hydrolysis models has been proposed so far in the literature (e.g. Kadam et al., 2004), few or none of the models were independently subject to experimental validation. This sets the aim of this contribution, which is to perform a validation of a dynamic cellulosic biomass model with particular focus on the validation of the hydrolysis and product inhibition mechanisms.

In order to carry out this analysis a number of dedicated experiments were performed. Avicel® and cellobiose were used as substrates. The end product glucose and/or intermediate cellobiose was added to the initial samples in different concentrations, to better identify their inhibition effect. The enzymes used in the experiments were Novozyme 188 and Celluclast 1.5L. Three sets of experiments were performed focusing exclusively on three different conversion reactions:

(i) conversion of cellulose to cellobiose (set 1),

(ii) conversion of cellulose to glucose (set 2) and

(iii) conversion of cellobiose to glucose (set 3).

A total of nine parameters (k1r, k2r, k3r, K1IG2, K2IG2, K3M, K1IG, K2IG and K3IG) involved in the conversion reaction of cellulose to glucose were estimated from the data set 2. The cellobiose to glucose conversion contains three parameters (k3r, K3M and K3IG) which were estimated from the dedicated experimental data (set 3). These three parameters were then used (assumed fixed) when estimating the parameters involved in the reaction for conversion of cellulose to glucose.

Experimental results show that the reaction from cellobiose to glucose goes much faster than initially assumed by the Kadam model (k3r; 3106.7 versus 285.5 and K3M; 2.377 versus 24.3; both result in a reaction that is ±20 times faster, depending on glucose and cellobiose concentrations). This means that inhibition of cellobiose is less critical than originally described by Kadam et al. (2004). Some calculated parameters (k1r, k2r, K1IG and K2IG2) are similar to the values proposed by Kadam et al. (2004) for cellulose to glucose reaction. However due to the fixed parameters of the cellobiose to glucose reaction (which were found faster), some of the model parameters change as well (e.g. faster cellobiose conversion to glucose results in more inhibition for the cellulose to cellobiose conversion). In general, simulation results (using the estimated parameters) fit well for all the experiments.

Overall the model of Kadam et al. (2004) was found to remain largely valid when describing experimental data collected under different initial concentrations of product inhibition (glucose and cellobiose respectively) as inhibitors. One caveat has to be mentioned, that is that some of the model parameters had to be fine-tuned to get a better fit to the experimental data. Nonetheless the results add credit to the reliability of the model and thereby provide confidence in utilising the model for process development and optimisation studies.

References

[1]. Sin, G., Meyer, A. S., Gernaey, K. V., 2010. Comput. Chem. Eng. doi:10.1016/j. compchemeng.2010.02.012.

[2]. Kadam, K.L. Rydholm, E.C. McMillan, J.D., 2004. Biotechnol. Prog. 20, 698-705.

See more of this Session: Reaction Kinetics and Transport Fundamentals for Biomass Conversion I
See more of this Group/Topical: Sustainable Engineering Forum

(617e) Syngas Fermentation: Reaction Kinetics and Pressure Dependencies of the Clostridial P11 Hydrogenase Enzyme

Bradley E. Skidmore, Douglas R. Tree, Jason M. Bray, Ryan A. Baker, Dila R. Banjade and **Randy S. Lewis**, Chemical Engineering, Brigham Young University, Provo, UT

Ethanol usage as a transportation fuel is rapidly increasing in the United States. Production of ethanol from cellulose feedstocks via gasification followed by syngas fermentation offers an approach that mitigates many of the adverse effects associated with production of ethanol from corn. In the syngas fermentation process, the hydrogenase enzyme of the fermentation bacterium supplies electrons to the metabolic pathway, facilitating ethanol production.

Though it is possible for the metabolic pathway to obtain electrons from CO oxidation during syngas fermentation, it is preferred to obtain electrons from H_2 as this leaves the CO available for incorporation into the desired organic chemicals. Processes that depend on CO oxidation to supply electrons greatly reduce their carbon-to-product yield. Hence, understanding the kinetics of H_2 oxidation via hydrogenase may facilitate process improvements resulting in greater H_2 utilization and, consequently, greater ethanol production.

Experiments were performed to measure the dependence of hydrogenase activity from Clostridium P11 on H_2 partial pressure. The data were fit with a variety of published hydrogenase rate laws and the Okura rapid-equilibrium rate law was found to best describe hydrogenase activity in Clostridium P11. A constant that regulates the effect of H_2 partial pressure on hydrogenase activity, K_{H2}, was determined to be independent of electron acceptor and to have a value of 0.31 atm, implying that H_2 must be supplied to syngas fermentation processes employing this bacterium at ~3 atm to maximize hydrogenase activity. K_{BV} and K_{MV}, constants that regulate the effect of benzyl viologen and methyl viologen (the electron acceptors used for the kinetic studies) on hydrogenase activity, were determined to be 1.7-2.4 mM and 10.6 mM, respectively.

See more of this Session: Reaction Kinetics and Transport Fundamentals for Biomass Conversion I
See more of this Group/Topical: Sustainable Engineering Forum

(617f) Secondary Hydrolysis of Pretreatment Liquor Obtained From Continuous High-Solids Dilute-Acid Pretreatment of Corn Stover

Suan Shi, Chemical Engineering, Auburn University, Auburn, AL, Y. Y. Lee, Department of Chemical Engineering, Auburn University, Auburn University, AL, Richard Elander, National Bioenergy Center, National Renewable Energy Laboratory, Golden, CO and Nick Nagle, National Renewable Energy Laboratory, Golden, CO

Dilute-acid pretreatment is one of the most promising process options applicable for various lignocellulosic biomass feedstocks. NREL has developed an efficient continuous pretreatment reactor that can be operated with high-solid charge and with short residence time. Using this reactor, NREL has investigated extensively on pretreatment of corn stover. Because of these unique features, the pretreatment liquors produced from this reactor contain high level of hemicellulose sugars, the concentration surpassing 10 wt. %. On the other hand, the short residence time and low-acid condition limits the hydrolysis of hemicellulose. The pretreatment liquor obtained from this reactor thus contains relatively high amount of xylose oligomers, typically in the range of 10-40% of the total sugar. In order to fully utilize the sugars in the pretreatment liquor, oligomers need to be hydrolyzed to monomers. However, the secondary hydrolysis of this liquor was found to be much more difficult than hydrolysis of xylose oligomers in clean environment. This investigation was undertaken to verify the factors inhibiting the acid-catalyzed oligomer hydrolysis and to establish the comprehensive kinetics recognizing all reactions occurring under the heavily contaminated environment. For this purpose, the secondary hydrolysis was experimentally studied using rapid-heating small-scale batch reactors. The experiments were done for pretreatment liquor and for clean xylose oligomers. The comparison of the results indicated that the a significant fraction oligomers may be bound to lignin fragments, and some of it may also exist in branched form. The kinetic model was set up recognizing different forms of oligomers (linear, branched, and lignin-oligomer complex), and additional reactions of xylose including decomposition, and reversion to xylose dimmers. Model verification was performed using batch hydrolysis data, xylose decomposition data, the data obtained from reversion reaction. The results of the model verification are presented and how the model can be used to determine the optimum operating conditions are discussed.

See more of this Session: Reaction Kinetics and Transport Fundamentals for Biomass Conversion I
See more of this Group/Topical: Sustainable Engineering Forum

() GENENTECH: A Rich History of Chemical Engineering Innovation

Jean Bender, Genentech, South San Francisco, CA and **Ann L. Lee**, Process Development, Roche, South San Francisco, CA

Throughout its history, Genentech scientists and engineers pioneered and then exploited the use of recombinant DNA technology to develop and manufacture proteins and monoclonal antibodies for therapeutic use. Today an entire industry uses these genetic engineering techniques in pursuit of novel medicines. Genentech is considered a leader in innovation and has developed and commercialized more than 16 novel medicines that address unmet medical needs. Much of the technology needed for process development, scale-up, and manufacturing of these medicines has relied on three decades of chemical engineering innovation.

Core to Genentech's success is the culture and values of the company. Innovative science is the basis of the company's existence and the foundation for its future. By building on its legacy of innovation, chemical engineers in the laboratory, in the pilot plants, in manufacturing, in engineering and in management will continue to play a key role in transforming the latest scientific breakthroughs into innovative medicines for our patients. This presentation will highlight the chemical engineering innovations that contributed to several breakthrough therapeutic medicines, and that inspire us to focus our imagination on what is possible for patients.

See more of this Session: 3rd AIChE Corporate Innovation Award Lecture
See more of this Group/Topical: Liaison Functions

(631a) Adsorption of Polyfunctional Molecules Onto Zeolites for the Development of Biorefinery Separation Processes

Elizabeth E. Mallon, Aditya Bhan and Michael Tsapatsis, Chemical Engineering and Materials Science, University of Minnesota, Minneapolis, MN

Adsorption isotherms have been developed for oxygenates with different carbon number and functionality on zeolites that vary in pore size and connectivity (FER, MWW, MFI, BEA, MOR, FAU) as well as on ordered microporous/mesoporous materials (MCM-36, 3DOm, and SBA-15) using liquid chromatography. Henry's constants (Kads) increases exponentially with carbon number for C2-C6 diol and triol adsorption on MFI, BEA, and MWW indicating that dispersion forces are critical for polyol adsorption; this conclusion is supported by the observation that Kads for propylene glycol adsorption decreases with increasing zeolite pore size and that propylene glycol preferentially adsorbs in the micropores of MCM-36 and 3DOm. It is also shown that the chemical composition of the zeolite only has a marginal effect on propylene glycol adsorption, which again demonstrates that confinement of the adsorbate in the zeolite pore is the major contributor to the energy of adsorption. Overall, initial correlations for oxygenate adsorption in zeolites are presented in this study which will have an impact on the separation and catalytic conversion of oxygenates in the processing of biomass to chemicals and fuels.

See more of this Session: Chemical and Catalytic Conversions and Processes for Renewable Feedstocks
See more of this Group/Topical: Sustainable Engineering Forum

(631b) Esterification of Free Fatty Acids with Methanol Over Metaloxides Supported ZrO2 Catalysts

Manhoe Kim[1], Craig DiMaggio[2], Shuli Yan[3], Steven O. Salley[4] and K. Y. Simon Ng[4], (1)National Biofuel Energy Lab, Next Nery, Detroit, MI, (2)National Biofuel Energy Laboratory, Detroit, MI, (3) Wayne State University, Detroit, MI, (4)Chemical Engineering and Materials Science, Wayne State University, Detroit, MI

Brown grease recovered from waste water is less expensive lipids and would make biodiesel much more competitive with petroleum-based fuels. About 80±5% of the brown grease is free fatty acids. In order to develop heterogeneous catalysts for brown grease conversion toward biodiesel, various mixed metal-oxides supported ZrO2 catalysts has been prepared via sol-gel method and impregnation method. In order to get the best catalyst composition for brown grease conversion, the performances of 35~70 mesh size of catalysts has been tested at 200°C with the mixture of brown grease, methanol, and tetrahydrofuran in batch reactors. Catalyst particles (size of 0.5~5mm) of three different formulas (ZnO2/ZrO2, ZnO2-Al2O3/ZrO2, and ZnO2-La2O3/ZrO2-La2O3) have been tested in continuous flow reactors with changes on temperatures (170, 180, 190, and 200°C) and retention times. A technical grade of oleic acid was used to continuous flow reactors as a model compound for brown grease. The FAME yields were measured with a GC-MS. Dissolved metal species (especially Zn) were measured with an Inductively Coupled Plasma Spectrometer. The ZnO2/ZrO2 catalyst showed very a high activity (98~99% yield) and durability for 54 days at 180°C with 5:1 ratio of methanol:FFA. Zr specie was very stable in the acidic stream of free fatty acids. However, Zn concentration in the product eluted from the continuous flow reactors were initially high as 807 ppm in the 1st day product, and decreased to below 10 ppm in 50th day product. The Zn dissolution can be prevented by using various metal-oxides combinations. In addition, the catalytic activities of the mixed metal-oxides supported catalysts for esterification and transesterification depended on the metal species and amounts of metals used.

See more of this Session: Chemical and Catalytic Conversions and Processes for Renewable Feedstocks
See more of this Group/Topical: Sustainable Engineering Forum

(631c) Aqueous Processing of Cellulosic Biomass to Reactive Intermediates for Biological and Catalytic Conversion to Liquid Fuels and Other Products

Deepti Tanjore, Center for Environmental Research and Technology, University of California-Riverside, Riverside, CA and Charles Wyman, College of Engineering-Center for Environmental Research and Technology, University of California, Riverside, Riverside, CA

Hot water and dilute acid pretreatments are industrially scalable processes that can break the lignin matrix of cellulosic biomass and release hemicellulose sugars with good yields. High severity treatments facilitate deconstruction of crystalline cellulosic structure but also lead to the dehydration of hemicellulose sugars to furfural and formation of other degradation products. As a result, moderate severity conditions are typically applied to ensure high recovery of xylose and dissolved xylooligomers. However, xylan along with furfural can be catalytically converted to alkanes that can be used to synthesize jet and other hydrocarbon fuels. Thus, higher severity pretreatment conditions than typically considered to feed fermentation pathways can improve yields of dissolved carbon (xylan, furfural, glucan, levulinic acid, formic acid, and HMF) and prove to be a more economical for catalytic routes to making biofuels. In the present study, release of dissolved compounds was followed for treatment of sorghum biomass with just hot water and with dilute sulfuric acid at temperatures ranging between 160 and 260°C and reaction times between 5 and 50 minutes. In addition, some of the volatile compounds were captured and analyzed with gas chromatography. Optimal severities were identified for (i) maximum recovery of dissolved xylose and (ii) maximum production of dissolved organic carbon. When using just hot water, 6.9g/l of xylose was extracted into the liquid phase when biomass was held at 180°C for 30 minutes, and a total organic carbon concentration of 14g/l was obtained from reaction at 200°C for 5 minutes. It was interesting to find that while the solid residue progressively lost organic carbon with increasing severity, the highest severity did not provide the highest extraction of organic carbon into the liquid phase. The energy yields at both the optimal conditions were calculated and compared along with the mass yields of the products, and a higher energy yield from single unit process was found to be more economical than higher mass yield from several processes.

See more of this Session: Chemical and Catalytic Conversions and Processes for Renewable Feedstocks
See more of this Group/Topical: Sustainable Engineering Forum

(631d) Catalytic Conversion of Hemicellulose Sugars to Furfural in Ionic Liquid Media

Akinwale Shittu[1], **Thehazhnan (Thihal) K. Ponnaiyan**[1], Bin Li[2], Wenwen Zhang[1], Sridhar Viamajala[1] and Sasidhar Varanasi[1], (1)Chemical & Environmental Engineering, University of Toledo, Toledo, OH, (2)Bioengineering, University of Toledo, Toledo, OH

In an attempt to provide alternatives to petroleum based building blocks used in petrochemical refining, Furan derivatives obtained from lignocellulosic biomass derived carbohydrates are viable substitutes. Preliminary work has been done by other researchers on converting glucose and fructose which are cellulose sugars into furanic compounds. But since lignocellulosic biomass contains cellulose and hemicellulose sugars, there is a need to develop a carbon neutral method of converting the hemicellulose sugars into furanic compounds.

Hemicellulose, which is a branched polymer that contains many different types of sugars accounts for a substantial part of lignocellulosic biomass as it is embedded in the cell walls of plants. One of the principal sugars obtained from Hemicellulose hydrolysis is xylose. In this paper, Chromium (II) chloride and Copper (II) chloride are combined with ionic liquids to facilitate the dehydration of xylose into furfural.

Furfural is a platform based chemical used in petrochemical refining, It has the potential to be used as a starting material for alternative liquid fuels and additives and it can be catalytically hydrogenated into 2-methyltetrahydrofuran (MTHF), which is an oxygenated fuel additive.

See more of this Session: Chemical and Catalytic Conversions and Processes for Renewable Feedstocks
See more of this Group/Topical: Sustainable Engineering Forum

(631e) Consolidated Acid Catalysis and Extraction of Fermentable Sugars From Switchgrass in Ionic Liquids

Timothy C. R. Brennan[1], Blake A. Simmons[1], Harvey Blanch[2] and **Bradley M. Holmes**[1], (1) Deconstruction Division, The Joint BioEnergy Institute, Emeryville, CA, (2)Department of Chemical Engineering, University of California, Berkeley, Berkeley, CA

Ionic liquids (ILs) continue to draw attention as a novel technology for the pretreatment of biomass. The realization of an economically-viable pretreatment technology requires the near complete conversion of biomass into its component sugars and a closed-loop use of the ionic liquid solvent. The IL 1-butyl-3-methyl imidazolium chloride ([C4mim]Cl) has been shown to greatly accelerate the rates of enzymatic hydrolysis of holocellulose into monomeric sugars, and has also been used to produce sugars in situ using mineral acids such as HCl. Recently we reported the use of boronic acids to recover sugars from an aqueous-IL phase into an organic phase. Using naphthalene-2-boronic acid (N2B) in solution with hexane and Aliquat 336™, more than 80% of the total sugars produced by the acid hydrolysis of switchgrass in a system of 1-ethyl-3-methyl imidazolium chloride ([C2mim]Cl) and hydrochloric acid could be extracted. Ionic liquids were recycled and used for subsequent treatment, hydrolysis and extraction with minimal loss of performance efficiency. This IL treatment process provides an alternative route to the production of monomeric sugars from biomass that may eliminate the need for saccharolytic enzymes, and could potentially provide an economical route to the production of biofuels and other biomass derived products.

See more of this Session: Chemical and Catalytic Conversions and Processes for Renewable Feedstocks
See more of this Group/Topical: Sustainable Engineering Forum

(631f) Dilute Sulfuric Acid Pretreatment of Corn Stover Assisted by Renewable Carbonaceous Catalysts

Erik M. Kuhn[1], Joseph Shekiro[2], Richard T. Elander[2], Nicholas J. Nagle[1] and Clare J. Dibble[3], (1) National Renewable Energy Laboratory, Golden, CO, (2)National Bioenergy Center, National Renewable Energy Lab, Golden, CO, (3)National Bioenergy Center, National Renewable Energy Laboratory, Golden, CO

A range of industrially important fuels and chemicals such as ethanol can be produced from sugars. Xylose and glucose can be effectively extracted from renewable ligno-cellulosic biomass via depolymerization of hemicellulose and cellulose. A cost competitive bioconversion typically involves thermo-chemical pretreatment the ligno-cellulosic substrate followed by enzymatic deconstruction of the remaining cellulose into glucose. Dilute sulfuric acid pretreatment solubilizes much of the xylan into xylose and xylo-oligomers. However, fermentations typically require monomeric sugars to effectively convert sugars to higher value products. Xylo-oligomer conversion into monomers has been demonstrated by a secondary, mild thermochemical hydrolysis step and/or the addition of xylanases and side chain debranching enzymes. Thermochemical hydrolysis suffers from degradation due to lack of product specificity while enzymes are feedback inhibited, which makes it difficult to drive the reaction to completion. We seek a heterogeneous catalyst to selectively assist the hydrolysis of xylo-oligomers to xylose on a process relevant time scale.

We developed a reproducible method of generating a heterogeneous acid catalyst from a renewable source. Literature suggests that the conversion of ligno-cellulosic biomass through heterogeneous catalysis via carbonaceous materials is possible. In order to obtain a range of catalyst activities and support geometries, different polysaccharide candidates were chosen for catalyst generation, including Avicel, starch, cotton, and corn stover. Conditions such as carbonization temperature and method of sulfonation were varied during catalyst generation. The amount of thermal degradation and degree of sulfonation of the catalyst was determined through gravimetric analysis. Catalyst structural information was characterized by scanning electron microscopy (SEM), X-ray diffraction (XRD), and Raman spectroscopy.

Fundamental engineering principles state that significant mass transfer limitations exist when a solid substrate interacts with a solid catalyst. Therefore this study focuses on the conversion of solubilized xylo-oligomers to monomers in the dilute sulfuric acid pretreatment hydrolyzate in bench scale batch and continuous packed bed reactors. Solid catalysts could serve as a solid-liquid separation media, thereby increasing their utility to the process. These results are compared with whole slurry batch catalysis and dilute sulfuric acid catalysis. Catalyst selectivity for xylo-oligomers is investigated, and the yield of xylo-oligomer conversion into monomers determined.

(669a) Mechanistic Models for High-Solids Loading Pretreatment and Enzymatic Hydrolysis of Lignocellulosic Biomass

Andrew J. Griggs, Erik M. Kuhn, Xiaowen Chen, James J. Lischeske, Melvin P. Tucker and Jonathan J. Stickel, National Renewable Energy Laboratory, Golden, CO

Conversion of lignocellulosic biomass to fermentable sugars for renewable fuel production remains a formidable challenge for the economically feasible operation of bio-refineries. An improved understanding of the conversion reactions and the development of predictive models will help in overcoming this challenge, especially for high-solids loadings, where mass transfer may limit overall conversion yields. In this work, mechanistically based kinetic models for dilute-acid pretreatment and enzymatic hydrolysis have been developed, which consider the various physical and chemical changes to the feedstock. For dilute-acid pretreatment at high-solids loadings, we have developed a population-balance model, which considers the spatial distribution of reactants and products. The effects of the initial polydisperse particle-size distribution, and its evolution, are captured using a fragmentation-abrasion model. Such a model can aid in interpreting experimental results, determining reaction mechanisms, scale-up of pretreatment processes, and identifying the chemical and structural changes that result in high-xylose yields and enhanced-enzymatic digestibility. For example, when sieved fractions of milled corn stover are subjected to high-solids, dilute-acid pretreatment, the fraction with the largest particle size showed the greatest amount of fragmentation and yielded the most enzymatically digestible substrate, relative to the other sieved fractions. This finding suggests that the amount of fragmentation during dilute-acid pretreatment may be a measure of an effective pretreatment. For enzymatic hydrolysis, the distinct modes of action of the cellulase enzymes on soluble and insoluble substrates are each described by separate rate expressions. Depolymerization of the insoluble substrate by adsorbed enzyme is calculated using distribution kinetics, which tracks the molecular-weight distribution of the cellulose chains, rather than the concentration of individual species, over time. When combined with the method-of-moments, computationally efficient calculations are obtained. Model results throughout enzymatic saccharification are compared to experimental results for cellulose substrates, over a range of solids loadings. These models are capable of examining many of the relevant phenomena occurring during pretreatment and enzymatic saccharification, and aim to guide process development and identify optimal process-operating conditions.

See more of this Session: Reaction Kinetics and Transport Fundamentals for Biomass Conversion II
See more of this Group/Topical: Sustainable Engineering Forum

(669b) Modeling for Identification of Operating Strategies for Continuous Simultaneous Saccharification and Fermentation (cSSF)

Yi Jin[1], Jiacheng Shen[2], Jian Shi[2], Mirvat Ebrik[3] and Charles Wyman[4], (1)Chemical and Environmental Engineering,Bourns College of Engineering, Center for Environmental Research and Technology (CE-CERT) University of California, Riverside, Riverside, CA, (2)Center for Environmental Research and Technology, Bourns College of Engineering, University of California, Riverside, Riverisde, CA, (3) College of Engineering-Center for Environmental Research and Technology, University of California, Riverside, Riverisde, CA, (4)College of Engineering-Center for Environmental Research and Technology, University of California, Riverside, Riverside, CA

Although many economic studies assume continuous Simultaneous Saccharification and Fermentation (cSSF) will be the preferred configuration for commercial applications to achieve higher productivity and eliminate the need for seed fermentors, little information is available to guide process design or optimization. Therefore, in this paper, we focused on developing kinetic models to predict cSSF performance and applying them to better understand potential tradeoffs among enzyme loadings, residence times, yields, organism washout, and organism fermentation features to inform our new experimental research approach. Comparisons are made between batch and continuous processing to understand how end product inhibition and other characteristics differ and their effects on performance. In addition, simplified kinetic models are devised that require fewer measurable parameters to more easily predict the threshold for organism washout and how enzyme loadings impact system performance. Experimental results are compared to model predictions, and opportunities are identified for improving model accuracy. The result will be applied to identify optimal enzyme loadings and residence times for different pretreated feedstocks based on their compositions.

See more of this Session: Reaction Kinetics and Transport Fundamentals for Biomass Conversion II
See more of this Group/Topical: Sustainable Engineering Forum

(669c) Molecular Modeling of Cellulose Pyrolysis Using Molecular Dynamics

Vishal Agarwal, Chemical Engineering, University of Massachusetts, Amherst, MA, George W. Huber, Chemical Engineering, University of Massachusetts-Amherst, Amherst, MA, William C. Conner, Department of Chemical Engineering, University of Massachusetts-Amherst, Amherst, MA and Scott M. Auerbach, Chemistry and Chemical Engineering, Univ. of Massachusetts, Amherst, MA

Biomass pyrolysis is one of the viable routes, which can be used for producing gases and liquids that can replace petroleum-derived products. However, molecular mechanisms that control the physical and chemical transformation of cellulose during pyrolysis is poorly understood. Understanding this can help to control degradation process and minimize unwanted coke formation. Low temperature degradation of cellulose is proposed to proceed via a "active cellulose" stage [1]. However, such transition process and the nature of so called "active cellulose" is poorly understood. We have applied NPT molecular dynamics (MD) using ReaxFF (a reactive force field) [2] to model this transition process. We have used cellulose Iβ as the starting material since it is the most abundant crystalline form in plant material. We have found that ReaxFF can accurately model the crystal structure of cellulose Iβ. The modeled crystal agreed reasonably well with experimental results, including crystal density, unit cell parameters, chain conformation parameters, hydrogen bonds and thermal expansion coefficient at room temperature. The transition from cellulose to "active cellulose" is a physical change involving breaking of only intermolecular hydrogen bonds. This transition occurs at 480 K agreeing well with experimental results [1]. We have computed the changes in cell parameters, hydrogen bonding network, crystal density and chain conformational parameters with temperature. The results have been found to be in good agreement with the experimental results.

References

1. Lin Y., Cho J., Tompsett G. A., Westmoreland P. R. and Huber G. W., "Kinetics and Mechanisms of Cellulose Pyrolysis", J. Phys. Chem. C 113, 2009, 20097-20107.

2. Chenoweth K., van Duin A. C. T. and Goddard A., "Reactive Force Field for Molecular Dynmaics Simulations of Hydrocarbon Oxygen", J. Phys. Chem. A 112, 2008, 1040-1053.

See more of this Session: Reaction Kinetics and Transport Fundamentals for Biomass Conversion II
See more of this Group/Topical: Sustainable Engineering Forum

(669d) Reaction Kinetics of Transesterification of Canola Oil to Biodiesel Catalyzed by Nanopowder Calcium Oxide

Lina Zhao, Zheyan Qiu and Susan M. Stagg-Williams, Chemical & Petroleum Engineering Department, The University of Kansas, Lawrence, KS

The reaction kinetics model of transesterification of canola oil to biodiesel catalyzed by nanopowder calcium oxide was expressed.The reaction mechanism based on Langmuir-hinshelwood model has been proposed. In this proposed mechanism the overall reaction rate is assumed to be controlled by triglyceride adsorption step in the initial period followed by chemical reaction-controlled period. The apparent reaction constants, apparent activation energy and pre-exponential factors for both periods have been calculated based on experimental data. The internal mass transfer does not limit the reaction due to the nanosize of the catalyst particle.

See more of this Session: Reaction Kinetics and Transport Fundamentals for Biomass Conversion II
See more of this Group/Topical: Sustainable Engineering Forum

(669e) Water-Only Flowthrough Pretreatment of Poplar and Birchwood Xylan

Heather L. McKenzie[1], Nancy L. Engle[2], Joshua F. Emory[3], Bruce A. Tomkins[3], Timothy J. Tschaplinski[2], Gary J. Van Berkel[3] and Charles E. Wyman[1], (1)Chemical and Environmental Engineering, Center for Environmental Research and Technology, University of California Riverside, Riverside, CA, (2)Environmental Sciences Division, Oak Ridge National Laboratory, Oak Ridge, TN, (3) Chemical Sciences Division, Oak Ridge National Laboratory, Oak Ridge, TN

The conversion of lignocellulosic biomass to ethanol requires recovery of the sugars contained in the hemicellulose and cellulose. High temperature water-only pretreatment solubilizes primarily sugars and oligomers in hemicellulose prior to the enzymatic hydrolysis of the cellulose in the remaining solids to glucose. The understanding and kinetic models of hemicellulose hydrolysis are generally simplistic and do not fully address the existence of xylooligomers, lignin, and the interaction of biomass components. Furthermore, pretreatment of cellulosic biomass in the batch reactors typically employed makes the collection of time-dependent data challenging. On the other hand, a fixed bed flowthrough reactor facilitates tracking the release of biomass components as a function of time, thus making it a valuable laboratory tool. Analyzing the flow behavior and release patterns of xylooligomers from poplar and xylan would improve the understanding of the relative importance of mass transfer and kinetics during pretreatment. Analysis of void volume and flow behavior of a fixed bed flowthrough reactor will be determined to help characterize mass transfer in the system. Although lignin morphology is known to change during pretreatment, the cause of these changes is not well-known. It has been suggested that lignin depolymerizes (Li et al., 2007) or melts (Donohoe et al., 2008) during pretreatment. In an attempt to clarify the differences between these mechanisms, the release of lignin and lignin derivatives from poplar are followed over a range of reaction times, temperatures, and flow rates. Finally, pretreatment results from poplar and purified xylan are compared to clarify xylan interactions with other poplar components, such as cellulose and lignin.

See more of this Session: Reaction Kinetics and Transport Fundamentals for Biomass Conversion II
See more of this Group/Topical: Sustainable Engineering Forum

(669f) Comparative Pyrolysis Kinetics for One Legume and Two Grass Hays

Joseph J. Biernacki[1], Jessica Murillo[1], Prejeeth Ambuken[1] and C. Pat Bagley[2], (1)Chemical Engineering, Tennessee Technological University, Cookeville, TN, (2)Agriculture and Human Ecology, Tennessee Technological University, Cookeville, TN

While the reactions involved in plant biomass are somewhat understood, there are still unknowns, particularly regarding the details of the kinetic pathways. Furthermore, there are a wide range of biomass feedstocks presently being considered for which there is no kinetic information available. This work compares the pyrolysis of one legume (alfalfa hay) and two grass hays (sudan grass hay and tall fescue). Alfalfa hay is a premium lifestock feed and would likely not be used in the future as a feed for production of biofuel, however, provides a basis upon which to compare the two grasses, which are of poor and very poor nutritional content and are likely candidates for biofuel conversion. Conventional pyrolysis using thermalgravimetric analysis (TGA) was performed on the dried, pulverized feeds. Heating rates of 5, 10, 30 and 50 oC/min were used as a basis. Mass spectroscopy was performed and and pyrolysis rate curves analyzed using a multi-process model wherein pyrolysis is assumed to be a series of discrete independent reactions. The results were used to infer alternative appraoches that account for the dramatic change in pyrolysis efficiency as a function of heating rate.

See more of this Session: Reaction Kinetics and Transport Fundamentals for Biomass Conversion II
See more of this Group/Topical: Sustainable Engineering Forum

(674a) Application of Top-Down Ecological Control Principles to Maximize Algal Productivity in Open Pond Systems

Belinda S.M. Sturm, Environmental Engineering, The University of Kansas, Lawrence, KS, Val H. Smith, Ecology and Evolutionary Biology, University of Kansas, Lawrence and Frank Jerry deNoyelles, Kansas Biological Survey, Lawrence

Of the biofuel feedstocks investigated to reduce the nation's dependence on foreign oil, algae are the most promising due to their high lipid content, rapid growth rate, and ability to sequester CO2. Algae do not compete with existing food commodities, can grow on marginal lands not suitable for agriculture, and do not require large volumes of fresh water. Moreover, algae can remove nitrogen and phosphorus from nutrient-rich wastewaters, thus reducing eutrophication of receiving water bodies that cause gross disruptions of ecosystems, such as the Gulf of Mexico hypoxia. Algal production has been estimated to yield 3,200 to 14,600 gallons of oil/acre/year, but mass culture efforts have yielded 10-20 times less oil than expected. Outdoor bioreactors that are subject to natural variations and function as ecosystems have not mimicked pure laboratory reactors, and sustainable production of algae will require an abundant and nutrient-rich water source, which is unlikely to support the mass cultivation of pure cultures. This project incorporates ecological and reactor engineering concepts to develop a sustainable algal growth system in wastewater-fed, outdoor bioreactors. Four 2600-gallon continuous-flow reactors with a ten day hydraulic residence time were operated at the Lawrence, KS Wastewater Treatment Plant for six months in 2009. Top-down ecological control practices were applied to two of the reactors in order to increase and stabilize algal production.

The growth of algae growing in mixed-species, outdoor community cultures will be regulated by their supplies of macronutrients (nitrogen and phosphorus), carbon dioxide, and incident light. These algal species will vary significantly in size and shape; some of them will be small and edible by algae-eating zooplankton, while some of them will be large and inedible. The amount of lipid that can be produced from these algae will be constrained by zooplankton grazing in the bioreactor. During the past 25 years, a very large and globally-accepted body of literature has been developed which indicates that in natural lakes, algal biomass losses to zooplankton grazing can be minimized by including zooplanktivorous fish in the food web (top-down ecological control; see Fig. 2 from Smith et al. 2010). Likewise, algal biomass in outdoor bioreactors can be strongly affected by food web structure. Fishless bioreactors can be expected to develop large populations of large-bodied herbivorous zooplankton, thus exhibiting significantly lower algal biomass than would be predicted from their nutrient supply, relative to bioreactors that contain zooplanktivorous fish and smaller, less-efficient grazers.

One of our primary research goals was to determine whether the ecological concept of top-down control applies to microalgae grown in outdoor bioreactors. If this concept indeed applies, then careful control of food-web structure will allow for the design and operation of future large-scale outdoor bioreactors or ponds for algal biomass production – without the danger of algal population crashes due to predation.

Highlights of Results • Algae may be successfully cultivated in open pond reactors using wastewater effluent as an N and P source. • Top-down ecological control practices can decrease losses of algae to zooplankton predation. • The % lipid content of algae cultivated in the field is consistent with laboratory studies.

References Smith VH, Sturm BSM, Billings S, deNoyelles FJ. 2010. "Ecological aspects of algal biodiesel production". Trends in Ecology & Evolution, 25(5), 301-309.

See more of this Session: Advances in Algal Biorefineries I
See more of this Group/Topical: Sustainable Engineering Forum

(674b) Maximizing Productivity in Batch Reactors of Microalgae Nannochloropsis Gaditana

Ming Ren, Department of Chemical and Environmental Engineering, University of Arizona, Tucson, AZ and Kimberly Ogden, University of Arizona, Tucson, AZ

Biodiesel from algae is receiving more attention. Just recently, the United States Department of Energy (DOE) announced to fund around $85 million in the development of algae-based biofuels and advanced infrastructure-compatible biofuels. Compared to other terrestrial crops, such as soybean, sunflower, canola, palm, jatropha and others, algae has many advantages. It has high oil productivity, a short growth cycle and survives in a wide variety of water sources including high salinity and waste water.

Nutrients, salinity, pH, temperature, light intensity and aging of the culture can all affect on both lipid content and fatty acid profile. Usually nutrient stress is the easiest way to manipulate lipid composition and increase lipid content. In our research, microalgae *Nannochloropsis Gaditana* has been shown to have high lipid content (more than 40%) and quick growth rate (around $0.018 hr^{-1}$). Current work focuses on how the harvest time and nitrogen sources affect on the lipid content and composition.

The cell growth, sodium nitrate concentration, and lipid content and profile are monitored as a function of time. Decreasing nitrogen concentration does stimulate lipid content increase with sacrificing growth rate. In contrast, increasing nitrogen concentration does not increase lipid content. Maximum lipid content is obtained in mid-exponential phase and is relatively constant from this point through stationary phase. Major lipid compounds are palmitic acid (16:0) and palmitoleic acid (16:1), and lipid profile does change significantly with time.

See more of this Session: Advances in Algal Biorefineries I
See more of this Group/Topical: Sustainable Engineering Forum

(674c) Photobioreactor Productivity Enhancement Offered by Oscillatory Flow Mixing

Benjamin J. Taylor and Malcolm Mackley, Department of Chemical Engineering and Biotechnology, University of Cambridge, Cambridge, United Kingdom

Closed photobioreactors offer certain advantages over open algal growth systems in terms of consistent product output and purity but still suffer from other drawbacks. Limitations of many tubular photobioreactors were identified, and a new reactor was constructed to meet these challenges by utilizing the proven processing advantages of oscillatory flow mixing (OFM). The unique mass and energy transfer characteristics of OFM offer excellent bulk-to-wall turnover and suspension capabilities similar to that provided by turbulent flow but at significantly reduced input energy requirements. Growth experiments with the algal strain *Scenedesmus obliquus* demonstrated improved growth kinetics and sustained productivity within the laboratory-scale OFM reactor surpassing those from other bioreactor types. Computational fluid dynamics accurately predicted the suspension flow patterns obtained experimentally with OFM and offer insight into proper process control variables tailored to a wide range of algal species and reactor scales.

See more of this Session: Advances in Algal Biorefineries I
See more of this Group/Topical: Sustainable Engineering Forum

Effects of Total Inorganic Carbon on Growth of *Chlorella vulgaris*

Jinsoo Kim [a], Joo-Youp Lee [a*], Kaniz F. Siddiqui [b]

[a] *Chemical Engineering Program, School of Energy, Environmental, Biological, and Medical Engineering, University of Cincinnati, Cincinnati, OH, 45221-0012, U.S.A.*

[b] *The Metropolitan Sewer District of Greater Cincinnati, Cincinnati, Ohio 45204-2022, U.S.A.*

1. Introduction

Diminishing petroleum resources along with increased demand for petroleum by emerging economies are driving our society to search for new renewable and carbon neutral liquid transportation fuels. Biofuels such as bioethanol and biodiesel derived from agricultural and forestry residues and oil crops have been considered to be candidates for renewable liquid fuels. Recently, microalgae-derived biodiesel production has received great attention due to their ability to produce substantial amounts of triacylglycerols (e.g., 20–50% dry cell weight) [1], as a feedstock for biodiesel [2, 3], and to utilize a large fraction of solar energy (up to 10%) [4], and carbon dioxide (CO_2), and to grow at much faster rates than terrestrial biomass [5]. In addition, sugars generated as metabolites have potential to produce bioalcohols [6-10].

Microalgae utilize carbon dioxide (CO_2) as their carbon source and energy primarily from sunlight. They are photosynthetic, like terrestrial plants, but lack the various structures that characterize plants. It is estimated that there are 200,000–1,000,000 different microalgal strains in nature, but only ~30,000 strains have been described [11]. Among the strains, *Chlorella vulgaris* belongs to a genus of single-celled green algae, and is known as one of the fastest growing microalgae, and typically includes 14–22 % of lipid, 51–58 % of protein, 12–17 % of carbohydrates, and 4–5 % of nucleic acid [12]. Several studies on lipid production [13] and growth kinetics of *Chlorella vulgaris* have been conducted in terms of initial CO_2 concentration [14] and pH level [15, 16] in order to find optimum growth conditions for large-scale biofuels production. However, these previous growth kinetic studies did not focus on actual initial total inorganic carbon concentration used for algae growth in the aqueous phase. The solubility of CO_2 gas in water is very limited, and some of microalgae have the capability to convert bicarbonate ion into CO_2 for their carbon uptake with the help of an enzyme called carbonic anhydrase [17-20].

In this study, the initial total inorganic carbon concentration in the culture medium was controlled by using sodium bicarbonate ($NaHCO_3$) and CO_2 gas as a carbon source for photosynthesis of *Chlorella vulgaris* under the similar initial pH condition. This method is allowed to be controlled the initial total inorganic carbon concentration in culture medium due to the high solubility of the sodium bicarbonate in the water. In addition, the bubbled CO_2 was used for the decreasing pH and injected CO_2 gas was captured by the sodium bicarbonate.

2. Experimental

2.1. Preparation of Culture Medium

A culture medium was prepared by following the shuisheng-4 method [21, 22], and included ammonium sulfate ($(NH_4)_2SO_4$, EM Industries, Inc., Gibbstown, N.J.), boric acid (Allied chemical), cupric sulfate (J. T. Baker), calcium phosphate monobasic monohydrate, magnesium sulfate hetahydrate, sodium bicarbonate, potassium chloride, iron (III) chloride, potassium phosphate dibasic, manganese chloride, zinc sulfate, sodium molybdate (Fisher Scientific Company). The medium was mixed together with 200 mg/L of ammonia sulfate ($(NH_4)_2SO_4$), 30 mg/L of calcium phosphate monobasic monohydrate ($Ca(H_2PO_4)_2 \cdot H_2O$), 100 mg/L of magnesium sulfate hetahydrate ($MgSO_4 \cdot 7H_2O_3$) 80 mg/L, 25 mg/L of potassium chloride (KCl), 1.5 mg/L of iron (III) chloride($FeCl_3$), 10 mg/L of potassium phosphate dibasic (K_2HPO_4), and 1 mL of A5 liquid. Herein, A5 liquid was prepared by mixing together with 2.86 g/L of boric acid (H_3BO_3), 1.81 g/L of manganese chloride ($MnCl_2 \cdot 4H_2O$), 0.222 g/L of zinc sulfate ($ZnSO_4 \cdot 7H_2O$), 0.391 g/L of sodium molybdate ($Na_2MO_4 \cdot 2H_2O$), 0.079 g/L of cupric sulfate ($CuSO_4 \cdot 5H_2O$). In addition, sodium bicarbonate ($NaHCO_3$) was not added in this step because the sodium bicarbonate was used as a inorganic carbon source in the next step.

2.2. Preparation of Different Initial Total Inorganic Carbon Concentration in Culture Medium

The initial total inorganic carbon in S1 culture medium was adjusted by adding sodium hydroxide (NaOH) until pH increased to 8.99 and bubbling CO_2 gas ($\geq$99.5%, Wright Brothers, Inc., Cincinnati, OH) until pH decreased to 6.99. Then, the initial total inorganic carbon in S2-S5 culture mediums were controlled by adding different amount of sodium bicarbonate ($NaHCO_3$) (e.g. 100 mg of $NaHCO_3$/ 1L H_2O (S2, pH=7.05), 400 mg of $NaHCO_3$/ 1L H_2O (S3, pH=7.36), 1600 mg of $NaHCO_3$/ 1L H_2O (S4, pH=7.68) and 3200 mg of $NaHCO_3$/ 1L H_2O and bubbling CO_2 gas until pH decreased to pH=7 and the final pH was 7.05 (S2), 6.82 (S3), 7.04 (S4), and 7.08 (S5), respectively. Here, in the case of S2 culture medium, extra CO_2 gas was not bubbled after adding sodium bicarbonate ($NaHCO_3$) because the pH was close to pH=7.

2.3. Determination of Cell Density of Chlorella vulgaris

The cell density of *Chlorella vulgaris* was determined by measuring the optical density of a 15-mL sample at 682 nm [13] for every 24 hours by using UV-vis spectrophotometer (UV-1800, Shimadzu Scientific Instruments). The absorbance of UV spectrophotometer at 682 nm was calibrated by measuring the weight of dried *Chlorella vulgaris*. Then, the weight of dried biomass was obtained from the prepared calibration curve.

2.4. *Determination of Total Inorganic Carbon Concentration*

An acid-base titration method [23] was used to determine the inorganic carbon species in the aqueous phase. This titration method determines total inorganic carbon concentration (carbonate (CO_3^{2-}), bicarbonate (HCO_3^-) and aqueous carbon dioxide (CO_2* (aq))) using a 0.01 N hydrochloric acid solution as a titrant.

3. Results and Discussion

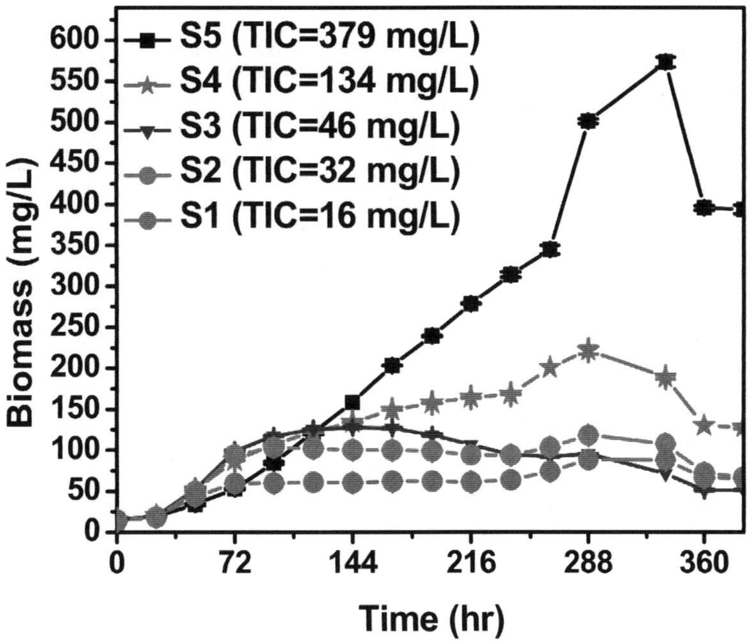

Figure 1. Growth kinetics of *Chlorella vulgaris* dependent on different initial total inorganic carbon (e.g. S1 (TIC=16 mg/L), S2 (TIC=32 mg/L), S3 (TIC=46 mg/L), S4 (TIC=134 mg/L), and S5 (TIC=379 mg/L).

The growth of *Chlorella vulgaris* and the consumption of total inorganic carbon are represented in Figures 1 and 2. At first, initial total inorganic carbon in S1-S4 culture mediums were 16 mg/L (S1), 32 mg/L (S2), 46 mg/L (S3), 134 mg/L (S4) and 379 mg/L (S5), respectively. The lag-phase of *Chlorella vulgaris* in S1 and S2 culture mediums was short (24 hrs) and the phase of *Chlorella vulgaris* in S3 and S4 culture mediums was modest (48 hrs) while a small amount of dissolved total inorganic carbon concentration in those mediums decreased during the period. Otherwise, the lag-phase in S5 culture medium showed the longest lag period (72 hrs) compared to other cases (S1-S4) because it might be derived from the high concentration of total inorganic carbon. During the exponential growth phase, the cell density of *Chlorella vulgaris* in S1-S5 culture mediums increased, and the dissolved total inorganic carbon dramatically decreased. Then, the growth of the algae in culture mediums stopped at 72 hrs (S1), 96 hrs (S2), 144 hrs (S3), 288 hrs (S4) and 336 hrs (S5), respectively. In addition, the maximum cell density decreased in the order of S5 (573.27±5.90 mg/L) > S4 (221.68±4.94 mg/L) > S3 (95.17±0.23) > S2 (118.06±0.23)> S1 (87.87±0.00). Therefore, it was noted that *Chlorella vulgaris* cultured in the medium with high total inorganic carbon concentrations can produce more algal biomass yield. However, the carbon consumption was delayed as its concentration increased, and too high a total inorganic carbon concentration resulted in longer lag-phase period. The pH of the culture medium continued to increase from the lag phase through the growth phase (Figure 2) as *Chlorella vulgaris* continued to produce hydroxyl ion (OH^-) during photosynthesis by the following reaction, $HCO_3^- \rightarrow CO_2 + OH^-$. For the culture mediums with moderate total inorganic carbon concentrations (i.e. S1–S3), as the uptake of total inorganic carbon levels off, the pH also reached a constant level.

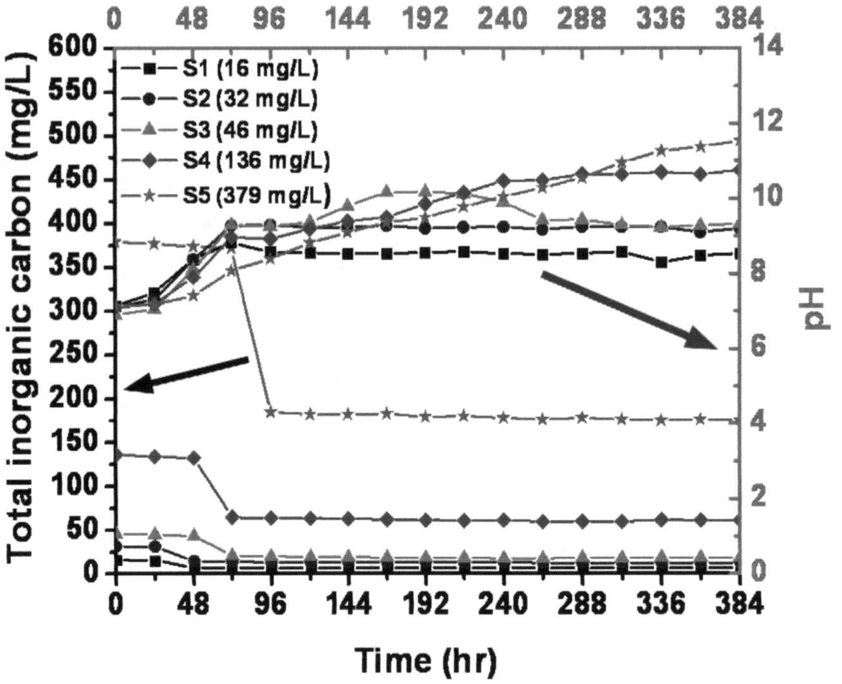

Figure 2. Change of total inorganic carbon concentration in culture mediums.

4. Conclusions

The effects of total inorganic carbon concentrations on the growth of *Chlorella vulgaris* have been studied using sodium bicarbonate and carbon dioxide gas as carbon sources for photosynthesis. The lag-phase period increased with an increase in inorganic carbon concentrations (i.e. 24 hrs for S1 and S2, 48 hrs for S3 and S4, and ~72 hrs for S5). The period of growth-phase increased with an increase in total inorganic carbon concentrations. However, a long lag-phase period was observed from the culture medium with an extremely high inorganic carbon concentration (i.e. S5). It was also observed that *Chlorella vulgaris* cultured in the medium with high total inorganic carbon concentrations can produce high algal biomass yield (i.e. S5 > S4> S3 > S2 >S1).

Acknowledgement

The This work was supported by the Metropolitan Sewer District of Greater Cincinnati (MSDGC) under Master Services Agreement No. 85X10431, Task Order No. 0210000209. The authors appreciate their financial support.

References

1. Hu, Q.; Sommerfeld, M.; Jarvis, E.; Ghirardi, M.; Posewitz, M.; Seibert, M.; Darzins, A., Microalgal triacylglycerols as feedstocks for biofuel production: perspectives and advances. *Plant Journal* **2008,** *54*, (4), 621-639.
2. Liu, Z.-Y.; Wang, G.-C.; Zhou, B.-C., Effect of iron on growth and lipid accumulation in Chlorella vulgaris. *Bioresource Technology* **2008,** *99*, (11), 4717-4722.
3. Hill, J.; Nelson, E.; Tilman, D.; Polasky, S.; Tiffany, D., Environmental, economic, and energetic costs and benefits of biodiesel and ethanol biofuels. *Proceedings of the National Academy of Sciences of the United States of America* **2006,** *103*, (30), 11206-11210.
4. Huber, G. W.; Iborra, S.; Corma, A., Synthesis of transportation fuels from biomass: chemistry, catalysts, and engineering. *Chemical Reviews* **2006,** *106*, (9), 4404-4498.
5. Chisti, Y., Biodiesel from microalgae. *Biotechnology Advances* **2007,** *25*.
6. Milne, T. A.; Evans, R. J.; Nagle, N., Catalytic conversion of microalgae and vegetable oils to premium gasoline, with shape-selective zeolites. *Biomass* **1990,** *21*, (3), 219-232.
7. Miao, X.; Wu, Q., Biodiesel production from heterotrophic microalgal oil. *Bioresource Technology* **2006,** *97*, (6), 841-846.
8. Dote, Y.; Sawayama, S.; Inoue, S.; Minowa, T.; Yokoyama, S.-y., Recovery of liquid fuel from hydrocarbon-rich microalgae by thermochemical liquefaction. *Fuel* **1994,** *73*, (12), 1855-1857.
9. Minowa, T.; Yokoyama, S.-y.; Kishimoto, M.; Okakura, T., Oil production from algal cells of Dunaliella tertiolecta by direct thermochemical liquefaction. *Fuel* **1995,** *74*, (12), 1735-1738.
10. Ginzburg, B.-Z., Liquid fuel (oil) from halophilic algae: A renewable source of non-polluting energy. *Renewable Energy* **1993,** *3*, (2-3), 249-252.
11. Sheehan, J.; Dunahay, T.; Benemann, J.; Roessler, P. *A look back at the U.S. Department of Energy's Aquatic Species Program—biodiesel from algae.*; National Renewable Energy Laboratory: 1998.

12. Becker, E. W., *Microalgae biotechnology and microbiology.* Cambridge University Press: New York, 1994.

13. Widjaja, A.; Chien, C.; Ju, Y., Study of increasing lipid production from fresh water microalgae Chlorella vulgaris. *Journal of the Taiwan Institute of Chemical Engineers* **2009,** *40,* (1), 13-20.

14. Azov, Y., Effect of pH on inorganic carbon uptake in algal cultures. *Applied and Environmental Microbiology* **1982,** *43,* (6), 1300-1306.

15. Goldman, J. C.; Azov, Y.; Riley, C. B.; Dennett, M. R., The effect of pH in intensive microalgal cultures. I. Biomass regulation. *Journal of Experimental Marine Biology and Ecology* **1982,** *57,* (1), 1-13.

16. Powell, E. E.; Mapiour, M. L.; Evitts, R. W.; Hill, G. A., Growth kinetics of Chlorella vulgaris and its use as a cathodic half cell. *Bioresource Technology* **2009,** *100,* (1), 269-274.

17. Badger, M. R.; Price, G. D., The role of carbonic anhydrase in photosynthesis. *Journal of Experimental Marine Biology and Ecology* **1994,** *45,* 369-392.

18. Kaplan, A.; Reinhold, L., CO2 concentrating mechanisms in photosynthetic microorganisms. *Annual Review of Plant Physiology and Plant Molecular Biology* **1999,** *50,* 539-570.

19. Moroney, J. V.; Somanchi, A., How do algae concentrate CO2 to increase the efficiency of photosynthetic carbon fixation? *Plant Physiology* **1999,** *119,* 9-16.

20. Giordano, M.; Beardall, J.; Reaven, J. A., CO2 concentrating mechanisms in algae: Mechanisms, environmental modulation, and evolution. *Annual Review of Plant Biology* **2005,** *56,* 99-131.

21. Zhou, Y. X.; Zhang, Z. S., *The toxicity methods in aquatic living.* Agriculture Press: Beijing, 1989.

22. Qian, H.; Chen, W.; Sheng, G. D.; Xu, X.; Liu, W.; Fu, Z., Effect of glufosinate on antioxidant enzymes, subcellular structure, and gene expression in the unicellular green alga Chlorella vulgaris. *Aquatic Toxicology* **2008,** *88,* (4), 301-307.

23. Fisher, R. B.; Peters, D. G., *Basic theory and practice of quantitative chemical analysis.* 3rd edition ed.; W. B. Saunders Company: Piladelphia, 1968.

(674e) Effects of CO_2 On South African Fresh Water Microalgae Growth

Edmore Kativu[1], Tonderayi S Matambo[2], Diane Hildebrandt[2] and David Glasser[2], (1)School of chemical and metallurgical engineering, University of the Witwatersrand, Johannesburg, Johannesburg, South Africa, (2)Chemical and Metallurgical Engineering, University of the Witwatersrand, Johannesburg, Johannesburg, South Africa

South Africa obtains over 80% of its energy needs from fossil fuels. Latest research indicate that South Africa emits over 222 000 000 tones of carbon dioxide (CO_2) per year and it is ranked number eight as the world's CO_2 emitter. Unabated or controlled emission of CO_2 and other green house gases has largely contributed to detrimental global warming and adverse climatic change. Microalgae are capable to convert the hazardous CO_2 into valuable biomass. A high CO_2 tolerating microalgae was collected from Johannesburg Zoo Lake. Batch cultures were grown on 1000 ml Erlenmeyer flasks. CO_2 was bubbled into microalgae culture media. The media contained optimal nutrients and under optimal photoperiod and light intensity and controlled pH. Concentrations of 100 %, 50 %, 25 %, 10 %, and 5 % CO_2 and total gas flow rates of 20, 50 and 100 ml/min were used. Final concentrations were made up to 100 % by mixing with nitrogen gas. The aim of this study is to find out the effective CO_2 concentrations that gives optimal microalgae growth. Microalgae have at 50% carbon in its dry mass and when all the other nutritional and environmental conditions are not limiting carbon is the main limiting factor. When atmospheric air was supplied at 50 ml/minute the growth rate was 0.1151 per day and it increased drastically by almost 5 folds when 100% CO_2 was supplied at 50 ml per minute. At a gas flow rate of 20 ml/min 10 % CO_2, growth rate optimal growth rate was 1.42 per day and a dry biomass yield of 809.96mg/l after 12 days. At a gas flow rate 50 ml/min 5% CO_2 and the highest growth rate was 1.993 per day and an overall biomass yield of 1200mg/l and the average pH was 6.36, at 100% CO_2 the growth rate was 1.265 per day and dry biomass yield of 469.81 mg/l and the average pH was 5.12. When gas flow rate was increased to 100 ml/minute 5 % CO_2 the growth rate slightly reduced to 1.30 per day and dry biomass yield of 1000 mg/l, while at 100% the growth rate was 0.34 and dry biomass yield of 279.47 mg/l and the average pH was 5.09. Higher flow rates and higher concentrations resulted in slightly reduced growth due to low pH. These investigations indicate that the species under study is CO_2 tolerant.

See more of this Session: Advances in Algal Biorefineries I
See more of this Group/Topical: Sustainable Engineering Forum

(674f) Investigation of Chlorella Microalgae Cultivation On Anaerobically Digested Manure

Amarjeet S. Bassi, Chemical & Biochemical Engineering, The University of Western Ontario, London, ON, Canada, Gureet Chandok, Chemical and Biochemical Engineering, The University of Western Ontario, London, ON, Canada, Shaikh Razzak, Chemical and Biochem Engg, The University of Western Ontario, London, ON, Canada and Peter Schnurr, Chemical and Biochemical Engg, The University of Western Ontario, London, ON, Canada

Laboratory and pilot scale studies were carried out to investigate the cultivation of the microalgae *Chlorella vulgaris* on anaerobically digested manure from a local farm. The effects of light, nutrient load and CO_2 to air ratios on cell growth and lipid accumulation were investigated. Both batch and continuous studies were carried out. Rates of CO_2 sequestration were also measured. The results showed that the photo-bioreactor configuration was appropriate to cultivate microalgae on a large on farm operation. Only a small fraction of CO_2 supplied was effectively taken up as microalgae biomass, indicating that using microalgae only for CO_2 uptake is not effective.

See more of this Session: Advances in Algal Biorefineries I
See more of this Group/Topical: Sustainable Engineering Forum

(675a) GLBRC Research Aimed at Improving Plant Biomass for Biofuels Generation

John C. Sedbrook, School Of Biological Sciences, Illinois State University, Great Lakes Bioenergy Research Center, Normal, IL

Plants use the energy from sunlight to fix carbon dioxide and generate a number of energy rich molecules that are incorporated into cell walls (polysaccharides and lignin) as well as used for energy storage (eg. triacylglycerides). Researchers at the Great Lakes Bioenergy Research Center (GLBRC) are pursuing fundamental research that will allow for the engineering and breeding of bioenergy crop plants with increased amounts of easily accessible and convertible polysaccharides, and that accumulate energy-dense oils in vegetative tissues. This presentation will highlight Center efforts to 1) engineer lignin to contain zips that can readily be chemically cleaved thereby releasing cell wall polysaccharides for saccharification and fermentation; 2) identify genes and gene variants affecting polysaccharide composition, accumulation, and enzymatic digestibility; and 3) engineer plants to synthesize and retain up to 20% oil in vegetative tissues such as leaves, stems, and tubers.

See more of this Session: Advances in Biofuels: DOE Bioenergy Research Centers I
See more of this Group/Topical: Sustainable Engineering Forum

(675b) Advances in Switchgrass (Panicum virgatum L.) Biotechnology towards Improved Feedstocks for Lignocellulosic Biomass Conversion

David G.J. Mann[1], C. Neal Stewart Jr.[1] and Paul Gilna[2], (1)The University of Tennessee, Knoxville, TN, (2)BioEnergy Science Center, Oak Ridge National Laboratory, Oak Ridge, TN

Switchgrass (Panicum virgatum L.) is a leading candidate feedstock for biofuels in the U.S. and is a crucial model feedstock component of the BioEnergy Science Center (BESC). Improvements in biotechnology of switchgrass are important for screening potential cell wall biosynthesis genes, and genetic transformation of switchgrass is being performed by five laboratories in three institutions within BESC. Within our lab, goals and milestones include improving tissue culture and transformation systems, isolating novel switchgrass promoters, developing a new versatile vector set for monocot transformation, and altering lignin and cellulose biosynthesis within switchgrass. Using a novel non-MS based tissue culture medium for increased production and maintenance of type II callus, the efficiency of stable transformation in switchgrass has increased from less than 3% to 10-15%, making it feasible for high throughput applications. Additionally, switchgrass cell suspension cultures have been produced and characterized for mutant selection, mass propagation, and gene transfer experiments via protoplast isolation for cell wall trait assessment. In order to coordinate gene expression within the BESC switchgrass transformation labs and to facilitate more rapid screening of genes, we have developed a Gateway-compatible transformation vector set (termed "pANIC") for overexpression and RNAi-mediated knockdown for use in switchgrass and other monocot species. Overall, BESC has facilitated the coordination of scientific expertise and research in switchgrass biotechnology that would have been otherwise impossible by one investigator with funding under a traditional grant.

See more of this Session: Advances in Biofuels: DOE Bioenergy Research Centers I
See more of this Group/Topical: Sustainable Engineering Forum

Chemical Transformations of Poplar Lignin during Dilute Acid Pretreatment

Yunqiao Pu[1,2], Shilin Cao[1,3], Michael Studer[1,4], Charles Wyman[1,4,5], and Arthur J. Ragauskas[1,3]

[1]BioEnergy Science Center, [2]Institute of Paper Science and Technology, [3]School of Chemistry and Biochemistry, Georgia Institute of Technology, Atlanta, GA; [4]CE-CERT, [5]Chemical and Environmental Engineering Department, Bourns College of Engineering, University of California, Riverside, CA;

Increasing concerns on diminishing fossil fuel resources and energy security, coupled with global climate change, have placed a renewed focus on the development of renewable energy and biofuels. The utilization of lignocellulosic biomass for biofuels production has been considered as one of promising alternative sources of renewable energy, providing a sustainable option for fuels production in an environmentally compatible manner. As such, the economical transformation of lignocellulosic biomass into biofuels has become a global theme.

While the first generation of biofuels utilizes readily processable bioresources as raw materials such as sucrose and starch from sugar cane and corn, or plant oils, the second generation of biofuels employs lignocellulosic biomass to produce liquid biofuel, suffering from intrinsic recalcitrance of biomass owing to the complicated structure of the plant cell walls. To-date, effective utilization of the biomass through biological routes is predicated on pretreatment technologies that can reduce biomass recalcitrance. The objective of pretreating lignocellulosics is to alter the structure of biomass and to make the polysaccharides more accessible/amenable to hydrolytic enzymes. The conversion of biomass to biofuels through biological conversion process typically consists of three main steps: pretreatment, enzymatic hydrolysis, and fermentation.

Dilute acid pretreatment has been shown to be a leading pretreatment technology that is effective on hardwoods. In this study, poplar was subjected to dilute acid pretreatment with a range of pretreatment time 0.3~27 min. After pretreatment, ball-milled lignin was isolated from pretreated poplar at varying pretreatment severities. Gel permeation chromatography (GPC) and nuclear magnetic resonance (NMR) techniques were utilized to elucidate structural transformations of poplar lignin during pretreatment. As revealed by NMR, the aryl-*O*-ether linkage (*β*-*O*-4) of lignin was extensively cleaved via homolysis and lignin condensation occurred during pretreatment. The lignin was also observed to have a decreased carboxylic group and methoxyl group content and these changes were accompanied with increases in condensed lignin. GPC analysis showed that dilute acid pretreatment resulted in nominal changes in the molecular weight of lignin. In summary, poplar lignin changes in structure was observed during dilute acid pretreatment and this influences plant recalcitrance and its use as a co-product in the biorefinery process.

(675d) Increase in 4-Coumaryl Alcohol (H) Units During Lignification in Alfalfa (Medicago sativa) Alters Extractability and Molecular Weight of Lignin

Angela Ziebell[1], Kristen Gracom[1], Rui Katahira[1], Fang Chen[2], Art J. Ragauskas[3], Yunqiao Pu[4], Richard A. Dixon[2] and Mark F. Davis[1], (1)National Renewable Energy Laboratory, Golden, CO, (2) Samuel Roberts Noble Foundation, Ardmore, OK, (3)School of Chemistry and Biochemistry, Georgia Institute of Technology, Atlanta, GA, (4)Georgia Institute of Technology, Atlanta, GA

The lignin content of biomass can impact the ease biomass processing and therefore genetic modification of the lignin biosynthetic pathway has the potential to reduce costs in biomass processing industries (e.g. pulp and paper, forage, and biofuels). We investigated compositional changes in two low lignin alfalfa (Medicago sativa) lines with antisense down-regulation of p-coumarate 3-hydroxylase (C3H) or hydroxycinnamoyl CoA: shikimate hydroxycinnamoyl transferase (HCT). Gel permeation chromatography (GPC) of isolated ball milled lignin indicated significant changes in the weight averaged molecular weight distribution of the control versus the transgenic lines. Sequential base extraction greatly reduced the lignin content of the transgenic lines leaving a residual H lignin component. We hypothesize the increased extractability of the lignin-like material is due to a decrease in molecular weight of the polymer. The decrease in molecular weight is explained by the reduced stability of the coumaryl alcohol monomeric building block during radical coupling compared to the more prevalent sinapyl and coniferyl alcohol. Our findings suggest that the lignin molecular weight, and as a result, the ease with which it can be removed during chemical processing, can be influenced by altering monomer ratio.

See more of this Session: Advances in Biofuels: DOE Bioenergy Research Centers I
See more of this Group/Topical: Sustainable Engineering Forum

ADVANCES IN THE USE OF NMR TO CHARACTERIZE BIOMASS IN AN EFFORT TO ELUCIDATE THE NATURE OF RECALCITRANCE

Dr. Marcus Foston and Dr. Arthur Ragauskas, BioEnergy Science Center, Georgia Institute of Technology, Atlanta, GA

Abstract

NMR spectroscopy has become one of the most important methods for biomass and biopolymer characterization. ^{1}H, ^{13}C and ^{31}P 1D and 2D NMR spectra have been shown to play a key role in characterizing major biomass constituent chemistry and in understanding mechanisms of plant cell wall growth and deconstruction. Because of the importance of understanding biomass recalcitrance to the biofuel industry, substantial efforts have been made in expansion of the use of NMR spectroscopy to characterize biomass. This work will review key issues related to biomass recalcitrance including the chemical, supramolecular and ultrastructural changes that can be achieved by pretreatment and enzymatic deconstruction as measured by new NMR methodologies including: dynamics and diffusion measurements, multidimensional NMR and rapid quantitative analytical techniques. The results of this study highlight key changes in biomass that reduce recalcitrance and need to be further optimized to reduce the overall complexity of 2nd and 3rd generation biofuels.

Introduction

There has been a recent sense of urgency and push towards the development of more sustainable, "green" technologies to meet global manufacturing and energy demands.[1,2] Biofuels, such as second generation bioethanol has been identified as a practical renewable energy source. The biochemical conversion of cellulose to ethanol is accomplished utilizing enzymatic hydrolysis generating glucose which is subsequently converted by fermentation to ethanol. The overall sugar yields, energy and material cost associated with biomass deconstruction dictate the economics of biofuel production which so far has hindered large-scale commercialization.[3] Biomass recalcitrance, the very properties that make lignocellulosics so useful in nature as structural biopolymers in the cell walls of plants also makes it difficult to deconstruct. Therefore, having a fundamental understanding of the mechanisms related to biomass recalcitrance is critical to any approaches designed to overcome it.

The structure of biomass occurs as a complex microstructural system composed of stereo-irregular and cross-linked lignin and hemicellulose encapsulating and supporting cellulose chains packed into bundles.[4] Due to its complex nature and lack of common solvents, chemical and structural analysis of biomass has always posed challenges. Traditionally, research has relied on methods to isolate, separate, or purify all components however there is evidence these methods change both the chemical and ultrastructural characteristics of individual biomass components.[5] In an effort to probe biomass mechanisms of recalcitrance and methods of deconstruction, the structure of native and pretreated lignocellulosics has been characterized before, during and after thermo-chemical and enzymatic treatments by a variety of analytical techniques. Nuclear magnetic resonance (NMR) however has become a vital tool in biomass characterization probing the chemical structure, conformation, and organization of cell wall polymers in their native and treated state.[5]

NMR has many advantages with respect to other similar spectroscopy methods. NMR is very versatile, detecting very fine chemical and structural components. NMR measurements can be either qualitative or quantitative, when conducted under the proper conditions the area under the resulting peak of a chemical component is linear and proportional to the concentration of that component. More importantly, typical NMR of different chemical groups have varying chemical shifts of high variance producing information rich spectra which can be interpreted for chemical and ultrastructural information.

Solution NMR is particularly useful in the analysis of the chemical characteristics and changes occurring in the three major biopolymers which constitute biomass: cellulose, hemicellulose, and lignin. The literature is full of studies with uses ^{1}H, ^{13}C and ^{31}P 1D and 2D NMR generate profiles containing information on the number and type of functional groups present such as alcohol, carboxylic acid, acetyl and aromatic groups, also determining inter-unit linkages, polysaccharide and monolignol distribution, etc.[6,7]

Solid-state NMR has also been used to probe the structure, conformation, and organization of cell wall polymers.[8-10] For example, ^{13}C cross polarization magic angle spinning (CP/MAS) NMR spectroscopy has been used to analyze the ultrastructure of cellulose for more than three decades. The first applications of ^{13}C CP/MAS NMR to investigate native cellulose were by Attala and Vanderhart.[11] Several chemically equivalent carbon resonances in the anhydroglucose units were shown to display magnetic non-equivalence by high resolution solid-state NMR. They proposed this was due to a rather rigid highly hydrogen-bonded system in which carbons within the anhydroglucose units reflected this non-equivalency in slight changes in chemical shift which could be related to secondary bonding patterns, and therefore solid state NMR could be used to analyze the ultrastructure of cellulose determining % crystallinity and the proportion of various cellulose crystalline allomorphs.[11-18]

In this study, we present use of ^{13}C, ^{1}H and ^{2}H solid state NMR to evaluate the supramolecular and ultrastructure of biomass and the changes occurring during dilute acid pretreatment in an effort elucidate the mechanisms of recalcitrance.

Materials and Methods

Baseline *Populus* (*Populus trichocarpa x deltoides*) samples were harvested between 2007-2008 from area 0800 at Oak Ridge National Laboratory, TN. The samples were then shipped to the National Renewable Energy Laboratory (NREL) in Golden, Co for room temperature air drying, de-barking and size-reduction. The samples were stored in a freezer to maintain the moisture content and shipped to Georgia Tech upon request. The biomass was sized-reduced in a Wiley mill using a 20-80 mesh screen. Extractives were subsequently removed by placing 5 g of biomass into an extraction thimble in a Soxhlet extraction apparatus. The extraction flask was filled with 1:2 ethanol/benzene mixture (~ 150 mL) and then refluxed at a boiling rate which cycled the biomass for at least 24 extractions over a 4 h period.

Lignocellulosic samples were first prepared by presoaking, which was done at 25 °C while continuously stirring in a ~ 0.10 mol/m^3 dilute sulfuric acid solution at 5 % dry extractive-free and size-reduced lignocellulosic solids (w of solids/w of suspension) for 4 h. The presoaked slurry was filtered to remove the solid material and washed with an excess of de-ionized (DI) filtered water. 2 g of the presoaked material (by dry mass) was transferred to a 4560 mini-Parr 300 mL pressure reactor with ~ .10 or .20 mol/m^3 H_2SO_4 solutions at 5 % dry solids and was then sealed under ambient atmospheric conditions. The impeller speed was set to about 1.67 Hz, and the vessel was heated to 160, 175 or 180 °C over ~ 25-30 min (at ~ 6 °C min^{-1}). The reactor was held at the pretreatment temperature ± 2 °C (~ 650-690 kPa) for the specified residence time ± 30 sec (2.5, 5, 10, 20 and 60 min). The reactor was then quenched in an ice bath (~ 5 min) and the steam/atmosphere vented. The pretreated slurry was filtered to remove the solid material and washed with an excess of DI filtered water. The pretreated lignocellulosic samples were dried in the fume hood overnight at room temperature. Paramagnetic impurities generated during pretreatment from the surface of the reactor were removed by washing the solids with a 2 wt. % aqueous solution of ethylenediaminetetraacetic acid (EDTA) and DI filtered water. All yields for biomass recovered after pretreatment ranged between 55-75 % by mass of dry extractive-free and size-reduced lignocellulosic solids.

The NMR samples were prepared with untreated and pretreated 80-20 mesh milled biomass pulp or on ground isolated cellulose from untreated and pretreated biomass. Various solid-state NMR measurements were carried out on a Bruker Avance-400 spectrometer operating at frequencies of 100.55 MHz for ^{13}C or 75.475 MHz for ^{2}H in a Bruker double-resonance MAS probehead at either non-spinning condition or spinning speeds of 10 kHz. A bench-top low-field magnetic resonance analyzer (Maran 23 ultra) was used to analyze ^{1}H resonances operating at a ^{1}H frequency of 23 MHz using a permanent magnet.

Results and Discussion

^{13}C CP/MAS NMR experiments on isolated cellulose were run to determine the relative intensity of the ultrastructure components in the pretreated *Populus* samples and how those relative intensities changed with pretreatment residence time. ^{13}C CP/MAS and non-linear line-fitting showed the crystallinity index increases with residence time as also shown in Figure 1. This finding was significant because it contradicts a widely held position that a critical function of pretreatment is the de-crystallization of cellulose. Another interesting result showed that while the relative intensity for the other crystalline allomorphs remain constant with residence time, the relative intensity of the crystalline cellulose I_{α} form decreases. This again suggests the crystal I_{α} form is susceptible to either degradation by acidic hydrolyzation or thermally induced transformation to other crystal polymorphs during DAP.[10]

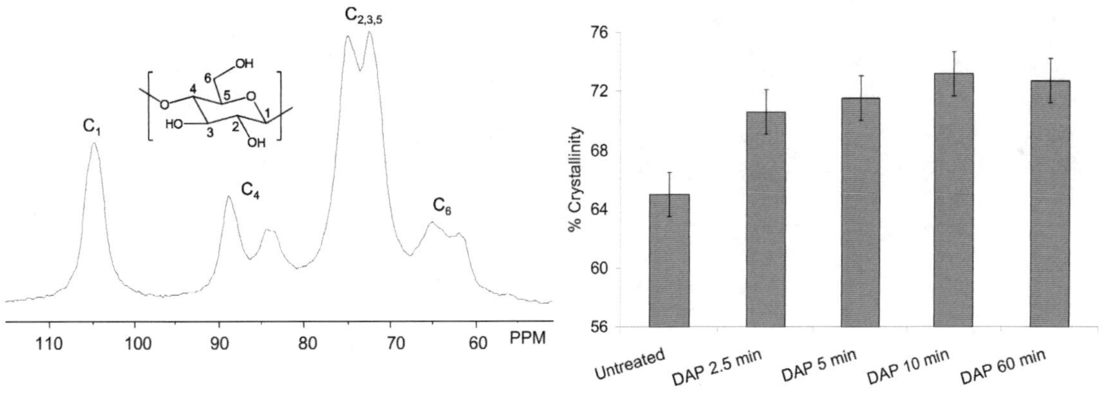

Figure 1. (left) A representation ^{13}C CP/MAS spectrum of isolated cellulose and (right) % crystallinity of untreated and pretreated *Populus* biomass (pretreated in 0.10 M H_2SO_4 at 160 °C for various residences time).

Changes occurring in the supramolecular and pore structure of lignocellulosics during dilute acid pretreatment can also be observed by NMR via a solvent probe, water. Water can be found spatially localized on and within cellulosic microfibrils, existing as capillary water in the lumen or between fibers and within the cellulose fibril matrix. The type and strength of this association, particularly in the cell wall material, is directly related to the ultrastructural and chemical state of the cellulose. These type of cellulose-water interactions exist in native wood pulps, however as pretreatment is performed, material is solublized and removed from the biomass, while also altering the ultrastructural state of the cellulose. By monitoring either the diffusive and/or relaxation behavior of the adsorbed water, information about changes in pore structure of lignocellulosic material can be inferred as shown in Figure 2.

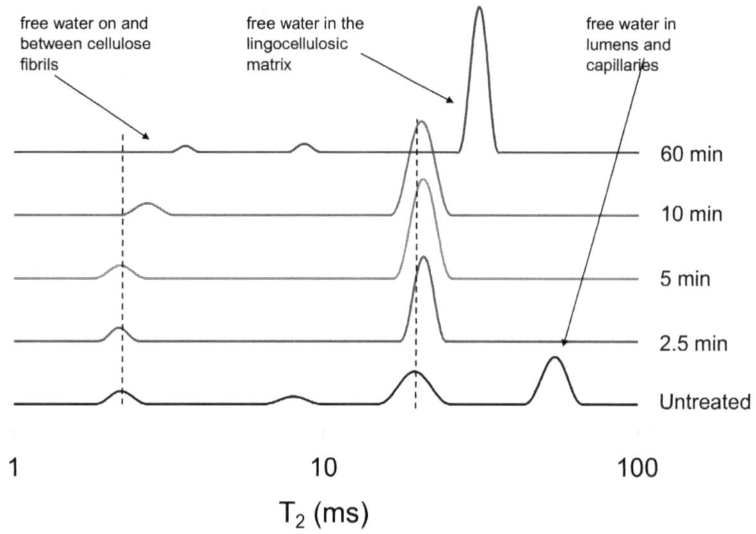

Figure 2. Distribution of spin-spin relaxation times of absorbed water within untreated and pretreated *Populus* biomass (pretreated in 0.10 M H_2SO_4 at 160 °C for various residences time) produced via ILTs of CMPG T_2 experiments at a moisture content of 60 ± 3 %. The vertical dotted lines serve to visually demonstrate shifts in peak position.

Spin-spin (T_2) relaxation times can be used to indicate the nature of cellulose–water interactions which change during DAP. Inverse Laplace distributions of the resulting T_2 decays from water adsorbed in pretreated biomass are shown in Figure 2, and demonstrate not only a shift in T_2 times to longer relaxation or a more mobile state but also indicate that the population of water with longer relaxation times increases. These changes indicate that pretreatment begins to breakdown and loosen the lignocellulosic ultrastructure within the biomass.[9]

^2H spin-lattice (T_1) times of deuterium oxide (D_2O) adsorbed within lignocellulosic biomass were also measured on untreated and pretreated populus, and an inverse Laplace transform of the T_1 decays was then used to generate pore size distributions shown in Figure 3. The resulting distributions indicate that substantial pore expansion within the cellulose fibril bundles occurs during pretreatment.[9]

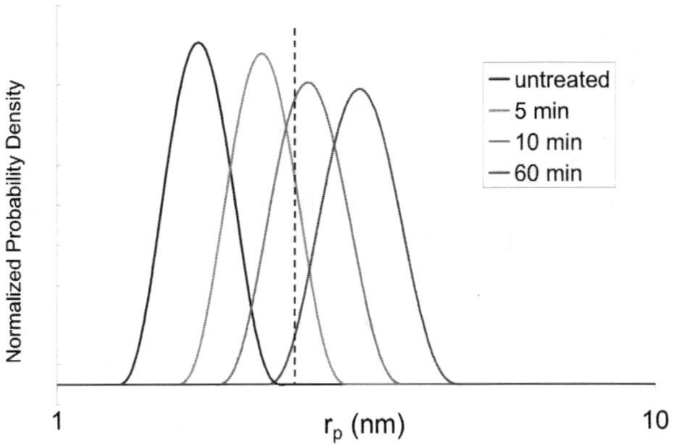

Figure 3. Pore size distribution based on D_2O adsorption and ^2H T_1 relaxation for untreated and pretreated *Populus* biomass (pretreated in 0.10 M H_2SO_4 at 160 °C for various residences times) at a moisture content of 24 ± 5 %. The vertical dotted line represents 26 Å, the minimal pore radius for cellulose.

Conclusions

Solid state NMR can be particularly useful in the analysis of biomass and biomass recalcitrance determining information about crystallinity, pore distributions and the relative organization of cell wall polymers. The ability to analyze biomass with more advanced NMR techniques may have large implications for the structural characterization of lignocellulose and large-scale biofuel production in the future.

References

1. Pu Y, Zhang D, Singh PM, Ragauskas AJ (2008), "The new forestry biofuels sector," *Biofuels, Bioprod. Bioref.*, **2** pp. 58-73.
2. Sierra R, Smith A, Granda C, Holtzapple M, *Producing fuels and chemicals from biomass*, in *Chemical Engineering Progress*. 2008, AIChE p. S10-S18.
3. Hsu TA (1980), "Alcohol from cellulose," *Chemical Technology*, **10**(5) pp. 315-319.
4. Fengel D, Wegener G, *Wood: Chemistry, Ultrastructure, Reaction*. 1984, New York: Walter de Gruyter.
5. Liitiä T, *Application of Modern NMR Spectroscopic Techniques to Structural Studies of Wood and Pulp Components*, in *Department of Chemistry*. 2002, University of Helsinki: Helsinki, Finland.
6. Samuel R, Pu Y, Raman B, Ragauskas AJ (2010), "Structural characterization and comparison of switchgrass ball-milled lignin before and after dilute acid pretreatment," *Applied Biochemistry and Biotechnology*, **162**(1) pp. 62-74.

7. Jiang N, Pu Y, Samuel R, Ragauskas A (2009), "Perdeuterated pyridinium molten salt (ionic liquid) for direct dissolution and NMR analysis of plant cell walls," *Green Chem.*, **11** pp. 1762-1756.
8. Foston M, Beckham H (2010), "Cyclic poly(dimethylsiloxane) from kinetically controlled cyclodepolymerization of linear precursors in dilute solution," *Polymer.*
9. Foston M, Ragauskas A (2010), "Changes in the structure of the cellulose fiber wall during dilute acid pretreatment in *Populus* studied by ^{1}H and ^{2}H NMR," *Energy and Fuels*, ***in press***.
10. Foston M, Ragauskas A (2010), "Changes in lignocellulosic supramolecular and ultrastructure during dilute acid pretreatment of *Populus* and switchgrass," *Biomass and Bioenergy*, ***in press***.
11. VanderHart DL, Atalla RH (1980), "^{13}C NMR Spectra of Cellulose Polymorphs," *J. Am. Chem. Soc.*, **109**(6) pp. 3249-3250.
12. VanderHart DL, Atalla RH (1981), "Observations by high-resolution carbon-13 nuclear magnetic resonance of cellulose I related to morphology and crystal structure," *Macromolecules*, **14**(3) pp. 570-574.
13. VanderHart DL, Atalla RH (1984), "Studies of microstructure in native celluloses using solid-state carbon-13 NMR," *Macromolecules*, **17**(8) pp. 1465-1472.
14. VanderHart DL, Atalla RH (1984), " A Composite of Two Distinct Crystalline Forms " *Science*, **223**(4633) pp. 283-285.
15. Larsson T, Westermark U, Iversen T (1995), "Determination of the cellulose Iα allomorph content in a tunicate cellulose by CP/MAS ^{13}C NMR spectroscopy," *Carbohydrate Research*, **278**.
16. Larsson T, Wickholm K, Iversen T (1997), "A CP/MAS ^{13}C NMR investigation of molecular ordering in celluloses," *Carbohydrate Research*, **302** pp. 19-25.
17. Larsson T, Hult EL, Wickholm K, Pettersson E, Iversen T (1999), "CP/MAS ^{13}C-NMR spectroscopy applied to structure and interaction studies on cellulose I," *Solid State Nuclear Magnetic Resonance*, **15** pp. 31-40.
18. Lennholm H, Larsson T, Iversen T (1994), "Determination of cellulose I_{α} and I_{β} in lignocellulosic materials," *Carbohydrate Research*, **261** pp. 119-131.

(675f) Feedstock Agnostic Pretreatment Approach for Efficient Hydrolysis of Bioenergy Crops

Seema Singh and Blake Simmons, Joint BioEnergy Institute, Emeryville, CA

Lignocellulosic biomass is resistant to microbial and enzymatic breakdown into monomeric sugars. There are several biophysical and chemical factors contributing to this recalcitrance including cellulose crystallinity and the presence of lignin-polysaccharide covalent linkages. Effective pretreatment of biomass is essential for breaking apart highly ordered and crystalline plant cell walls and disrupting the lignin-carbohydrate complex, thereby facilitating enzyme accessibility and reducing the costs of downstream saccharification processes. Pretreatment technologies such as dilute acid, ammonia fiber expansion (AFEX), and ionic liquid (IL) show great promise, yet pretreatement still remains one of the most costly steps in lignocellulosic biofuel production. It is therefore critically important to fundamentally understand the origin of biomass recalcitrance to enable facile deconstruction and the design of less expensive and practical approaches for next-generation biofuel commercialization.

We are utilizing ionic liquid pretreatment on a variety of potential biofuel feedstocks to understand the origins of biomass recalcitrance and the mechanism of IL pretreatment. In addition to employing molecular dynamics simulation for understanding cellulose solubility in IL compared with traditional solvents, we are utilizing advanced imaging techniques (hyperspectral confocal fluorescence and Raman) to track dynamic events during pretreatment. We will present the results (TGA, saccharification, dynamic light scattering, porosimetry, XRD, SANS, pyro-GC-MS) of a series of experiments including the study of rheological behavior during ionic liquid pretreatment.

See more of this Session: Advances in Biofuels: DOE Bioenergy Research Centers I
See more of this Group/Topical: Sustainable Engineering Forum

The Production Of Alkanes From Algae

Liane S. Carlson, Michael Y. Lee, Chukuemeka A.E. Oje, Arthur Xu, Stuart W. Churchill
Department of Chemical and Biomolecular Engineering, University of Pennsylvania

The potential of farmed algae as a biofuel has been recognized for some time because 1) its cultivation does not directly affect the food supply, 2) its productivity of lipids per acre far exceeds that of any other agricultural commodity, 3) it grows by absorbing CO_2, and if converted to a fuel and burned, is CO_2-neutral, and 4) the lipids can be converted to biodiesel or conventional fuel-range hydrocarbons. The purpose of our investigation has been to evaluate the technical and economic feasibility of an integrated algae-to-fuel process that incorporates several recently proposed technologies. This evaluation is based on the production of 20,000 BPD (226,000 lbs/hr) of n-alkanes since this is comparable to the daily output of a small refinery hydrotreater unit.

This analysis differs from prior ones in two major respects. First, it incorporates a new method of lipid extraction under development by OriginOil, Inc. rather than the conventional process of extraction with hexane. Second, the processing of the lipid has the objective of producing n-alkanes from triglycerides, as opposed to producing fatty acid methyl ester (FAME) biodiesel. Since the n-alkane product from this process is comparable to petroleum-based diesel fuels, it can be readily incorporated into existing energy infrastructures as a diesel blending stock or as a feedstock for other processing units in the refinery.

The conversion of algae to useful biofuels by means of this new process and in accord with this new concept can be divided into three modules: 1) algae cultivation, 2) lipid extraction, and 3) lipid conversion to alkanes. For each of the three modules, there are a number of possible approaches, each with a high degree of uncertainty. We attempted to evaluate the several methodologies for each module separately, and to formulate an integrated process of the most promising options.

Location. The selection of a site for an algae-to-alkanes plant is critical because of the need for 1) isolation from a population center, 2) a large relatively flat area with sufficient and relatively constant insolation, 3) a moderate year-around temperature, 4) a nearby coal-fired power plant whose stack gas contains a sufficient quantity of CO_2, 5) a sufficient source of saline water, 6) a nearby petroleum refinery as a market for the alkanes and as a source of hydrogen, and 7) a means of transportation of the lipids or alkanes thereto.

Sites that might meet these requirements were identified in Arizona and California, but East Texas appeared to be more promising overall. As a representative location that meets all of the listed requirements we propose an undeveloped area near the small town of Thompsons in East Texas. Thompsons, which is about 25 miles southwest from Houston, 30, 60, and 60 miles northward from three petroleum refineries on the Gulf Coast, and on a branch of the BNSF railway, receives insolation varying from the equivalent of 112 W/m^2 in January to 250W/m^2 in June, so the plant can be operated year-round [1]. A near-by coal-fired power-plant generates up to 2500 MW and produces sufficient CO_2 in its stack gas to meet the needs for algae cultivation. In addition, saline water for use in the cultivation process can be drawn from an underground aquifer and reused.

Cultivation. Three promising approaches for cultivation were identified. The first, which utilizes open-air "raceway" ponds, does not require much energy or attention but is hampered by contamination and low-growth rates. The second, which utilizes a compact photo-bioreactor with cultivation of the algae inside transparent tubes enclosing complex internals designed to optimize the rate of growth, requires an accounting for the concentration of carbon dioxide, free exposure to light, an expeditious rate of mass transfer from the gas-phase to the liquid phase, and the possible use of nitrogen starvation to maximize lipid production at the expense of proteins. The third is a hybrid that utilizes a simple infrastructure modeled after agricultural practices combined with the ability to control certain parameters of the environment by means of cultivating the algae inside polyethylene tubes. The Simgae™ Algal Biomass Production System developed by the Diversified Energy Corporation was chosen as a prototype of this methodology. *Nannochloropsis sp.*, a marine species with a high lipid content of 31- 68% (dry wt.) and other favorable qualities, was chosen as the strain of algae [2].

With a 50% recycle, a residence time of nearly 48 hours on average is required to double the amount of algae under the chosen conditions [3]. The corresponding length of the 6-in tubes is 1250 ft with a flow rate of 0.023ft/s. Each field consists of 100 reactor beds, each with 16 tubes, and, together with manifolding and paths, occupies 33 acres [4]. 1120 such fields, totaling 36,960 acres, were required for the chosen rate of production of algae.

Nutrients in suspension and solution in the saline water are combined with the recycled algae at the inlet. Stack gas pumped from the power plant by hydraulic induction and mixed with air to reduce its CO_2 content from 32 to 6 wt% is introduced every 300 ft along the tubing. The traces of SO_2 and NO_x in the stack gas enhance the growth of the algae. The O_2 produced by the growth of the algae is released by relief valves every 300 ft.

An overview of the proposed process is shown in Figure 1, the layout of the cultivation field in Figure 2, and of a single element thereof in Figure 3.

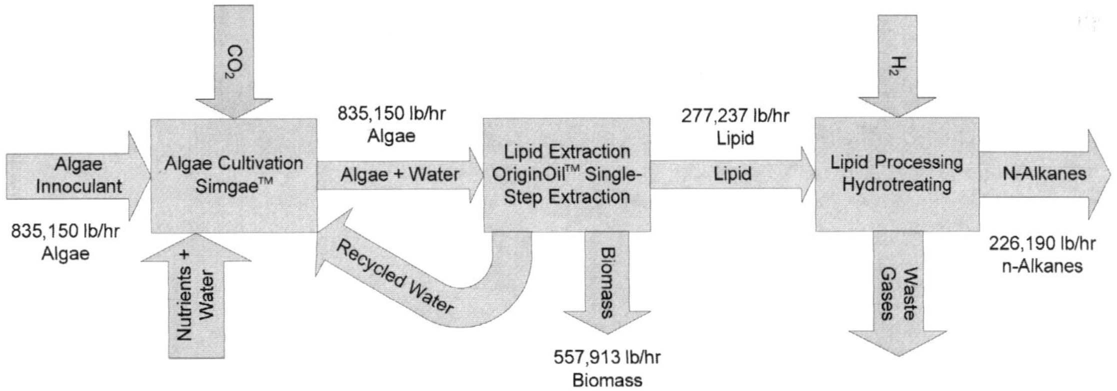

Figure 1. Overview of proposed integrated algae-to-fuel venture.

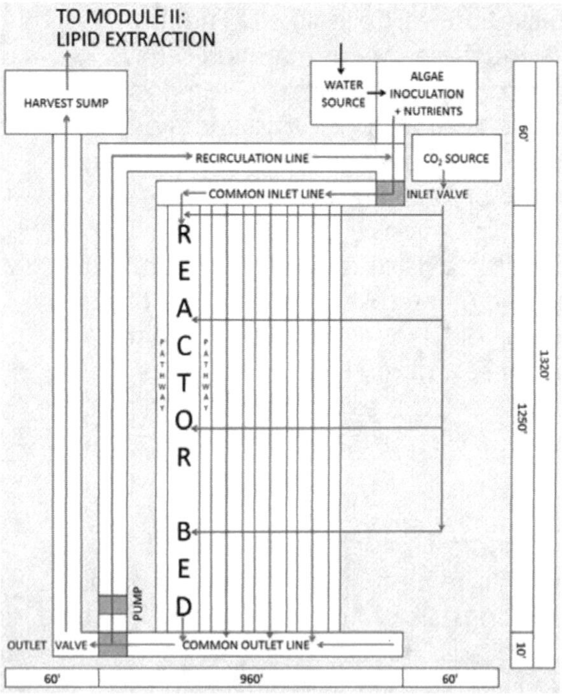

Figure 1. Schematic of cultivation field.

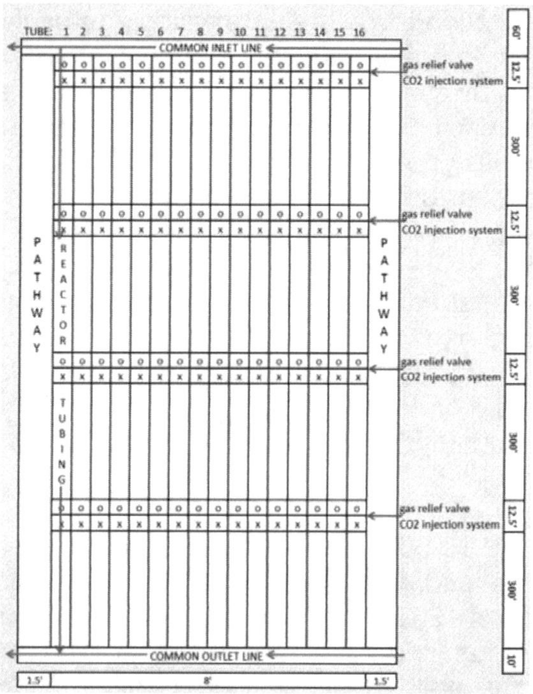

Figure 2. Schematic of a single reactor bed.

Lipid extraction. Historically, the desirable triglyceride fraction has been separated from the rest of the raw algae (proteins, carbohydrates, phospholipids, nucleic acids, and water) through extraction with hexane. A new process called Quantum Fracturing^{TM} by OriginOil, Inc., for which a patent has been applied but not yet issued, claims a combination of pH modification, microwaves, and ultrasonic pulses can rupture the wall of the algae cells while in slurry form, and allow the lipids to escape, resulting in a 90% reduction in energy consumption [5]. The resulting slurry can then be separated through gravitational settling into a lipid layer, an aqueous layer, and a wet layer of biomass. The lipids are transported to a petroleum refinery for conversion into alkanes, the aqueous fluid is recycled, and the biomass is centrifuged and drum-dried to reduce the residual water below 10 wt% [5]. It may then be sent to a nearby market for livestock feed or other uses. This process is shown schematically in Figure 4, the conjectured characteristics are compared with those for the conventional process in Table 1, and the estimated electrical consumption in Figure 5.

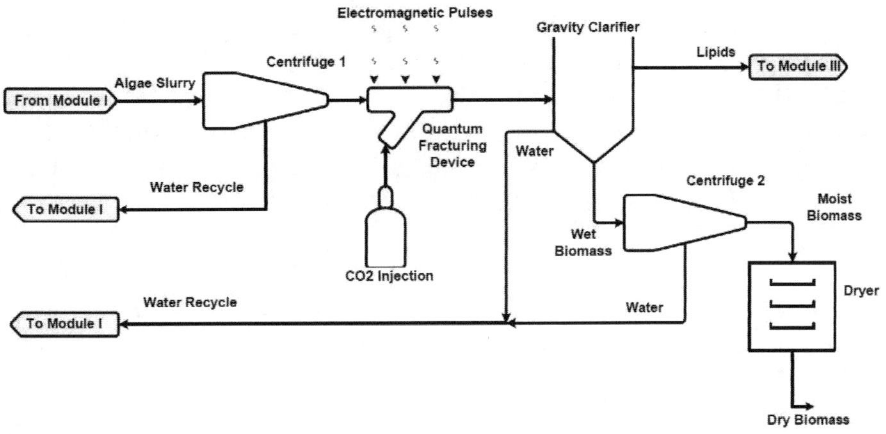

Figure 3. Schematic of proposed lipid extraction process based on Quantum Fracturing^{TM} by OriginOil

Table 1. Comparison between conventional hexane extraction and Quantum Fracturing^{TM} by OriginOil

	Conventional hexane extraction	OriginOil process
Separate by	Hexane	Gravity
Dewatering	Before separation	After separation
Lipid Yield	70 – 95%	85 – 97%
Energy Use (per lb lipid)	8.05 kWh	1.30 kWh

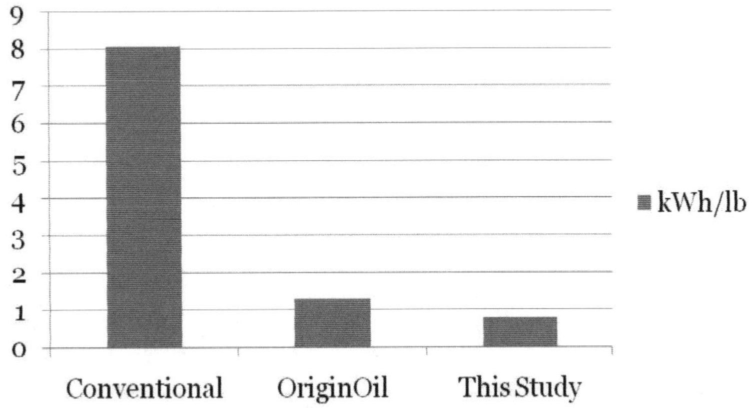

Figure 4. Comparison of estimated electrical consumption per pound of algae produced.

Lipid processing. Because the catalytic hydrotreating process required to convert the lipids to alkanes is similar to those employed in hydrocarbon processing and requires a source of hydrogen, it is proposed to carry out this step at a petroleum refinery rather than at the site of cultivation and extraction. The most feasible form of transportation to the refinery is by rail in tank cars. Since railway shipments are intermittent while lipid production and processing are continuous, it is proposed to provide for seven days of storage in tanks at the site of production and for two days of storage in tanks at the refinery.

The lipids react with hydrogen at 350°C and 50 bars in downward flow through packed beds of granules impregnated with a $NiMo/Al_2O_3$ catalyst [6]. In the first step, the triglycerides are hydrogenated and broken down into C_3 hydrocarbons and components of free fatty-acid.

These intermediates are then converted into straight-chain alkanes by three competing reaction pathways: decarboxylation, decarbonylation, and hydrodeoxygenation [7]. Figure 6 is a simplified schematic of the catalytic hydrotreating process. The product is separated by means of a high-temperature flash drum, a low-temperature, three-phase flash drum, a steam-stripper, and an amine scrubber with intermediate heat exchange. The alkane product from this process is comparable to petroleum-based diesel fuels. Thus it can be readily incorporated into the existing energy infrastructure as a diesel blending-stock or as a feedstock for other processing units in the refinery.

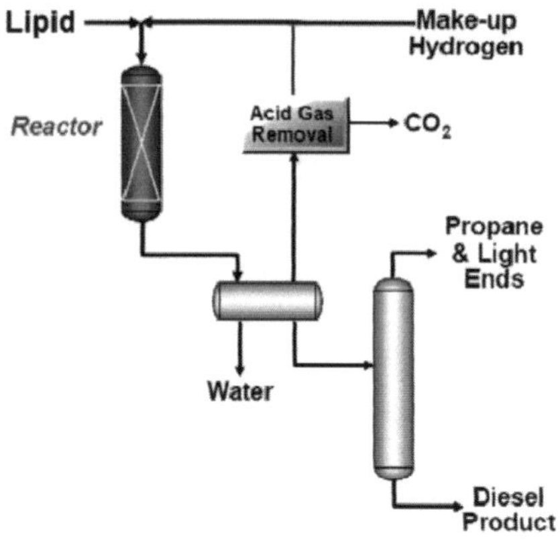

Figure 5. Simplified schematic of catalytic hydrotreating process.

Economic analysis. The profitability of the algae-to-fuel process described herein depends on a more accurate determination of costs and sources of revenues than is possible at the moment. Figure 7 shows a crude estimation of feedstock and utility costs per gallon of n-alkane, and Figure 8 a crude estimate of potential revenue. It is particularly difficult to estimate the utility costs because of great uncertainty of the electricity requirements in the Simgae™ and OriginOil™ processes.

Algae-to-fuel economics can be improved by developing higher-value uses of the algae biomass. Biomass which constitutes 55% of the algae, has possible use in pharmaceuticals and chemicals, and as a source of energy. These products are more valuable than livestock feed. As indicated by Figure 8, the sale of byproducts might exceed the revenue from the fuel oil. Carbon credits from potential cap-and-trade programs could be an additional source of revenue.

The bad news has been saved for last. The capital cost, which is primarily for the purchase of the 36,960 acres required for cultivation, is estimated to be roughly $2.6 billion. Despite the encouraging aspects of the potential income relative to the operating costs, the huge outlay for land suggests that consideration should be given to all possible means of increasing the rate of growth of algae and thereby reducing the required acreage before proceeding.

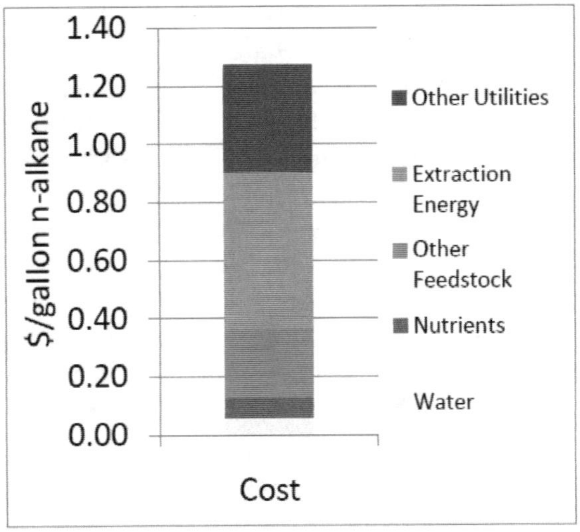

Figure 6. Estimated feedstock and utility costs of the proposed integrated algae-to-fuel venture.

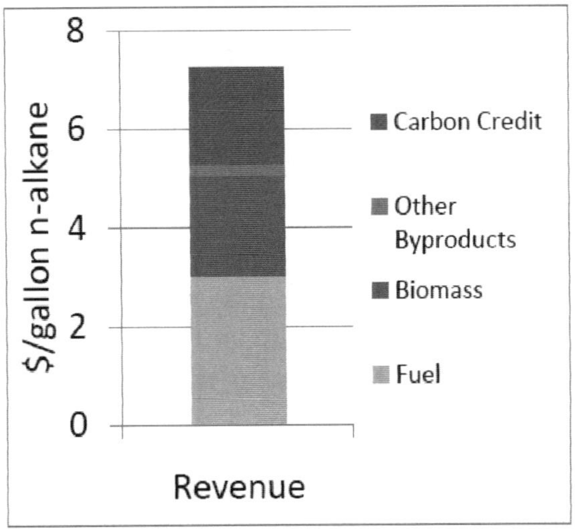

Figure 7. Projected revenue from the proposed integrated algae-to-fuel venture.

Conclusions. This manuscript describes a possibly improved process for the production of fuel from algae. Although encouraging both technically and economically, the analysis identifies several challenges that must be overcome before it merits commercialization. Future work should focus on these challenges.

Some of the challenges posed by the use of the hybrid SIMGAE™ design are maintenance of the proper temperature in the cultivation system, makeup water, the resistance of algae strain to invasion from other species, the environment impact of the algae farm, and most importantly, containing capital and operating costs. Nutrient consumption and costs, electrical requirements, CO_2-injection requirements, and most of all, land requirements are unresolved.

Some of the challenges in terms of oil (lipid) recovery from the algae include dewatering methods, lipid purification, energy costs, and optimizing the return from the residual biomass. The OriginOil™ lipid extraction process claims to reduce significantly the energy costs compared to those for conventional extraction processes. The most innovative aspect of this process is that dewatering is not necessary before lipid separation. However, the Quantum Fracturing™ technology from OriginOil™ is still in the development stage and is not a proven technology and is apparently not yet available for licensing.

The lipid processing step is probably the least risky of the three steps of the integrated algae-to-fuel process. Although its use to break up triglycerides is innovative, the fundamental technology behind catalytic hydrotreating is practiced in petroleum refineries worldwide. When adapting this technology for a lipid extraction process, it is important to focus on the catalyst selection and hydrotreater design for a lipid feedstock as opposed to a crude-oil feedstock.

It is important to optimize the value of biomass byproducts in order to maximize overall revenue, but the key challenge for this venture is to increase the rate of growth of the algae per acre. The possibilities include improvements by means of genetic engineering of the algae and by means of the physical and optical design of the reactors.

References.

1. Solar Radiation Data Manual for Flat-Plate and Concentrating Collectors. NREL. Web. 4 Apr. 2010. http://rredc.nrel.gov/solar/pubs/redbook/PDFs/TX.PDF.
2. Zittelli, G. Chini, et al. "Production of eicosapentaenoic acid by Nannochloropsis sp. cultures in outdoor tubular photobioreactors." Journal of Biotechnology (1999) 299-312.
3. Hassania, J. (Diversified Energy). "Simgae – Low Cost, Broad Application – Algal Biomass Production System." NETL 2008 Conference Proceedings. 18 Sept 2008.
4. Cloud, G. et al., US2008/0311649 A1. "Pressurized Flexible Tubing System for Producing Algae." 18 Dec 2008. XL Renewables.
5. OriginOil presentation at the World Biofuel Market Conference, 15-17 March 2010
6. UOP/Eni Ecofining™ Process for Green Diesel Fuel. Web. 20 April 2010. http://www.uop.com/renewables/10010.html
7. Huber, George W., et al. "Processing biomass in conventional oil refineries: Production of high quality diesel by hydrotreating vegetable oils in heavy vacuum oil mixtures." Applied Catalysis A: General, Volume 329, (2007), 120-129.

(718b) An Economic Evaluation of Lipid Production by Cultivation of Algae – A Critical Path Analysis

Ramalingam Subramaniam[1], Cunwen Wang[1], **Stephen Dufreche**[1], Mark Zappi[1], Barbara C. Benson[2] and Rakesh Bajpai[1], (1)Chemical Engineering, University of Louisiana at Lafayette, Lafayette, LA, (2) Bioprocessing Research Laboratory, University of Louisiana, Lafayette, LA

This presentation will deal with evaluation of heterotrophic growth of oleaginous algae on carbohydrates available from different sources and compare its economics of lipid production from this route with that from autotrophically growing algae. Energy efficiency during the different processes, light delivery issues, cell harvesting and recovery of lipids from the cells will be considered in identifying the effect of different processing steps on the unit cost of production of lipids.

See more of this Session: Advances in Algal Biorefineries II
See more of this Group/Topical: Sustainable Engineering Forum

(718c) An Integrated Algal Culture System for Biofuel Production Based On Mixotrophic and Heterotrophic Growth Modes

Shulin Chen, Zhanyou Chi, Yubin Zheng and Ben Lucker, Biological Systems Engineering, Washington State University, Pullman, WA

The commercial reality of algae fuel production depends upon innovations ingeniously incorporating engineering and algal biology and efficient utilization of resources. The Bioprocessing and Bioproduct Engineering Laboratory (BBEL) at Washington State University has developed a novel algal culture system that (1) incorporates the dual-trophism capability of some algae species into a two-stage culture process, and (2) integrates heterotrophic and phototrophic processes for year around fuel production in cold climate regions. In this system, the first stage utilizes organic carbon as substrate for algae fermentation, growing seed cells at high efficiency for inoculating to open ponds for phototrophic culture. This system can significantly improve productivity and offers a strategy for contaminant control. In parallel to the mixotrophic process, additional algae production can be accomplished using organic carbon derived from waste materials and agricultural residues. Systematic research efforts have been made in BBEL to further the science discovery and technology development for refining the system. This paper will present a summary of the related progresses. The information will provide new insights for biomass production for algae-based biofuel systems.

See more of this Session: Advances in Algal Biorefineries II
See more of this Group/Topical: Sustainable Engineering Forum

(718d) Biofuel Feedstock Production From Microalgae Grown in Municipal Wastewaters

Sage R. Hiibel, Mark S. Lemos and John C. Cushman, Biochemistry & Molecular Biology, University of Nevada Reno, Reno, NV

Microalgae offer great potential as a second-generation biofuel feedstock. Microalgae are attractive because they require minimal nutrient inputs, can capture atmospheric CO2, and can be grown in fresh, brackish, or salt water. Additionally, they do not compete with current food, feed or fiber feedstock needs, and can be grown on non-arable lands. However, limited water supplies and nutrient sources can significantly add to microalgae production costs.

Currently, municipal wastewater treatment facilities spend millions of dollars each year to remove abundant nitrogen and phosphorus from wastewater. In lieu of traditional treatments, microalgae can be used to effectively and efficiently remove these compounds, while simultaneously producing biofuel feedstocks. Preliminary research has demonstrated that some microalgal species can be grown successfully on centrate, the liquid fraction remaining after dewatering anaerobically digested wastewater sludge.

This project evaluates the potential for growth on centrate of over forty microalgae strains from fresh, brackish, or salt water genera. Generally, freshwater strains were able to grow in higher centrate concentrations, with two species, *Neochloris pseudostigmata* and *N. conjuncta*, demonstrating growth with up to 40 *vol%* centrate. Nine of eighteen salt water *Dunaliella* spp. also demonstrated growth in centrate levels exceeding 25 *vol%*. The use of UV sterilization for centrate sterilization was also evaluated. UV dosage requirements were significantly higher than those used in traditional drinking water treatment plants due to coloration and high TSS/TDS levels.

The lipid profiles of strains with high centrate tolerance were determined and compared against growth on defined maintenance media using GC-MS. The major lipid compounds were found to be palmitic (16:0), oleic (18:1), and linoleic (18:2) acids for all strains grown on either centrate or their respective maintenance medium.

See more of this Session: Advances in Algal Biorefineries II
See more of this Group/Topical: Sustainable Engineering Forum

(718e) Two-Step, Catalyst-Free Biodiesel Production From Wet Algal Biomass

Robert Levine, Tanawan Pinnarat and Phillip E. Savage, Chemical Engineering, University of Michigan, Ann Arbor, MI

Worldwide energy demand, particularly for high-density liquid transportation fuels, continues to rise as populations grow and become more affluent. Concurrently, concern over climate change, declining petroleum reserves, and national security has encouraged the use of biodiesel, a mixture of fatty acid alkyl esters commonly derived from the transesterification of vegetable oils. Recent interest in using oleaginous microalgae as a non-edible biodiesel feedstock has grown considerably, largely on the promise of high oil yields (1,000 to 100,000 L ha-1 y-1), the opportunity to capture CO2 from flue gases, and the ability to cultivate it on abandoned or unproductive land using brackish, salt or wastewaters instead of freshwater. Despite recent advancements and substantial public and private investment, large-scale algal biofuel production has not reached cost parity with the petroleum products they seek to replace. While energy inputs and costs must be reduced at each stage of algal biofuel production, harvesting, dewatering, and lipid processing remain significant challenges.

Instead of attempting to remove lipids from cells in an aqueous suspension or convert them directly to biodiesel in the presence of water, we propose a novel two-step, catalyst-free biodiesel production process involving intracellular lipid hydrolysis coupled with supercritical in-situ (trans)esterification (SC-IST/E). In the first step, wet algal biomass (ca. 80% moisture) is reacted at subcritical water conditions to hydrolyze intracellular lipids, conglomerate cells into an easily filterable solid which retains the lipids, and produce a sterile, nutrient-rich aqueous phase. In the second step, the wet fatty acid (FA)-rich solids are subjected to SC-IST/E with ethanol to produce biodiesel in the form of fatty acid ethyl esters (FAEEs). In this paper, we demonstrate that it is possible to carry out in-situ hydrolysis of cellular lipids in wet algal biomass, retain those lipids within a filterable solid, and then produce biodiesel using supercritical ethanol. In an effort to maximize lipid productivity and efficient carbon substrate utilization, Chlorella vulgaris was grown phototrophically in a bubble column reactor and then heterotrophically. Wet algal biomass contained about 50% total lipids as FAEEs and was used in hydrolysis reaction as a fresh paste or dried and rehydrated. The effects of hydrolysis processing time and temperature were investigated and lipid conversion to FA and smaller glycerides was quantified. Esterification reaction conditions and alcohol loading were investigated as predictors of biodiesel yield and biodiesel was analyzed for impurities in accordance with standard methods (ASTM 6584/EN14105 and EN14103).

See more of this Session: Advances in Algal Biorefineries II
See more of this Group/Topical: Sustainable Engineering Forum

(718f) The Sustainable Conversion of Algal Sugars Into Butanol

Jamie Hestekin[1], **Thomas M. Potts**[2], Jianjun Du[1], Robert R. Beitle[1] and Edgar C. Clausen[1], (1)Ralph E. Martin Department of Chemical Engineering, University of Arkansas, Fayetteville, AR, (2)Chemical Engineering, University of Arkansas, Fayettville, AR

Algae is an attractive feedstock since it can grow on marginal land, has a very high growth rate, and takes up nutrients as it grows so it has the potential to clean wastewater while making fuel feedstock. However, although much research is taking place on converting the algal lipids to biodiesel, there are few studies on converting algal carbohydrates to alcohols. The algal carbohydrates are mainly starches and simple sugars so they are much easier to pretreat and ferment than cellulosic sugars. Our group is working on taking a two step butanol process and continually converting algal sugars into butanol. The first step does an acidgenesis, with Clostridium Tyrobutyricum, for the conversion of sugars to organic acids, specifically butyric acid. The second step takes the butyric acid and converts it into butanol via Clostridium Beijerinckii. Although this is an anaerobic fermentation, we have used membrane processes to make both steps in the two-step process continuous and have modeled the reaction steps in the first and the second step. The yield of this process can go from 6:3:1 butanol, acetone, and ethanol, respecitively to greater than 90% butanol by optimizing both steps.

See more of this Session: Advances in Algal Biorefineries II
See more of this Group/Topical: Sustainable Engineering Forum

(731a) A Rigorous Graphic Approach to Modeling Transportation Fuel Lifecycle: Axioms and Theorems

Jiahong Liu, Energy Systems, Argonne National Lab, Argonne, IL

An innovative approach based on graph theory has been proposed for facilitating the lifecycle assessment of transportation fuels in general and renewable feedstock-based transportation fuels in specific. The complexity of the fuel lifecycle system is analyzed. The proposed approach is designed to tackle the specificities of a transportation fuel in its lifecycle: it represents the structure of the lifecycle by a unique graph, termed a Lifecycle graph, or an L-graph for short. This graph captures not only the syntactic but also the semantic contents of the system. A novel concept of "material state in its lifecycle" is proposed to help describe the pathway of a transportation fuel in its lifecycle and its complexity together with the concepts of material and location. An axiom system underlying the approach has been established to define the rigorous lifecycle structure. This axiom system is based on a given set of specifications for the lifecycle assessment problem. Such specifications include the types, input and output materials, origins and destinations of processes, and a variety of parameters such as efficiencies and losses associated with these processes, natural energy sources, energy products, co-products and their geographical specificities. All pathways in the lifecycle of a specific transportation fuel are embedded in a maximal structure, from which individual pathways can be extracted. Various theorems have been derived from the axiom system to ensure that this approach is mathematically rigorous, thereby rendering it possible to develop efficient transportation fuel lifecycle pathway generation and management methods on the basis of a rigorous mathematical foundation. Examples of multiple renewable feedstock-based transportation fuels are presented to highlight the significance and efficacy of the present approach.

See more of this Session: Life Cycle Analysis of Renewable Feedstock-Based Processes and Products
See more of this Group/Topical: Sustainable Engineering Forum

(731b) A Wiki-Based Platform for Technoeconomic Analysis of Lignocellulosic Ethanol Biorefineries

Daniel Klein-Marcuschamer, Lawrence Berkeley National Laboratory / Joint BioEnergy Institute, Berkeley, CA, Blake Simmons, Joint BioEnergy Institute, Emeryville, CA and Harvey Blanch, Department of Chemical Engineering, University of California, Berkeley, Berkeley, CA

Biofuels, and ethanol in particular, have been the target of both misguided criticism and blind fervor. While the environmental and economic performance of biofuels is currently debated, the scientific community continues to research technological advances toward the commercialization of renewable transportation fuels from non-food resources. The technical and economic performance of these advanced biofuel production processes must be repeatedly evaluated to determine their potential. To make different analyses comparable, we have constructed the first open, transparent, and collaborative process model of a lignocellulosic ethanol biorefinery. The model has been deposited on a public wiki and is available free of charge for non-commercial use. All of the assumptions made in the various unit operations have been described on the wiki, and experts in different areas of the process can discuss them and suggest changes as experimental data are gathered and incorporated into the model. We expect that, with the input of the biofuels scientific and professional community, this tool will help accelerate research on the environmental, energetic, carbon footprint, and economic potential of lignocellulosic ethanol.

See more of this Session: Life Cycle Analysis of Renewable Feedstock-Based Processes and Products
See more of this Group/Topical: Sustainable Engineering Forum

Life Cycle Cost of Greenhouse Gas Reduction through Lignin-Land Application

Ghaisdeh Pourhashem[1], Paul R. Adler[2], Andrew McAloon[3], Sabrina Spatari[1]

[1] **Civil, Architectural and Environmental Engineering Department, Drexel University, Philadelphia, PA**

[2] **USDA-ARS, Pasture Systems and Watershed Management Research Unit, University Park, PA**

[3] **USDA-ARS, Eastern Regional Research Center, Wyndmoor, PA**

1-Introduction

Energy security and greenhouse gas abatement have become driving forces for developing low carbon, renewable and sustainable energy technologies. Biofuels are one of the most prominent technical options in light duty vehicle transportation due to the possibility of blending with fossil fuels, and fitting with existing fuel distribution infrastructure and light duty vehicles without significant adaptation (Gnansounou, 2010).

Lignocellulose is an important and potentially sustainable feedstock for conversion to low-carbon liquid fuel (biofuel) alternatives to gasoline, and ones that do not compete for prime agricultural land, disrupt food markets, or induce land use change . Using lignocelulosic feedstocks for biofuel production has been investigated by many (Wyman 1994, Szczodrak and Fiedurek, 1996, McAloon et al., 2000, Aden et al., 2002, Sheehan et al., 2004, Aden et al., 2007, Hendriks and Zeeman, 2008, Laser et al., 2009a, 2009b, 2009c, Margeot et al., 2009, Uihlein and schebek, 2009, Hsu et al 2010, Luo et al., 2010, Singh et al., 2010, Spatari et al., 2010) and the resulting environmental performance of the biofuel differs due to the different processes and system boundaries.

Cornstover is the nongrain agricultural residue left above ground on the field after corn grain harvest (Wilhelm et al., 2004). Studies show that 80% of the residue left on land will return to the atmosphere within 2 years (Buyanovsky and Wagner, 1997). Removing stover from land and converting it to biofuel could result in a reduced net CO_2 release to the atmosphere relative to simply leaving it on the ground, but at the same time other impacts of removing cornstover must be considered. In general for the percentage of stover removal from land, soil moisture, water and wind erosion, crop rotation, and irrigation and tillage practices must be taken into consideration. In a three year study Blanco-Canqui and Lal (2008) showed that soil aggregates were weaker, less stable, and retained less water when cornstover was removed which was most likely due to the decrease in stover-derived organic binding agents.

In addition, removing cornstover from land removes essential soil organic carbon (SOC). Wilhelm et al. (2007) deduced that the quantity of corn stover required to maintain SOC is greater than the amount needed to control water and wind erosion. On the other hand, covering the soil with crop residue can create weed control problems, excessive wet and cold soils, and poor seed placement (Wilhelm et al., 2004). Thus considering cornstover removal to a point of just avoiding land erosion could be an option while trying to return organic carbon to soil in the form of bioethanol byproduct. Johnson et al. (2004) have proposed amendment of soil with the by-product of cornstover fermentation, which consists of about 70% lignin. The authors reported that during 118 days about 50% to 60% of the cornstover and 22% to 25% of the by-product was decomposed. Bahri et al. (2008) state that while the majority of lignin is rapidly mineralized, a small fraction becomes much longer lived in soil and estimated the turnover time of lignin molecules is quite long, on the order of about 25 years. These findings and the characteristics of lignin which can provide carbon for the soil while having a longer life time (sequestered) compared to cornstover lead us to consider returning the by-product of the bioconversion process to land as a soil amendment and providing power for the biorefinery from another source which here would be natural gas and purchased electricity from the regional electricity grid.

Another goal of this study is to make the biorefinery more economically favorable while having environmental benefits so as to make this industry more sustainable. The energy in the combustible by-products of the process can be recovered to a certain extent by the power generation. Hamelinck et al. (2005) state that using a gasifier combined cycle rather than a boiler and steam cycle would improve the power generation efficiency but would increase ethanol production costs due to a larger initial investment cost. Natural gas having a higher lower heating value (LHV) than the lignin by-product of the ethanol production process and less air pollution problems (Thornley, 2008, Tiwary and Colls, 2010), represents a potentially better substitute for byproduct combustion. Investigating this possible strategy would help us have a better understanding of a system which is not at full-scale commercial production. Thus this study involves economic analysis of the new design and its environmental impacts through investigation of the life cycle greenhouse gases (GHGs) and costs of the bioconversion of cornstover to ethanol and returning the lignin byproduct to land.

2-Method

The goal of this study was to conduct a cost and environmental life cycle inventory analysis of GHG emissions for cornstover conversion to ethanol, considering lignin return to land and electricity purchase from the grid as well as NG Boiler to provide plant's steam need. The functional unit chosen is 1 MJ of ethanol production.

System boundaries and steps of the life cycle for the proposed bioethanol production include (1) production and collection of cornstover, (2) transport of the cornstover feedstock from the farm to a processing facility (3) pretreatment and conditioning of cornstover, saccharification and fermentation, natural gas boiler steam generation, electricity delivery to plant, by-product dewatering (4) lignin return to the land. Life cycle stages of natural gas from extraction to its use in the biorefinery as well as electricity production are also included.

We will compare the option of producing steam and electricity for process energy and by-products from the lignin fraction of the feedstock with the option of land application as noted (Figure 1).

In this study we assumed that 50% of stover is collected from the land and the cornstover feed rate for the biorefinery is 98,039 kg/hr at 15% moisture as described by Aden et al. (2002).

The process considered in this study can be described as using co-current dilute acid prehydrolysis of the lignocellulosic biomass with simultaneous enzymatic saccharification of the remaining cellulose and co-fermentation of the resulting glucose and xylose to ethanol (Aden et al., 2002, McAloon, 2000). In addition to these unit operations, the process involves feedstock handling and storage, product purification, wastewater treatment, enzyme production, by-product dewatering, natural gas combustion, product storage, and other utilities.

All the steps for the process of ethanol production are the same as Aden et al. (2002) except the power production and byproduct handling part. The scenario we develop assumes that two fractions of the boiler feed (biogas and evaporator syrup) in the Aspen Plus model design by Aden et al. (2002) model will be sent to a new boiler to be combusted along with natural gas. The other fraction (lignin by-product) of the Aden et al. (2002) model boiler, as described before will be dewatered using a pressure filter and transferred to land. For the new scenario, biorefinery electricity requirements are provided by the purchase from the grid.

For the environmental performance of the refinery, we estimate life cycle global warming potential (GWP) for the GHGs CO_2, CH_4, and N_2O. Processes and steps are modeled using Aspen Plus and SimaPro software.

3-Results and Discussion

3-1-Environmental impact assessment

Natural gas production and delivery to the biorefinery to produce steam for the plant generates 2.2gCO_2e/MJ ethanol (Spath and Mann, 2000). The natural gas boiler and air compressor for the scenario developed were simulated in Aspen Plus. Aspen results show that power generation section produces 34,700kgCO_2e/hr, almost half the emissions from Aden et al.'s original design (65,600 kg CO_2e/hr).

Steam generation contributes greatest to GHG emissions in the whole process with 52 gCO_2e/MJ ethanol. Natural gas production and delivery and electricity production generate 13gCO_2e/MJ ethanol and 15gCO_2e/MJ ethanol, respectively.

By-product output of the pressure filter which is to be sent back to land is 44,729 kg/hr at 10% moisture. SimaPro simulation gives about 1gCO_2e/MJ ethanol for the transportation of the by-product to land.

Figure 2 shows the GHG emissions for the Aden et al. (2002) and the lignin-land application pathways. For fertilizers, feedstock transportation and chemicals and enzymes added to the process, GHG emissions of 8, 2 and 6gCO_2e/MJ of ethanol was assumed (Spatari et al., 2010). Figure 2 also shows the net GHG emission which is higher for the Aden et al. (2002) model.

Most of the difference in GHG emissions between the two scenarios is the result of returning the by-product, which has a high carbon content, to land, instead of burning it in the boiler which produces a high amount of GHG. As a result and considering that natural gas is a low carbon fossil fuel, the new boiler GHG emissions are less than the boiler fed with by-product. Another source of difference in GHG emission of two scenarios is the carbon being sequestered by return of byproduct to the land, which replaces the carbon lost by soil microbial activity.

3-2-Economic Analysis

Total fixed cost of the new system

Using Aspen Plus the simulation for the lignin land scenario showed that to produce the necessary steam for the plant a smaller boiler followed by an even smaller burner would be needed. This is due to the higher lower heating value (LHV) of the natural gas (20,745 Btu/lb, Spath and Mann, 2000) compared to the lignin by-product (4,179 Btu/lb, Aden et al., 2000), which results in less natural gas needed for producing the same amount steam needed. The total cost of the new equipments is $5.96 million. Considering no other changes compared to the Aden et al. (2002) (the plant operates on the same amount of feedstock and the same technology of ethanol production), the total fixed cost of plant in Aden et al. (2002) model is $153 million dollars, while it is $102 million dollars for the new scenario. This difference in cost as mentioned is mostly due to the smaller boiler.

Table 1 summarizes the costs difference in Aden et al. (2002) and the scenario investigated in this study.

6-Conclusion

Environmental and economic analyses show that the biorefinery designed with the aim of applying the lignin by-product to land has certain advantages in terms of reduced GHG emissions compared to burning the lignin for energy recovery. Results show that the lignin-land amendment scenario not only has less GHG emissions compared to using the lignin for energy as modeled in Aden et al. (2002), but may also be able to reduce SOC loss in agricultural soils that results from tillage, microbial activity and cornstover removal for fuel. This is one of the biggest advantages of the lignin-land amendment scenario we explore: reducing GHG emission of the whole scenario while facilitating soil fertility and soil aggregation.

In addition, economic analysis results showed that equipment installed costs for this study will be lower than Aden et al. (2002) which results in a much lower capital investment too making it potentially a more favorable option to investors.

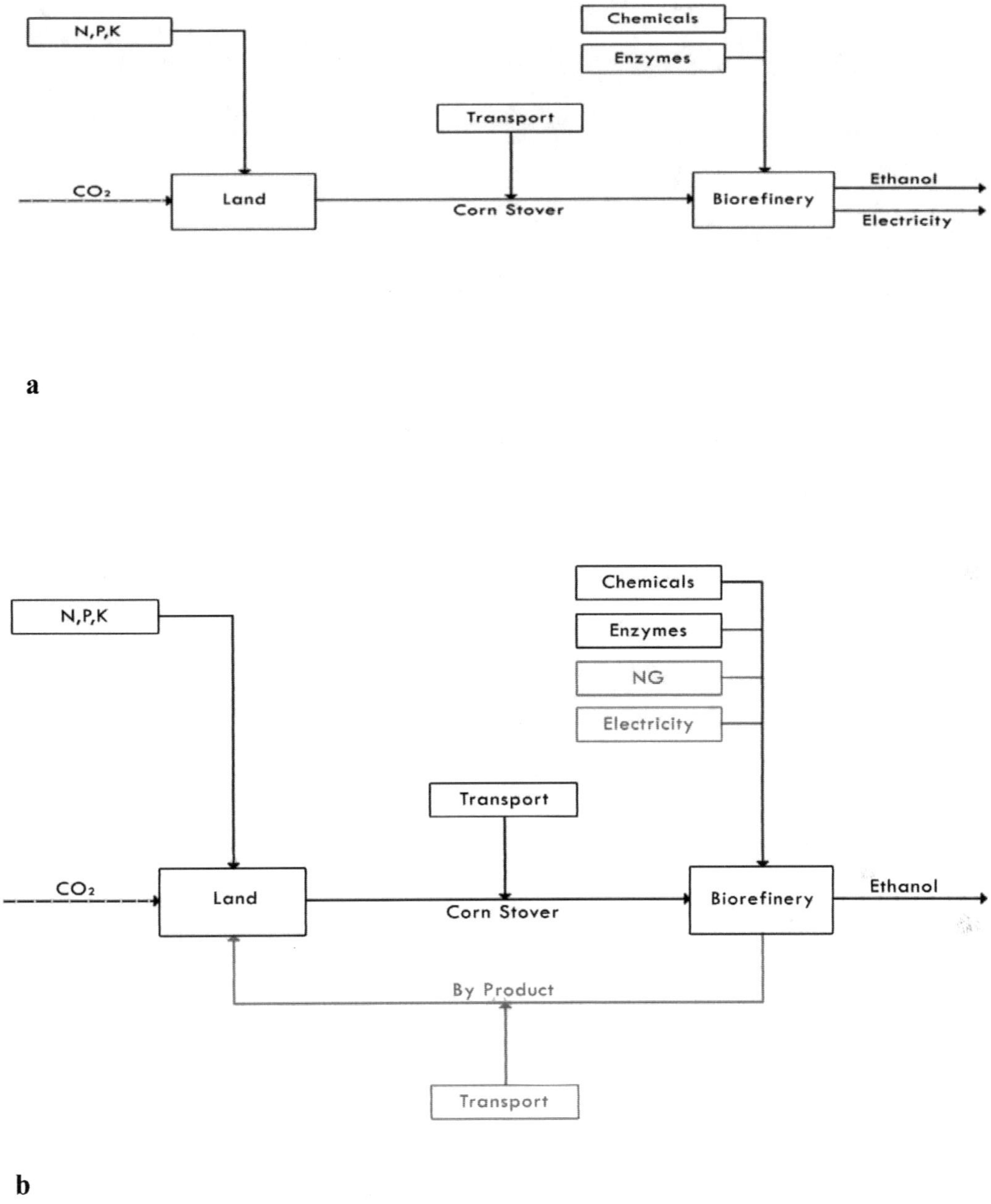

Figure 1. System Boundary (a) Ethanol Production using byproduct for power production, (b) Ethanol production using NG for Power production and by-product return to land

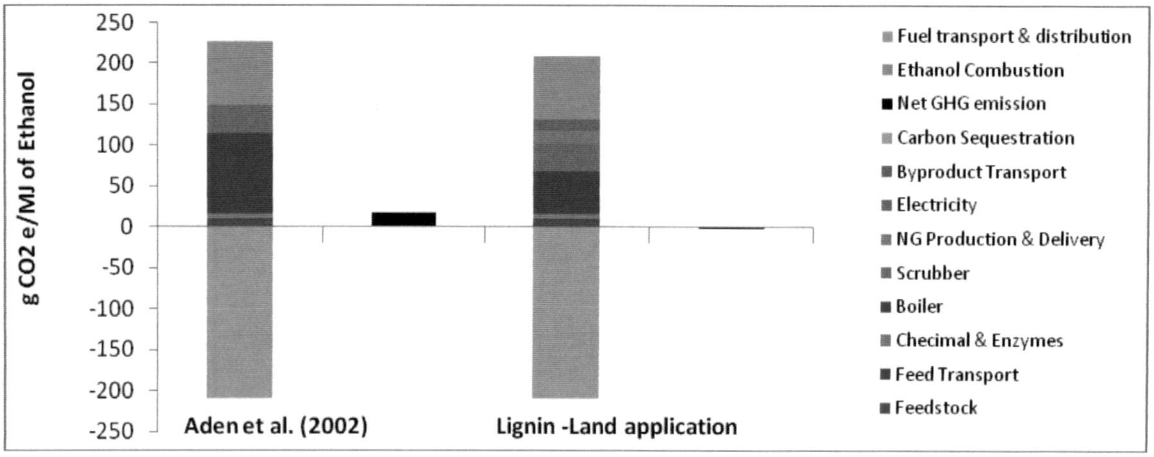

Figure 2. GHG emissions for the Aden et al. (2002) and lignin-land application scenario

Table 1. Cost summary comparison of Aden et al. (2002) and the proposed Scenario*

	Aden et al.	By-product return to land
Equipment installed cost		
Boiler&Compressor	$29,400,000	$3,000,000
Pressure Filter	N/A†	$1,608,000
Total equipment installed cost of the plant	$153,000,000	$102,000,000
Capital investment	$241,000,000	$160,000,000

*** Costs are for year 2010**
†Not applicable

References

Aden A., Ruth M., Ibsen K., Jechura J., Neeves K., Sheehan J., Wallace B., 2002, Lignocellulosic Biomass to Ethanol Process Design and Economics Utilizing Co-Current Dilute Acid Prehydrolysis and Enzymatic Hydrolysis for Corn Stover, NREL/TP-510-32438

Bahri B., Dignac M.F., Rumpel C., Rasse D.P., Chenu C., Mariotti A., 2008, Lignin turnover kinetics in an agricultural soil is monomer specific, Soil Biology and Biochemistry, 38, 1977–1988

Blanco-Canqui B. , Lal R., 2008, Corn stover removal impacts on micro-scale soil physical properties, Geoderma 145, 335–346

Farrell A.E., Plevin R.J., Turner B.T., Jones A.D., O'Hare M., Kammen D.M., 2006. Ethanol Can Contribute to Energy and Environmental Goals, Science, 311, 506-508

Gnansounou E., 2010, Production and use of lignocellulosic bioethanol in Europe: Current situation and perspectives, Bioresource Technology, 101, 4842–4850

Hamelinck C.N., Van Hooijdonk G., PC Faaij A., 2005, Ethanol from lignocellulosic biomass: techno-economic performance in short-, middle- and long-term, Biomass and Bioenergy, 28, 384–410

Hendriks A.T.W.M., Zeeman G., 2009. Pretreatments to enhance the digestibility of lignocellulosic biomass, Bioresource Technology 100, 10–18

Jaramillo P., Griffin W. M., Matthews H.S., 2007, Comparative Life-Cycle Air Emissions of Coal, Domestic Natural Gas, LNG, and SNG for Electricity Generation, Environmental Science and Technology, 41, 6290-6296

Johnson J. M. F., Reicosky D., Sharratt B., Lindstrom M., Voorhees W., Carpenter-Boggs L., 2004, Characterization of Soil Amended with the By-Product of Corn Stover Fermentation, Soil Science Society of America, 68, 139–147

Larson E.D., Fiorese G., Liu G., Williams R.H., Kreutz T.G., Consonni S., 2010. Co-production of decarbonized synfuels and electricity from coal + biomass with CO2 capture and storage: an Illinois case study, Energy & Environmental Science, 3, 28–42

Margeot A., Hahn-Hagerdal B., Edlund M., Slade R., Monot F., New improvements for lignocellulosic ethanol, Current Opinion in Biotechnology 2009, 20:372–380

McAloon A., Taylor F., Yee W., Ibsen K., Wooley R., 2000, Determining the Cost of Producing Ethanol from Corn Starch and Lignocellulosic Feedstocks, NREL/TP-580-28893

Reijnders L., 2008, Ethanol production from crop residues and soil organic carbon, Resources, Conservation and Recycling 52, 653–658

Sheehan J., Aden A., Paustian K., Killian K., Brenner J., Walsh M., Nelson R., 2004, Energy and Environmental Aspects of Using Corn Stover for Fuel Ethanol, Journal of Industrial Ecology, 7, 118-146

Sparling G. P., Wheeler D., Vesely E.T., Schipper L. A., 2006, What is Soil Organic Matter Worth?, Journal of Environmental Quality, 35, 548–557

Spatari S., Bagley D.M., McLean H.L., 2010, Life cycle evaluation of emerging lignocellulosic ethanol conversion technologies, Bioresource Technology, Volume 101, Issue 2, 654-667

Spath P.L. and Mann M.K., 2000, Life Cycle Assessment of a Natural Gas Combined-Cycle Power Generation System, NREL/TP-570-27715

Szczodrak J., Fiedurek J., 1996. Technology for Conversion of Lignocellulosic Biomass to Ethanol, Biomass and Bioenergy Vol. 10, Nos S/6, 367-375

Thonley P., 2008, Airborne emissions from biomass based power generation systems, Environmental research letters, 3, 014004

Tiwary A., Colls J., 2010, Mitigating secondary aerosol generation potentials from biofuel use in the energy sector, Science of the total environment, 408, 607-616

Uihlein A., Schebek L., 2010, Environmental impacts of a lignocellulose feedstock biorefinery system: An assessment, Biomass and Bioenergy, 33, 7 9 3 – 8 0 2

Walter A., Ensinas A.V., 2010, Combined production of second-generation biofuels and electricity from sugarcane residues, Energy 35, 874–879

Wilhelm W. W., Johnson J. M. F., Hatfield J.L., Voorhees W.B., Linden D.R., 2004, Crop and soil productivity response to corn residue removal: a literature review, Agronomy Journal, 96, 1–17

Wilhelm W. W., Johnson J. M. F., Karlen D. L., Lightle D.T., 2007, Corn Stover to Sustain Soil Organic Carbon Further Constrains Biomass Supply, Agronomy Journal, 99, 1665–1667

Wyman C.E., 1994. Ethanol from Lignocellulosic Biomass: Technology, Economics, AND Opportunities, Bioresource Technology 50, 3-16

(731d) Land Allocation in a Sustainable Landscape and a Comparative Life Cycle Assessment of Distributed Vs. Centralized Biorefinery Pre-Processing Systems

Pragnya L. Eranki, Department of Chemical Engineering, Michigan State University, Lansing, MI and **Bruce E. Dale**, Department of Chemical Engineering and Material Science, Michigan State University, East Lansing, MI

Cellulosic biofuels are a potentially large scale, environmentally superior alternative to petroleum-derived fuels. However, in order to produce large quantities of cellulosic biofuels, biorefineries handling enormous tonnages of a large variety of feedstocks are required and in such conditions a consistent and safe supply of feedstocks becomes critical. Under current thinking, biorefineries are fully integrated operations in which all processes (e.g. size reduction, pretreatment, hydrolysis, fermentation, distillation) of biomass conversion to fuels are present in a single centralized location. This implies contracting with literally thousands of individual farmers, interrupted feedstock supplies, large transport and storage costs of feedstock and other business and market power issues. We believe these and other drawbacks of fully integrated biorefineries can be addressed by network of facilities called "Regional Biomass Processing Depots" (RBPDs) which could bridge the gap between feedstock suppliers and biorefineries. In its simplest configuration, an RBPD would procure feedstock, pre-process/pre-treat it, densify it and deliver it to the biorefinery in an arrangement that satisfies the biorefinery requirements and return by-products to the appropriate end-user. We envision RBPDs that are configured to provide benefits in all three major sustainability criteria: environmental, economic and social. Specifically, the aim of this study is to allocate land usage in a landscape to various feedstocks based on their sustainability and ethanol/animal feed-yield characteristics, to estimate the total energy consumption and emissions generated due to the operation of the system that consists of feedstock suppliers, animal farms, the RBPD and all related transportation activities and to assess the sustainability of this distributed processing system by performing a comparative Life Cycle Assessment and weighing it against a fully integrated and centralized biorefinery. The study also includes several sensitivity analyses using parameters like yields of grasses, forms of densification and transport energy and emission comparisons using different transport models like GREET and IBSAL. Multiple landscape scales and locations are incorporated in order to evaluate the effect of site-selection and size on the entire system.

See more of this Session: Life Cycle Analysis of Renewable Feedstock-Based Processes and Products
See more of this Group/Topical: Sustainable Engineering Forum

(731e) Camelina-Derived Jet Fuel and Diesel: Sustainable Advanced Biofuels

David R. Shonnard[1], Jiqing Fan[1], Larry Williams[2] and T.N. Kalnes[3], (1)Chemical Engineering, Michigan Technological University, Houghton, MI, (2)Targeted Growth, Inc, Seattle, WA, (3)Research and Development, UOP LLC, Des Plaines, IL

Abstract

Recently, an isoparaffin-rich jet fuel derived from camelina, a low-input non-food oilseed crop, was flight-tested by a commercial airline. To date, all test results indicate that this hydrotreated renewable jet fuel (HRJ) not only meets stringent engine fuel and performance specifications but also reduces environmental emissions. The goal of this study is to determine the life cycle greenhouse gas (GHG) emissions, cumulative energy demand, and fossil energy demand of camelina-derived HRJ as well as green diesel (GD) produced by a UOP process, and compute the GHG savings per MJ of fuel compared to petroleum-based jet fuels and diesel. The scope of this study encompasses the entire life cycle from camelina cultivation and raw materials acquisition through the production and use of the fuels in vehicles and aircraft operations. In the life cycle assessment (LCA) model, camelina is grown in Montana on either marginal lands or as a rotation crop during fallow periods on existing lands already in food crop production, thus avoiding the conflict with food cultivation and concerns with indirect land use change impacts. After harvesting, the camelina seeds are transported to a processing facility, where the oils can be extracted by a crusher and solvent based process, and with seed meal being extracted as co-product. The oil is processed to remove impurities, and then transported by rail (from Montana to Seattle) to a hydroprocessing plant for GD and HRJ production. Inventory data for camelina cultivation, transportation, and oil extraction were provided by Targeted Growth, Inc. Data for conversion of oil to HRJ and GD were obtained from engineering design data, and was supplied by UOP. All the inventory data were assembled based on energy content of 1 MJ of final fuel product, which was the functional unit of this LCA study. Energy allocation was applied to distribute environmental impacts among various products and co-products along the life cycle of each biofuel. The life cycle GHG emissions for GD and HRJ are 18.0 and 22.4 g CO_2 eq./MJ fuel respectively, which represent savings relative to petroleum fuel counterparts of 80% and 75% respectively. Total energy consumption for each camelina biofuels product is comparable to petroleum fuels, but most energy is renewable biomass as opposed to non renewable fossil. Scenario analyses were also conducted to determine response to model assumptions and data uncertainty, including allocation methodology, N fertilizer application rate, N_2O emission factor, source of H_2, and farm diesel consumption. The results show that camelina derived biofuels qualify as advanced biofuels by the Renewable Fuels Standard (RFS2). With expected future gains in yield, camelina oil production and hydroprocessing has the potential to provide the US an estimated 800 million gallons per year of high quality renewable fuel.

See more of this Session: Life Cycle Analysis of Renewable Feedstock-Based Processes and Products

(731f) Optimization of Water Consumption in Bioethanol Plants

Mariano Martin[1], Elvis Ahmetovic[2] and Ignacio E. Grossmann[1], (1)Department of Chemical Engineering, Carnegie Mellon University, Pittsburgh, PA, (2)Department of Process Engineering, University of Tuzla, Tuzla, Bosnia

Current industrial practices and population demands are placing extra pressure on natural resources. Concerns related to overconsumption and depletion of these resources have focused on energy sources like crude oil, carbon and natural gas [1] while water has been overlooked from that list mainly due its wide availability in many regions of the world. As a result, water is inexpensive compared to any other raw material, in spite of an average annual increase of 6.7% over the last years according to the 2008 GWI/OECD[2] However, the industrial growth and the development suggests that by 2025 the industrial water usage (including utility cooling and heating, processing, transportation, air conditioning, cleaning, etc.) will account for about 11% of the total world water consumption [3,4] which can cause water stress in several regions of the world. The increasing concern towards water resources[5,6] and the green policies supported by many governments are making the management of water consumption and wastewater treatment an important topic with new economic incentives for implementing technologies that are more environmentally friendly, and that can ensure efficient use of water resources including the treatment and recycling of wastewater [7]

In this work we address the water consumption optimization of first and second generation bioethanol production plants. Corn-based ethanol is considered the base case to compare the water consumption with the one resulting from the production of ethanol from switchgrass following thermo-chemical, thermo-biochemical or biochemical routes. In order to optimize the water consumption a three stage method is used. First, energy consumption is optimized in the production processes, which reduces the cooling needs of the processes. We consider the optimal processes described by Karuppiah et al, Martín & Grossmann.[8-10]

Next a number of technologies are implemented to substitute the use of water as cooling agent so that is possible to reduce the water to be cooled down reducing the evaporation and drift losses in the cooling tower[11]. Finally, in order to synthesize the optimal water networks for the bioethanol production plants together with the cooling water and stream loops, we use the general superstructure of integrated process water networks, which has been proposed recently by Ahmetovi• and Grossmann [12]. We define the processes units that generate contaminants like washing stages, cooling tower and boiler, sources of water like distillation columns or gas-liquid flash separators, sinks like fermentation processes, cooling towers, boilers, pretreatment tanks or gasifiers. In this way it is possible to indentify the main contaminants. We assume that total suspended solids, organics and total dissolved solids are the most important contaminants generated. Thus, we propose a number of treatment units to handle them such as screens, combined aerobic and anaerobic treatment and reverse osmosis, respectively. With this information we design the optimal water network for each of the energetically optimized production processes of ethanol.

By energy optimization of the flowsheet together with the implementation of multieffect columns and the design of optimal heat exchanger networks, it is not only possible to reduce the energy consumption by one half but also to reduce the cooling needs by two thirds. As a result, the water to be treated in the cooling tower is reduced and so are the water loses by evaporation and drift. Furthermore, the use of air cooling to reduce the use of water as cooling agent has an impact only in case of thermo-chemical processes, where the intercooling of the compression stages has a large contribution in the cooling needs. Finally, by coupling energy optimization, air cooling and the optimal design of water networks, all the ethanol production processes considered consume less than 3 gal water per gal ethanol in the range of the needs for the production of gasoline [13]. Moreover, it turns out that corn ethanol is not as water intense as has been reported so far and even consumes lees freshwater than any other process (1.46 gal/gal). Zero discharge is not achieved for all the production processes but small values are reported.

References

(1) Hayward, T. (2008) BP Statistical Review of World Energy June **2008**

(2) GWI/OCDE (2008) http://www.globalwaterintel.com/archive/9/9/analysis/world-water-prices-rise-by-67.html

(3) Rosegrant, M.W. Cai, X. Cline, (2002) Global Water Outlook to 2025 – Averting an Impending Crisis. International Food Policy Research institute, Washington

(4) Sparks Companies Inc. (2003) Global water and food outlook (**2003**)

(5) Elcock, D. (2008) Baseline and Projected Water Demand Data for Energy and Competing Water Use Sectors, ANL/EVS/TM/08-8,

(6) Chiu, Yi-Wen, Walseth‡, B., Suh, S. Water Embodied in Bioethanol in the United States *Environ. Sci. Technol.*, 2009, 43 (8), pp 2688–2692

(7) Petrakis S. Reduce cooling water consumption. New closed loop cooling method improves process cooling tower operations. Hydrocarbon Processing, **2008**, December, 95-98

(8) Karuppiah et al. 2008 AICHE J. 54, 1499-1525

(9) Martin, M., Grossmann, I.E: Process Optimization of Bioethanol Production via Gasification of Switchgrass to be submitted AIChE J. **2010**

(10) Martin, M., Grossmann, I.E: Process Optimization of Bioethanol Production via Hydrolysis of Switchgrass to be submitted AIChE J. (**2010b)**

(11) Phillips, S., Aden, A., Jechura, J. and Dayton, D., Eggeman, T (2007) Thermochemical Ethanol via Indirect Gasification and Mixed Alcohol Synthesis of Lignocellulosic Biomass *Technical Report,* NREL/ TP-510-41168, April **2007**

(12) Ahmetovi•, E., & Grossmann, I. E. General superstructure and global optimization for the design of integrated process water networks Submitted to AIChE J. **2010**

(13) Andy Aden Water Usage for Current and Future Ethanol Production. Southwest Hydrology, September/October **2007**. 22-23

See more of this Session: Life Cycle Analysis of Renewable Feedstock-Based Processes and Products
See more of this Group/Topical: Sustainable Engineering Forum

AUTHOR INDEX

AUTHOR INDEX

AUTHOR INDEX

AUTHOR INDEX

AUTHOR INDEX

American Institute of Chemical Engineers
3 Park Avenue
New York, New York 10016-5991

CURRAN ASSOCIATES INC
proceedings.com